Advanced Communication and Networking Technologies for Vehicular Ad Hoc Networks (VANETs)

Advanced Communication and Networking Technologies for Vehicular Ad Hoc Networks (VANETs)

Guest Editors

Qiong Wu
Pingyi Fan

Basel • Beijing • Wuhan • Barcelona • Belgrade • Novi Sad • Cluj • Manchester

Guest Editors

Qiong Wu
Jiangnan University
Wuxi
China

Pingyi Fan
Tsinghua University
Beijing
China

Editorial Office
MDPI AG
Grosspeteranlage 5
4052 Basel, Switzerland

This is a reprint of the Special Issue, published open access by the journal *Sensors* (ISSN 1424-8220), freely accessible at: https://www.mdpi.com/journal/sensors/special_issues/4F6N1XA61K.

For citation purposes, cite each article independently as indicated on the article page online and as indicated below:

Lastname, A.A.; Lastname, B.B. Article Title. *Journal Name* **Year**, *Volume Number*, Page Range.

ISBN 978-3-7258-2961-3 (Hbk)
ISBN 978-3-7258-2962-0 (PDF)
https://doi.org/10.3390/books978-3-7258-2962-0

© 2024 by the authors. Articles in this book are Open Access and distributed under the Creative Commons Attribution (CC BY) license. The book as a whole is distributed by MDPI under the terms and conditions of the Creative Commons Attribution-NonCommercial-NoDerivs (CC BY-NC-ND) license (https://creativecommons.org/licenses/by-nc-nd/4.0/).

Contents

Qiong Wu, Zheng Zhang, Hongbiao Zhu, Pingyi Fan, Qiang Fan, Huiling Zhu and Jiangzhou Wang
Deep Reinforcement Learning-Based Power Allocation for Minimizing Age of Information and Energy Consumption in Multi-Input Multi-Output and Non-Orthogonal Multiple Access Internet of Things Systems
Reprinted from: *Sensors* **2023**, 23, 9687, https://doi.org/10.3390/s23249687 1

Zheng Zhang, Qiong Wu, Pingyi Fan and Qiang Fan
Age of Information Analysis for Multi-Priority Queue and Non-Orthoganal Multiple Access (NOMA)-Enabled Cellular Vehicle-to-Everything in Internet of Vehicles
Reprinted from: *Sensors* **2024**, 24, 7966, https://doi.org/10.3390/s24247966 21

Lianyou Lai, Zhongzhe Song and Weijian Xu
A Novel Framed Slotted Aloha Medium Access Control Protocol Based on Capture Effect in Vehicular Ad Hoc Networks
Reprinted from: *Sensors* **2024**, 24, 992, https://doi.org/10.3390/s24030992 34

Ping Sun, Haibo Dai and Baoyun Wang
Integrated Sensing and Secure Communication with XL-MIMO
Reprinted from: *Sensors* **2024**, 24, 295, https://doi.org/10.3390/s24010295 54

Yang Wang, Jianghong Shi, Zhiyuan Fang and Lingyu Chen
A Novel Analytical Model for the IEEE 802.11p/bd Medium Access Control, with Consideration of the Capture Effect in the Internet of Vehicles
Reprinted from: *Sensors* **2023**, 23, 9589, https://doi.org/10.3390/s23239589 66

Lei Nie, Junjie Zhang, Haizhou Bao and Yiming Huo
Heuristic Path Search and Multi-Attribute Decision-Making-Based Routing Method for Vehicular Safety Messages
Reprinted from: *Sensors* **2023**, 23, 9506, https://doi.org/10.3390/s23239506 87

Tianqing Zhou, Ming Xu, Dong Qin, Xuefang Nie, Xuan Li and Chunguo Li
Computing Offloading Based on TD3 Algorithm in Cache-Assisted Vehicular NOMA–MEC Networks
Reprinted from: *Sensors* **2023**, 23, 9064, https://doi.org/10.3390/s23229064 107

Weijian Xu, Zhongzhe Song, Yanglong Sun, Yang Wang and Lianyou Lai
Capture-Aware Dense Tag Identification Using RFID Systems in Vehicular Networks
Reprinted from: *Sensors* **2023**, 23, 6792, https://doi.org/10.3390/s23156792 129

Yuan-Xu Fu, Tao Shen and Jiang-Ling Dou
Mutual Coupling Reduction of a Multiple-Input Multiple-Output Antenna Using an Absorber Wall and a Combline Filter for V2X Communication
Reprinted from: *Sensors* **2023**, 23, 6355, https://doi.org/10.3390/s23146355 145

Sixing Ma, Meng Li, Ruizhe Yang, Yang Sun, Zhuwei Wang and Pengbo Si
Next-Hop Relay Selection for Ad Hoc Network- Assisted Train-to-Train Communications in the CBTC System
Reprinted from: *Sensors* **2023**, 23, 5883, https://doi.org/10.3390/s23135883 160

YuanXu Fu, Tao Shen, JiangLing Dou and Zhe Chen
Absorbing Material of Button Antenna with Directional Radiation of High Gain for
P2V Communication
Reprinted from: *Sensors* **2023**, *23*, 5195, https://doi.org/10.3390/s23115195 . . . . **181**

Yuanxu Fu, Tao Shen, Jiangling Dou and Zhe Chen
Stereoscopic UWB Yagi–Uda Antenna with Stable Gain by Metamaterial for Vehicular
5G Communication
Reprinted from: *Sensors* **2023**, *23*, 4534, https://doi.org/10.3390/s23094534 . . . . **199**

Article

Deep Reinforcement Learning-Based Power Allocation for Minimizing Age of Information and Energy Consumption in Multi-Input Multi-Output and Non-Orthogonal Multiple Access Internet of Things Systems

Qiong Wu [1,2,*,†], Zheng Zhang [1,2,†], Hongbiao Zhu [1,2,†], Pingyi Fan [3,*], Qiang Fan [4], Huiling Zhu [5] and Jiangzhou Wang [5]

[1] School of Internet of Things Engineering, Jiangnan University, Wuxi 214122, China; zhengzhang@stu.jiangnan.edu.cn (Z.Z.); hongbiaozhu@stu.jiangnan.edu.cn (H.Z.)
[2] State Key Laboratory of Integrated Services Networks, Xidian University, Xi'an 710071, China
[3] Department of Electronic Engineering, Beijing National Research Center for Information Science and Technology, Tsinghua University, Beijing 100084, China
[4] Qualcomm, San Jose, CA 95110, USA; qf9898@gmail.com
[5] School of Engineering, University of Kent, Canterbury CT2 7NT, UK; h.zhu@kent.ac.uk (H.Z.); j.z.wang@kent.ac.uk (J.W.)
* Correspondence: qiongwu@jiangnan.edu.cn (Q.W.); fpy@tsinghua.edu.cn (P.F.); Tel.: +86-0510-8591-0633 (Q.W.); +86-010-6279-6973 (P.F.)
† These authors contributed equally to this work.

Abstract: Multi-input multi-output and non-orthogonal multiple access (MIMO-NOMA) Internet-of-Things (IoT) systems can improve channel capacity and spectrum efficiency distinctly to support real-time applications. Age of information (AoI) plays a crucial role in real-time applications as it determines the timeliness of the extracted information. In MIMO-NOMA IoT systems, the base station (BS) determines the sample collection commands and allocates the transmit power for each IoT device. Each device determines whether to sample data according to the sample collection commands and adopts the allocated power to transmit the sampled data to the BS over the MIMO-NOMA channel. Afterwards, the BS employs the successive interference cancellation (SIC) technique to decode the signal of the data transmitted by each device. The sample collection commands and power allocation may affect the AoI and energy consumption of the system. Optimizing the sample collection commands and power allocation is essential for minimizing both AoI and energy consumption in MIMO-NOMA IoT systems. In this paper, we propose the optimal power allocation to achieve it based on deep reinforcement learning (DRL). Simulations have demonstrated that the optimal power allocation effectively achieves lower AoI and energy consumption compared to other algorithms. Overall, the reward is reduced by 6.44% and 11.78% compared to the to GA algorithm and random algorithm, respectively.

Keywords: deep reinforcement learning; age of information; MIMO-NOMA; Internet of Things

1. Introduction

With the development of the Internet of Things (IoT), the base station (BS) can support the real-time applications such as disaster management, information recommendation, vehicle network, smart city, connected health and smart manufacturing by collecting the data sampled by IoT devices [1,2]. However, the amount of sampled data is enormous and the number of IoT devices is usually high; thus, the realization of these IoT applications requires a large bandwidth spectrum [3]. The multi-input multi-output and non-orthogonal multiple access (MIMO-NOMA) IoT can transmit data through the MIMO-NOMA channel to solve these problems, wherein multiple antennas are deployed at the BS to improve the channel capacity and multiple IoT devices access the common bandwidth simultaneously to improve the spectrum efficiency.

The BS collects data during discrete slots in the MIMO-NOMA IoT system. In each slot, a BS first determines the sample collection commands and allocates the transmit power for each IoT device and then sends the corresponding sample collection commands and transmission power to each IoT device. Afterwards, each IoT device determines whether to sample data from the physical world according to their sample collection commands. Then, each IoT device adopts its allocated power to transmit the sampled data to the BS over the MIMO-NOMA channel. In the transmission process, multiple IoT devices transmit the signals of the data by using the same spectrum, and therefore interference exists between different IoT devices. To eliminate the interference, the BS adopts the successive interference cancellation (SIC) technique to decode the signals from each device [4]. Specifically, the BS sorts the power of all received signals in descending order and decodes the signal with the highest received power by considering other signals as interferences. Then, the BS removes the decoded signal from the received signals and resorts the received signals to decode the next signal. The process is repeated until all signals are decoded.

The age of information (AoI) is a metric to measure the freshness of the data, which is defined as the time from the data sampling to the time when the sampled data are received [5]. In the MIMO-NOMA IoT system, the BS needs to receive data, i.e., decode the signals of the data, in a timely manner after they are sampled to provide the real-time applications; thus, a low AoI is critical in MIMO-NOMA IoT systems [6]. Furthermore, the IoT devices are energy-limited. Thus, the MIMO-NOMA IoT system should also keep its energy consumption low to prolong the working time of the IoT devices [7]. Hence, the AoI and energy consumption are two important performance metrics of the MIMO-NOMA IoT system [8]. The sample collection commands and power allocation may affect the AoI and energy consumption of the system. Specifically, for the sample collection commands, if the BS selects more IoT devices to sample, the system will consume more energy because more IoT devices consume energy to sample data. However, if the BS selects less IoT devices to sample, the data transmitted from the unselected IoT devices become obsolete, which may increase the AoI of the system. Hence, the sample collection commands affect both the AoI and energy consumption of the MIMO-NOMA IoT system. For the power allocation, if an IoT device transmits with high power, the signal transmitted by the IoT device will be decoded wherein a significant amount of signals with lower power act upon the interferences in the SIC process, which may lead to a low signal-to-interference-plus-noise ratio (SINR). Otherwise, if an IoT device transmits data with low power, the SINR may also be deteriorated due to the low transmission power. The low SINR causes a low transmission rate, which may cause a long transmission delay and a high AoI of the MIMO-NOMA IoT system. Hence, the power allocation affects the AoI of the MIMO-NOMA IoT system. Moreover, the power allocation affects the energy consumption directly. Thus, the transmission power affects both the AoI and energy consumption of the MIMO-NOMA IoT system. As mentioned above, it is critical to determine the optimal policy including sample collection commands and power allocation to minimize the AoI and energy consumption of the MIMO-NOMA IoT system. To the best of our knowledge, there is no work to minimize the AoI in the MIMO-NOMA IoT system, which motivates us to conduct this work. In the MIMO-NOMA IoT system, the allocation of transmission powers has a direct impact on the transmission rate during the SIC process. Additionally, the MIMO-NOMA channel is inherently affected by stochastic noise. Model-based algorithms struggle to construct an accurate model to describe this process, which causes the traditional model-based algorithms unsuitable to solve the problem. Deep reinforcement learning (DRL) is a type of model-free-based method that enables an agent to learn how to make sequential decisions in a complex environment to achieve a specific goal. DRL can learn the near-optimal policy by learning from the interaction between action and the environment (i.e., dynamic stochastic MIMO-NOMA IoT system) [9]. There are some existing studies on DRL-based optimization frameworks in similar systems. In [10], Zhao et al. formalized the joint optimization problem of video frame resolution selection, computation offloading and resource allocation strategy, and proposed a hierarchical reward function based on the DRL algorithm that minimizes energy consumption, maximizes quality of experience (QoE) delay and analyzes the accuracy in the IoT system. In [11], Chen et al. considered a marginalized IoT system and studied the joint caching and computing service deployment (JCCSP) problem for IoT applications driven by perceptual data. An improved method based on twin-delayed (TD) deep deterministic policy gradient (DDPG) was proposed, which achieved significant convergence performance compared to benchmarks. In general, the DRL algorithm is used to solve problems with either continuous or discrete action spaces separately. However, we focus on simplifying the joint optimization problem when the space of the sample collection commands is discrete while the space of the transmission power is continuous, to make it applicable to DDPG. We achieve this goal by establishing the relationship between sample collection commands and transmission power, and then propose a DRL-based power allocation to minimize the AoI and energy consumption of the MIMO-NOMA IoT system (The source code has been released at: https://github.com/qiongwu86/MIMO-NOMA_AoI_GA.git (7 March 2023)). The main contributions are summarized as follows:

(1) We formulated the joint optimization problem to minimize the AoI and energy consumption of the MIMO-NOMA IoT system by determining the sample selection and power allocation. Specifically, we constructed an MIMO-NOMA channel model and an AoI model to find the relationship between transmission rate and AoI of each device under the SIC mode. Additionally, we constructed an energy consumption model. Then, the joint optimization problem was formulated based on the constructed models.

(2) Then, we simplified the formulated optimization problem to make it suitable for DRL algorithms. In the formulated optimization problem, the sample selection is discrete and power allocation is continuous, which cannot be solved by the traditional DRL method and results in a challenge for optimization. We substituted the energy model and AoI model by the formulated optimization problem, merged the homogeneous terms containing sample selection and simplified the formulated problem to make it suitable to be solved by the traditional continuous-control DRL algorithm.

(3) To solve the formulated optimization problem, we first designed a DRL framework which included the state, action and reward function, and then adopted the DDPG algorithm to obtain the optimal power allocation to minimize the AoI and energy consumption of the MIMO-NOMA IoT system.

(4) Extensive simulations were carried out to demonstrate that the DDPG algorithm successfully optimizes both the AoI and energy consumption compared with other baseline algorithms.

The rest of this paper is organized as follows. Section 2 reviews the related work. Section 3 introduces the system model and formulates the optimization problem. Section 4 simplifies the formulated optimization problem and presents the near-optimal solution by DRL. We carry out some simulation to demonstrate the effectiveness of our proposed DRL method in Section 5, and conclude this paper in Section 6.

2. Related Work

In this section, we first review the studies about the AoI in the IoT system, and then survey the state of the arts on the MIMO-NOMA IoT system.

2.1. AoI in IoT

In [12], Grybosi et al. proposed the SIC-aided age-independent random access (AIRA-SIC) scheme (i.e., a slotted ALOHA fashion) for the IoT system, wherein the receiver operates SIC to reconstruct the collisions of various devices. In [13], Wang et al. focused on the problem that minimizes the weighted sum of AoI cost and energy consumption in the IoT systems by adjusting the sample policy, and proposed a distributed DRL algorithm based on the local observation of each device. In [14], Elmagid et al. aimed to minimize the AoI at the BS and the energy consumption of the generate status for the IoT devices, formulated an optimization problem based on the Markov decision process (MDP) and then proved the monotonicity property of the value function associated with the MDP. In [15], Li et al. designed a resource block (RB) allocation, modulation-selecting and coding-selecting scheme for each IoT device based on its channel condition to minimize the long-term AoI of the IoT system. In [16], Hatami et al. employed the reinforcement learning to minimize the average AoI for users in an IoT system consisting of users, energy harvesting sensors and a cache-enabled edge node. In [17], Sun et al. aimed to minimize the weighted sum of the expected average AoI of all IoT devices, propulsion energy of unmanned aerial vehicle (UAV) and transmission energy of IoT devices by determining the UAV flight speed, UAV placement and channel resource allocation in the UAV-assisted IoT system. In [18], Hu et al. considered an IoT system wherein the UAVs take off from a data center to deliver energy and collect data from sensor nodes, and then fly back to the data center. They minimized the AoI of the collected data by dynamic programming (DP) and ant colony (AC) heuristic algorithms. In [19], Emara et al. developed a spatio-temporal framework to evaluate the peak AoI (PAoI) of the IoT system, and compared the PAoI under the time-triggered traffic with event-triggered traffic. In [20], Lyu et al. considered a marine IoT scenario, wherein the AoI is utilized to represent the impact of the packet loss and transmission delay. They investigated the relationship between AoI and state estimation error, and minimized the state estimation error by the decomposition method. In [21], Wang et al. investigated the impact of AoI on the system cost which consists of control cost and communication energy consumption of the industrial-Internet-of-Things (IIoT) system. They proved that the upper bound of cost is affected by the AoI. In [22], Hao et al. maximized the sum of the energy efficiency of the IoT devices under the constraints of AoI by optimizing the transmission power and channel allocation in a cognitive radio-based IoT system. However, none of these works have taken the MIMO-NOMA channel into account.

2.2. MIMO-NOMA IoT System

In [23], Yilmaz et al. proposed a user selection algorithm for the MIMO-NOMA IoT system to improve the sum data rate, and adopted the physical layer network coding (PNC) to improve the spectral efficiency. In [24], Shi et al. considered the downlink of the MIMO-NOMA IoT networks and studied the outage probability and goodput of the system with the Kronecker model. In [25], Wang et al. proposed that the resource allocation problem consists of the optimal beamforming strategy and power allocation in the MIMO-NOMA IoT system, wherein the beamforming optimization is solved by the zero-forcing method, and after that the power allocation is solved by the convex functions. In [26], Han et al. proposed a novel millimeter wave (mmWave) positioning MIMO-NOMA IoT system and proposed the position error bound (PEB) as a novel performance evaluation metric. In [27], Zhang et al. considered the massive MIMO and NOMA to study the performance of the IoT system, and calculated the closed-form function for spectral and energy efficiencies. In [28], Chinnadurai et al. considered the heterogeneous cellular network and formulated a problem to maximize the energy efficiency of the MIMO-NOMA IoT system, wherein the non-convex problem was solved based on the branch and reduced-bound (BRB) approach. In [29], Gao et al. considered the mmWave massive MIMO and NOMA IoT system to maximize the weighted sum transmission rate by optimizing the power allocation, and then solved the problem by the convex method. In [30], Feng et al. considered an UAV-aided MIMO-NOMA IoT system and regarded an UAV as the BS. They formulated the problem to maximize the sum transmission rate of the downlink by optimizing the placement of UAVs, beam pattern and transmission power, and then solved the problem by convex methods. In [31], Ding et al. designed a novel MIMO-NOMA system consisting of two different users, wherein user one should be served with strict quality-of-service (QoS) requirement, and user two accesses the channel by the non-orthogonal way opportunistically; thus, the requirement that small packets of user one in the IoT system should be transmitted in time can be met. In [32], Bulut et al. proposed the water cycle algorithm (WCA) based on the energy allocation method for MIMO-NOMA IoT systems. Their simulation results demonstrated that the proposed method performs better than empirical search algorithm (ESA) and genetic algorithm (GA). In [33], Ullah et al. proposed a power allocation algorithm based on DDPG to maximize energy efficiency in MIMO-NOMA next-generation Internet-of-Things (NG-IoT) networks. Their simulation results demonstrated that the proposed method achieved better performance compared with random algorithms and greedy algorithms. However, these works have not considered the AoI of the MIMO-NOMA IoT system.

As mentioned above, there is no work considering the joint optimization problem of age of information and energy in the MIMO-NOMA IoT system, which motivates us to conduct this work. The comparison of the related works is shown in Table 1.

Table 1. The comparison between related works.

Related Work	MIMO-NOMA	AoI Minimization	Energy Optimization
[12,15,17–19]	×	✓	×
[13,14,16,22]	×	✓	✓
[25,28,29,32,33]	✓	×	✓
[23,24,26,27]	✓	×	×

3. System Model And Problem Formulation

3.1. Scenario Description

The network scenario is illustrated in Figure 1. We consider a MIMO-NOMA IoT system consisting of a BS with K antennas and a set $\mathcal{M} = \{1, \ldots, m, \ldots, M\}$ of the single-antenna IoT devices. Here, each IoT device is embedded with a sensor and a transmitter. The time duration is divided into T slots, each of which is τ. The set of slots is denoted as $\mathcal{T} = \{1, \ldots, t, \ldots, T\}$. At the beginning of each slot t, the BS determines the policy (including the sample collection commands of each device m, denoted as $s_{m,t}$, and the transmission power of each device m, denoted as $p_{m,t}$) and then sends $s_{m,t}$ and $p_{m,t}$ to each device m. If $s_{m,t} = 1$, device m will sample data in slot t, and transmit the data to the BS with transmission power $p_{m,t}$ over the MIMO-NOMA channel. This action reduces the AoI, but also incurs a cost in terms of energy consumption. Otherwise, it does not sample data in slot t; therefore, it does not consume energy for sampling and transmission, while increasing the AoI due to a lack of updates. The key notations are listed in Table 2. Next, we will construct the MIMO-NOMA channel model.

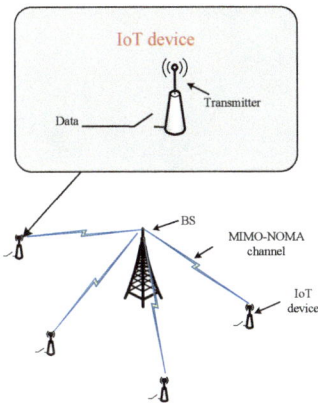

Figure 1. MIMO-NOMA IoT system.

Table 2. The summary of the notations.

Notation	Description	Notation	Description
B	Population size of genetic algorithm.	C_s	The energy consumption to sample fresh information and generate upload packet.
$c_{m,t}$	Complex data symbol with 1 as variance.	d_m	The communication distance between device m and BS.
E	Number of episodes.	F_c/F_m	Probability of offspring in genetic algorithm for crossover/mutation.
G_P/U_P	Complexity of the primary networks for computing gradients/updating parameters.	$h_m(t)$	The channel vector between device m and BS in slot t.
i	Index of transition tuples in mini-batch.	I	The number of transition tuples in a mini-batch.
$\mathcal{I}_m$	The set of devices in which the received power is weaker than device m.	$J(\mu)$	The long-term discounted reward under policy μ.
K	The number of antennas equipped in BS.	$l_{m,t}$	The transmission delay of device m in slot t.
L	Loss function.	$n(t)$	Additive white Gaussian noise.
N_{GA}	Evolution times of genetic algorithm	$m/M/\mathcal{M}$	Index/number/set of devices.
$o_t/o_{m,t}$	State in slot t of all devices/device m	$p_t/p_{m,t}$	Transmission power of all devices/device m.
$P_{m,t}$	Maximum transmission power device m.	$Q(o_t, p_t)$	Action-value function under o_t and p_t.
$Q(o_t, p_t)$	Action-value function under o_t and p_t.	Q	Packet size.
r_t	Reward function.	$s_t/s_{m,t}$	Indicator of sample or not for all devices/device m.
$s_t/s_{m,t}$	Indicator of sample or not for all devices/device m.	S_d	Complexity of calculating sample decisions based on power allocation.
S_d	Complexity of calculating sample decisions based on power allocation.	$t/\mathcal{T}$	Index/set of slot.
$\mathcal{U}$	The set of undecoded received power of BS.	$u_t/u_{m,t}$	Indicator of transmission success for all devices/device m.
W	Bandwidth of system.	α_a/α_c	Learning rate of actor network/critic network.
β	Discounting factor.	γ_a, γ_c	Weighted factors of reward function.
$\Gamma_{m,t}$	Received power of BS for device m in slot t.	Δ_t	Exploration noise.
$\varepsilon_{m,t}$	The energy consumed by device m in slot t.	$\bar{\varepsilon}$	The average sum energy consumption in slot t.
ζ/ζ'	Parameters of critic-network/target critic-network.	$\theta/\theta'/\theta^*$	Parameters of actor-network/target actor-network/optimal policy.
κ	The constant for the update of target networks.	μ_θ	Policy approximated by actor-network with θ.
$\pi_{m,t}$	Transmission rate of device m in slot t.	ρ_m	Normalized channel correlation coefficient.
σ_R^2	Variance of received signal's noise.	$\phi_{m,t}/\Phi_{m,t}$	AoI of device m in slot t on device/BS.
$\bar{\Phi}$	The average sum AoI.		

3.2. MIMO-NOMA Channel Model

Let $c_{m,t}$ be the data symbol of device m in slot t with 1 as variance; thus, the signal of the data transmitted by device m is $\sqrt{p_{m,t}}c_{m,t}$. Let $h_m(t) \in \mathbb{C}^{K \times 1}$ be the channel power gain between the BS and device m in slot t; thus, the corresponding signal received by the BS is $h_m(t)\sqrt{p_{m,t}}c_{m,t}$. Note that $c_{m,t}$ is unknown for the BS, so that it is difficult for the BS to calculate the received signal. Hence, the BS needs to adopt the SIC technology to decode the received signal transmitted by each device, which is expressed as

$$y(t) = \sum_{m \in \mathcal{M}} h_m(t)\sqrt{p_{m,t}}c_{m,t} + n(t)$$
$$p_{m,t} \in [0, P_{m,max}], \forall m \in \mathcal{M}, \forall t \in \mathcal{T}$$
(1)

where $n(t) \in \mathbb{C}^{K \times 1}$ is the complex additive white Gaussian noise (AWGN) with variance σ_R^2 and $P_{m,max}$ is the maximum transmission power of device m.

In [34,35], the authors adopted $\mathbf{h}_m(t)$ estimated by the deep neural network or minimum mean square error method in the SIC process and demonstrated its efficiency. In addition, the BS also knows $p_{m,t}$; thus, the BS can calculate the power of the received signal transmitted by device m as

$$\Gamma_{m,t} = p_{m,t}||\mathbf{h}_m(t)||^2. \tag{2}$$

Then, the BS decodes the received signal transmitted by each device sequentially. For one iteration, the BS decodes the signal with the highest received power from $\mathbf{y}(t)$ while considering the other signals as interference, and then removes the decoded signal from $\mathbf{y}(t)$ and starts the next iteration until all signals are decoded.

For instance, in an iteration, the received power of the signal transmitted by device m is the highest among the signals without being decoded. Denote $\mathcal{I}_m = \{k \in \mathcal{M} \mid \Gamma_{k,t} < \Gamma_{m,t}\}$ as the set of devices whose signals' received powers is less than device m. Thus, the signal transmitted by each device $k \in \mathcal{I}_m$ is deemed as the interference. In this case, $\mathbf{y}(t)$ is rewritten as

$$\mathbf{y}(t) = \mathbf{h}_m(t)\sqrt{p_{m,t}}c_{m,t} + \sum_{k \in \mathcal{I}_m} \mathbf{h}_k(t)\sqrt{p_{k,t}}c_{k,t} + \mathbf{n}(t), \tag{3}$$

where $\sum_{k \in \mathcal{I}_m} \mathbf{h}_k(t)\sqrt{p_{k,t}}c_{k,t}$ indicates the interference; thus, the signal-to-interference-plus-noise ratio (SINR) of device m is calculated as

$$\gamma_{m,t} = \frac{p_{m,t}||\mathbf{h}_m(t)||^2}{\sum\limits_{k \in \mathcal{I}_m} p_{k,t}||\mathbf{h}_k(t)||^2 + \sigma_R^2} = \frac{\Gamma_{m,t}}{\sum\limits_{k \in \mathcal{I}_m} \Gamma_{k,t} + \sigma_R^2}. \tag{4}$$

The transmission rate of device m in slot t can be derived according to Shannon capacity formula, i.e.,

$$\pi_{m,t} = W \log_2(1 + \gamma_{m,t}), \tag{5}$$

where W is the bandwidth of the MIMO-NOMA channel.

3.3. AoI Model

Denote $\phi_{m,t}$ as the AoI at device m in slot t, which can be calculated as

$$\phi_{m,t} = \begin{cases} 0, & s_{m,t} = 1 \\ \phi_{m,t-1} + \tau, & \text{otherwise} \end{cases}. \tag{6}$$

According to Equation (6), at the beginning of slot t, if device m samples data, i.e., $s_{m,t} = 1$, $\phi_{m,t}$ will be reset to 0. Otherwise, $\phi_{m,t}$ will be increased by τ.

Device m will transmit data with transmission power $p_{m,t}$ after sampling data. If the data volume transmitted within a slot is larger than the packet size Q, i.e., $\pi_{m,t} \cdot \tau \geq Q$, device m will transmit the data successfully; otherwise, the transmission fails. Denoting $u_{m,t} = 1$ as a successful transmission by device m in slot t and $u_{m,t} = 0$ as an unsuccessful transmission, we have

$$u_{m,t} = \begin{cases} 1, & \pi_{m,t} \cdot \tau \geq Q \\ 0, & \text{otherwise} \end{cases}. \tag{7}$$

According to [36], if a transmission from device m is successful, the AoI at the BS equals the aggregation of AoI at device m and the transmission delay. Otherwise, the AoI at the BS is increased by a slot; therefore, we have

$$\Phi_{m,t} = \begin{cases} \phi_{m,t} + l_{m,t}, & u_{m,t} = 1 \\ \Phi_{m,t-1} + \tau, & \text{otherwise} \end{cases}, \tag{8}$$

where $l_{m,t}$ is the transmission delay of device m in slot t, which is calculated as

$$l_{m,t} = \frac{Q}{\pi_{m,t}}. \qquad (9)$$

The AoI of the MIMO-NOMA IoT system is measured by averaging the AoI of all devices at the BS, i.e.,

$$\overline{\Phi} = \frac{1}{T} \sum_{t \in \mathcal{T}} \sum_{m \in \mathcal{M}} \Phi_{m,t}. \qquad (10)$$

3.4. Energy Consumption Model

Since each device consumes energy in data sampling and transmission, the energy consumption of device m in slot t can be calculated as

$$\varepsilon_{m,t} = s_{m,t} C_s + p_{m,t} l_{m,t}, \qquad (11)$$

where C_s is the energy consumption for data sampling [13], and $p_{m,t} l_{m,t}$ is the energy consumption for transmission.

The BS has a stable power supply; hence, the energy consumption of the BS is sufficient and thus it is not taken into account in the system. Hence, the energy consumption of the MIMO-NOMA IoT system is measured by averaging the energy consumption of all devices, i.e.,

$$\overline{\varepsilon} = \frac{1}{T} \sum_{t \in \mathcal{T}} \sum_{m \in \mathcal{M}} \varepsilon_{m,t}. \qquad (12)$$

3.5. Problem Formulation

In this work, our target is to minimize the AoI and energy consumption of the MIMO-NOMA IoT system, which is impacted by $p_{m,t}$ and $s_{m,t}$. Therefore, the optimization problem is formulated as

$$\min_{s_t, p_t} \left[\gamma_a \overline{\Phi} + \gamma_e \overline{\varepsilon} \right] \qquad (13)$$

$$\text{s.t.} \quad p_{m,t} \in [0, P_{m,max}], \forall m \in \mathcal{M}, \forall t \in \mathcal{T}, \qquad (13a)$$

$$s_{m,t} \in \{0, 1\}, \forall m \in \mathcal{M}, \forall t \in \mathcal{T}, \qquad (13b)$$

where $s_t = \{s_{1,t}, \ldots, s_{m,t}, \ldots, s_{M,t}\}$ and $p_t = \{p_{1,t}, \ldots, p_{m,t}, \ldots, p_{M,t}\}$, γ_a and γ_e are the non-negative weighted factors. Next, we will present a solution to the problem based on DRL.

4. DRL Method for Optimization of Power Allocation

In this section, we solve the optimization problem based on the DRL. First, we design the DRL framework including the state, action and reward function, wherein the relationship between the sample collection commands and transmission power is derived to facilitate the DRL algorithm in solving the problem. Then, we obtain a near-optimal power allocation based on the DRL algorithm.

4.1. DRL Framework

The DRL framework consists of three significant elements: state, action and reward function. For each slot, the agent observes the current state and takes the current action according to policy μ, where policy μ yields the action based on the state. Then, the agent calculates its corresponding reward under the current state and action according to the reward function, while the current state in the environment transits to the next state. Next, we will design agent, state action and reward function based on DRL [37], respectively.

- **Agent:** In each slot, the BS determines the transmission power and sample collection commands of each device based on its observation; thus, we consider the BS as the agent.
- **State:** In the system model, the state o_t observed by the BS in slot t is defined as

$$o_t = [o_{1,t}, \ldots, o_{m,t}, \ldots, o_{M,t}], \quad (14)$$

where $o_{m,t}$ represents the observation of device m, which is designed as

$$o_{m,t} = [u_{m,t-1}, \gamma_{m,t-1}, \Phi_{m,t-1}]. \quad (15)$$

Here, $u_{m,t-1}, \gamma_{m,t-1}$ and $\Phi_{m,t-1}$ can be calculated by the BS from the historical data in slot $t-1$.

- **Action:** According to the problem formulated in Equation (13), the action in slot t is set as

$$a_t = [s_t, p_t]. \quad (16)$$

The two traditional DRL algorithms, namely DDPG and Deep Q-Learning (DQN), are suitable for continuous and discrete action space, respectively. However, $s_{m,t} \in \{0,1\}$ and $p_{m,t} \in [0, P_{m,max}]$ in Equation (16); thus, the space of s_t is discrete while the space of p_t is continuous. Hence, the optimization problem can neither be solved by DQN nor DDPG. Next, we will investigate the relationship between $p_{m,t}$ and $s_{m,t}$ to handle this dilemma.

Substituting Equations (10) and (12) by Equation (13), the optimization objective is rewritten as Equation (17a). Then, substituting Equations (8) and (11) by Equation (17a), we can obtain Equation (17b), where $\phi_{m,t}$ is denoted as $\phi_{m,t}(s_{m,t})$ to indicate that it is the function of $s_{m,t}$. Then, by reorganizing Equation (17b), we have Equation (17c). The first term of Equation (17c) is related with $s_{m,t}$; next, we rewrite the first term of Equation (17c) as Equation (18) to investigate the relationship between $s_{m,t}$ and $p_{m,t}$. Substituting Equation (6) by Equation (18), we have Equation (18a). Then, by merging the homogeneous terms containing $s_{m,t}$ and γ_a in Equation (18a), respectively, we have Equation (18b). Let $C_{m,t,1} = \gamma_e C_s - \gamma_a u_{m,t}(\phi_{m,t-1} + \tau)$ and $C_{m,t,2} = \gamma_a[u_{m,t}(\phi_{m,t-1} + \tau) + (1 - u_{m,t})(\Phi_{m,t-1} + \tau) + u_{m,t} l_{m,t}] + \gamma_e p_{m,t} l_{m,t}$; thus, Equation (18b) is rewritten as Equation (18c), where $C_{m,t,1}$ is the coefficient for homogeneous terms containing $s_{m,t}$ in Equation (18b), and $C_{m,t,2}$ contains all terms without $s_{m,t}$ in Equation (18b).

$$\gamma_a \overline{\Phi} + \gamma_e \overline{\varepsilon} \quad (17)$$

$$= \frac{1}{T} \sum_{t \in \mathcal{T}} \sum_{m \in \mathcal{M}} \left[\gamma_a \Phi_{m,t} + \gamma_e \varepsilon_{m,t} \right] \quad (17a)$$

$$= \frac{1}{T} \sum_{t \in \mathcal{T}} \sum_{m \in \mathcal{M}} \left[\gamma_a [(1 - u_{m,t})(\Phi_{m,t-1} + \tau) + u_{m,t}(\phi_{m,t}(s_{m,t}) + l_{m,t})] + \gamma_e(s_{m,t} C_s + p_{m,t} l_{m,t}) \right] \quad (17b)$$

$$= \frac{1}{T} \sum_{t \in \mathcal{T}} \sum_{m \in \mathcal{M}} \left[[\gamma_a u_{m,t} \phi_{m,t}(s_{m,t}) + \gamma_e s_{m,t} C_s] + \gamma_a[(1 - u_{m,t})(\Phi_{m,t-1} + \tau) + u_{m,t} l_{m,t}] + \gamma_e p_{m,t} l_{m,t} \right] \quad (17c)$$

$$\gamma_a \Phi_{m,t}(s_{m,t}, p_{m,t}) + \gamma_e \varepsilon_{m,t}(s_{m,t}, p_{m,t}) \quad (18)$$
$$= \gamma_a u_{m,t}(1 - s_{m,t})(\phi_{m,t-1} + \tau) + \gamma_e s_{m,t} C_s + \gamma_a[(1 - u_{m,t})(\Phi_{m,t-1} + \tau) + u_{m,t} l_{m,t}] + \gamma_e p_{m,t} l_{m,t} \quad (18a)$$
$$= s_{m,t}[\gamma_e C_s - \gamma_a u_{m,t}(\phi_{m,t-1} + \tau)] + \gamma_a[u_{m,t}(\phi_{m,t-1} + \tau) + (1 - u_{m,t})(\Phi_{m,t-1} + \tau) + u_{m,t} l_{m,t}] + \gamma_e p_{m,t} l_{m,t} \quad (18b)$$
$$= s_{m,t} C_{m,t,1} + C_{m,t,2} \quad (18c)$$

In $C_{m,t,1}$ and $C_{m,t,2}$, $\phi_{m,t-1}$ can be calculated by the BS based on the historical data in slot $t-1$ [13] and $\Phi_{m,t-1}$ is known for the BS. In addition, the BS can calculate $\gamma_{m,t}$ according to Equations (4) and (5); thus, $u_{m,t}$ and $l_{m,t}$ can be further calculated according to Equations (7) and (9) given $p_{m,t}$, which means that $C_{m,t,1}$ and $C_{m,t,2}$ depend on $p_{m,t}$ and are independent of $s_{m,t}$. Hence, the optimal sample collection commands to minimize $s_{m,t}C_{m,t,1} + C_{m,t,2}$, denoted as $s^*_{m,t}$, are achieved when the term $s_{m,t}C_{m,t,1}$ is at its minimum; thus, we have

$$s^*_{m,t} = \begin{cases} 1, & C_{m,t,1} < 0 \\ 0, & otherwise \end{cases}. \quad (19)$$

Hence, the optimal sample collection commands can be determined according to Equation (19) when $p_{m,t}$ is given and Equation (13) can be rewritten as

$$\min_{p_t} [\gamma_a \overline{\Phi} + \gamma_e \overline{\varepsilon}] \quad (20)$$

$$s.t. \quad p_{m,t} \in [0, P_{m,max}], \forall m \in \mathcal{M}, \forall t \in \mathcal{T}, \quad (20a)$$

$$s^*_{m,t} = \begin{cases} 1, & C_{m,t,1} < 0 \\ 0, & otherwise \end{cases}. \quad (20b)$$

According to Equation (20), the action a_t is only reflected by p_t. Therefore, DDPG, which is suitable for the continuous action space, can be employed as the desired algorithm to solve the optimization problem in Equation (20).

- **Reward function:** The BS aims to minimize the AoI and energy consumption of the MIMO-NOMA IoT system, and the target of the DDPG algorithm is to maximize the reward function. Therefore, the reward function in slot t can be defined as

$$r_t(o_t, p_t) = - \sum_{m \in \mathcal{M}} [\gamma_a \Phi_{m,t} + \gamma_e \varepsilon_{m,t}]. \quad (21)$$

Furthermore, the expected long-term discounted reward of the system can be defined as

$$J(\mu) = \mathbb{E}\left[\sum_{t=1}^{T} \beta^{t-1} r_t(o_t, p_t) |_{p_t = \mu(o_t)} \right], \quad (22)$$

where $\beta \in [0,1]$ is the discounting factor, $p_t = \mu(o_t)$ indicates the action under the state o_t, which is derived through policy μ. Thus, our objective in this paper becomes finding the optimal policy to minimize $J(\mu)$.

4.2. Optimizing Power Allocation Based on DDPG

In this subsection, we will introduce the architecture of the DDPG algorithm including primary networks (an actor network and a critic network) and target networks (a target actor network and a target critic network) [38], wherein the actor network is adopted for policy approximation and improvement, the critic network is adopted for policy evaluation and the target networks are adopted to improve the stability of the algorithm. Both primary and target networks are neural networks (DNNs). The flow diagram is shown in Figure 2. Denote θ, ζ, θ' and ζ' as parameters of the actor network, critic network, target actor network and target critic network, respectively, μ_θ as the policy approximated by actor network and Δ_t as the noise added upon action for the exploration in slot t. Next, we will present the training stage of the DDPG algorithm in detail.

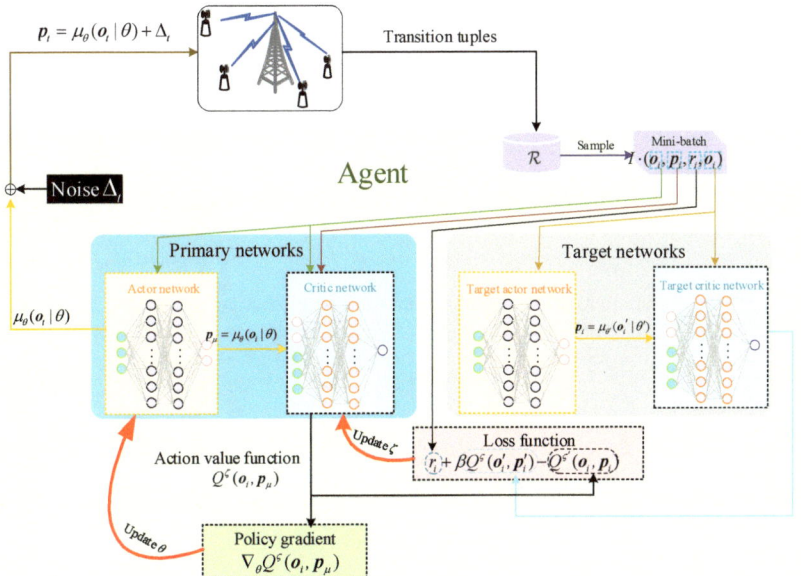

Figure 2. Flow diagram of DDPG.

The parameters θ and ζ are first initialized randomly, θ' and ζ' are set as θ and ζ, respectively. In addition, a replay experience buffer $\mathcal{R}$ is built to cache the state transitions (lines 1–3).

Next, the algorithm loops for E episodes. At the beginning of each episode, the simulation parameters of the system model are reset as $u_{m,0} = 0$, $p_{m,0} = 1$ and $\Phi_{m,0} = 0$ for each device m, $h_m(0)$ is initialized randomly. Given $p_{m,0}$ and $h_m(0)$, the SINR $\gamma_{m,0}$ is calculated according to Equations (2)–(4); then, the state of each device m, i.e., $o_{m,1} = [u_{m,0}, \gamma_{m,0}, \Phi_{m,0}]$ is observed by the agent (lines 4–6).

Afterwards, the algorithm iterates from slot 1 to T. For slot t, the actor network yields the output $\mu_\theta(o_t|\theta)$ under the observed state o_t and policy μ with parameters θ. Then, a noise Δ_t is generated and the agent calculates the transmission powers of all devices according to $p_t = \mu_\theta(o_t|\theta) + \Delta_t$. After that, the agent calculates the $u_{m,t}$, $s_{m,t}$ and $\gamma_{m,t}$ of each device m according to Equations (4), (7) and (19), respectively. Afterwards, the agent calculates $\Phi_{m,t}$ and $\varepsilon_{m,t}$ according to Equations (8) and (11), respectively, and thus obtains the state of slot t, i.e., o_{t+1}, and then calculates r_t according to Equation (21). The above tuple $[o_t, p_t, r_t, o_{t+1}]$ in the replay buffer. Then, the agent inputs o_{t+1} into the actor network and starts the next iteration if the number of samples in the replay buffer is not larger than I (lines 7–10).

If the number of tuples in the replay buffer exceeds I, the parameters θ, θ', ζ and ζ' will be updated to maximize $J(\mu_\theta)$. Here, θ is updated toward the direction of the gradient $\nabla_\theta J(\mu_\theta)$. Specifically, the agent uniformly retrieves a mini-batch consisting of I tuples from the replay buffer. For each tuple i, i.e., (o_i, p_i, r_i, o'_i) ($i \in \{1, 2, \ldots, I\}$), the agent inputs o'_i into the target actor network and outputs $p'_i = \mu_{\theta'}(o'_i|\theta')$, inputs o'_i and p'_i into the target critic network and outputs $Q^{\zeta'}(o'_i, p'_i)$ and then calculates the target value as

$$y_i = r_i + \beta Q^{\zeta'}(o'_i, p'_i)|_{p'_i = \mu_{\theta'}(o'_i|\theta')}. \tag{23}$$

While o_i and p_i are the input and $Q^\zeta(o_i, p_i)$ is the output of the critic network, the loss function can be expressed as

$$L(\zeta) = \frac{1}{I} \sum_{i=1}^{I} \left[y_i - Q^\zeta(o_i, p_i) \right]^2. \tag{24}$$

Then, the critic network is updated by the gradient descending method with the gradient of loss function $\nabla_\zeta L(\zeta)$ [39] (lines 11–13), i.e.,

$$\zeta \leftarrow \zeta - \alpha_c \nabla_\zeta L(\zeta), \tag{25}$$

where α_c is the learning rate of the critic network.

After that, the agent calculates the gradient $\nabla_\theta J(\mu_\theta)$ as [40]

$$\begin{aligned}
&\nabla_\theta J(\mu_\theta) \\
&\approx \frac{1}{I} \sum_{i=1}^{I} \nabla_\theta Q^\zeta(o_i, p_\mu)|_{p_\mu = \mu_\theta(o_i|\theta)} \\
&= \frac{1}{I} \sum_{i=1}^{I} \nabla_\theta \mu_\theta(o_i|\theta) \cdot \nabla_{p_\mu} Q^\zeta(o_i, p_\mu)|_{p_\mu = \mu_\theta(o_i|\theta)}
\end{aligned} \tag{26}$$

where the chain rule is applied to derive the gradient of $Q^\zeta(o_i, p_\mu)$ with respect to θ [40]. Given $\nabla_\theta J(\mu_\theta)$, the actor network can be updated by gradient ascending to maximize $J(\mu_\theta)$, i.e.,

$$\theta \leftarrow \theta + \alpha_a \nabla_\theta J(\mu_\theta), \tag{27}$$

where α_a is the learning rate of the actor network.

After the parameters of the primary networks are updated, the parameters of the target networks are updated based on the parameters of primary networks, i.e.,

$$\begin{aligned}
\zeta' &\leftarrow \kappa \zeta + (1-\kappa)\zeta' \\
\theta' &\leftarrow \kappa \theta + (1-\kappa)\theta'
\end{aligned}, \tag{28}$$

where κ is a constant much smaller than 1, i.e., $\kappa \ll 1$ (line 15).

Up to now, the iteration for slot t is finished and the agent starts the next iteration until the number of slots reaches T. Then, the agent starts the next episode. When the number of episodes reaches E, the training stage is finished and outputs the near-optimal policy. The pseudo-code of the training stage is described in Algorithm 1.

Next, the testing stage is initialized to test the performance under the near-optimal policy. Compared with the training stage, the parameter-updating process is omitted in the testing process and actions in each slot are generated by the near-optimal policy. The corresponding pseudo-code is shown in Algorithm 2, where θ^* is the parameter to achieve the near-optimal policy in the training stage.

Algorithm 1: Training stage of the DDPG algorithm

Input: $\gamma, \tau, \theta, \zeta$
Output: optimized DNNs
Randomly initialize the θ, ζ;
Initialize target networks by $\zeta' \leftarrow \zeta, \theta' \leftarrow \theta$;
Initialize replay experience buffer $\mathcal{R}$;
for *episode from* 1 *to* E **do**
    Reset simulation parameters for the system model;
    Receive initial observation state o_1;
    for *slot t from* 1 *to* T **do**
        Generate the transmission power of all devices according to the current policy, state and exploration noise $p_t = \mu_\theta(o_t|\theta) + \Delta_t$;
        Execute action p_t, observe reward r_t and new state o_{t+1} from the system model;
        Store transition tuple (o_t, p_t, r_t, o_{t+1}) in $\mathcal{R}$;
        if *number of tuples in $\mathcal{R}$ is larger than I* **then**
            Randomly sample a mini-batch of I transitions tuples from $\mathcal{R}$;
            Update the critic network by minimizing the loss function according to Equation (25);
            Update the actor network according to Equation (27);
            Update target networks according to Equations (28).
        end
    end
end

Algorithm 2: Testing stage of the DDPG algorithm

for *episode from* 1 *to* E **do**
    Reset simulation parameters for the system model;
    Receive initial observation state o_1;
    for *slot t from* 1 *to* T **do**
        Generate the transmission power of all devices according to the near-optimal policy and current state $p_t = \mu_\theta(o_t|\theta^{m*})$;
        Execute the action p_t;
        Observe reward the r_t and new state o_{t+1}.
    end
end

4.3. Complexity Investigation

In this subsection, we investigate the complexity of the proposed algorithm. Denote G_P and U_p as the computational complexity for computing gradients and updating parameters of the primary networks, respectively. Since the architecture of the target networks is the same as that for the primary networks, the computational complexity for updating the parameters of the target networks is also the same as the one for the primary networks. The complexity of the proposed algorithm is related to the number of slots in the training process. To be specific, during each slot, the primary networks calculate the gradients and updating parameters, while the target networks update the parameters with the parameters of the primary networks according to Equation (28). Moreover, we denote the complexity of calculating sample decisions based on power allocation as S_d. Thus, the complexity of the proposed algorithm in a slot is $O(G_P + 2U_P + S_d)$. Note that the gradients calculation and parameters updating will be processed until the number of tuples cached in the replay buffer exceeds I. The proposed algorithm will loop for E episodes, each of which contains T slots. Thus, the complexity of the proposed algorithm can be expressed as $O((E \cdot T - I)(G_P + 2U_P + S_d))$.

5. Simulation Results and Analysis

In this section, we provide simulation results to verify the effectiveness of the proposed power allocation strategy. The scenario is described in the system model. The experiments were conducted during the training and testing phases. The simulation tool was Python 3.6. In the simulation, both the actor network and critic network are the four-layer fully connected DNN with two hidden layers which are equipped with 400 and 300 neurons, respectively. The Adam optimization method [41] is adopted to update the parameters of the critic network and actor network. The noise Δ_t (for exploration) follows the Ornstein–Uhlenbeck (OU) process with decay-rate 0.15 and variation 0.004, respectively [42]. The small-scale fading of each device is initialized by white Gaussian noise, and the Rayleigh block fading model is employed to simulate the stochastic small-scale fading [43]. The reference channel gain of each device is -30 dB when the communications distance is 1 m, the path-loss exponent is 2 and the communication distances is randomly set within a range of $[50, 100]$ meters. The parameters of the measurement setup and DDPG algorithm are set according to [13] and [39], respectively, which are shown in Table 3.

Table 3. Values of the parameters in the experiments.

Parameters of System Model [13]			
Parameter	Value	Parameter	Value
τ	0.1 s	K	4
W	18 kHz	C_s	0.5 J
$P_{m,max}$	2 W	T	500
Parameters of Agent [39]			
Parameter	Value	Parameter	Value
κ	0.001	I	64
E	800	β	0.99
$|\mathcal{R}|$	2.5×10^5	γ_e	0.5
γ_a	0.5	α_a	10^{-3}
α_c	10^{-4}	F_c/F_m	0.8/0.5
B	10	N_{GA}	50

5.1. Training Stage

Figure 3 shows the learning curves in the training stage, i.e., rewards in different episodes, for different numbers of IoT devices. It can be seen that the rewards of different curves rise and fluctuate from episode 0 to 150, which reflects that the agent is learning the policy to maximize the average reward. After that, the learning curves turn out to be stable, which indicates that the near-optimal policy has been learned by the agent. Note that there is a litter jitters after episode 150, which is due to the fact that the agent is adjusting slightly since the exploration noise prevents the agent from converging into the local optima. It can also be seen that the large number of devices incurs a low reward. This is attributed to the fact that each device will be affected by more interference as the number of devices in the system increases, which leads to the lower transmission rate. This will prolong the transmission delay and further increase the AoI of the system. Then, the BS may inform the devices to consume more energy to sample more frequently and transmit faster; thus, the lower AoI can be guaranteed.

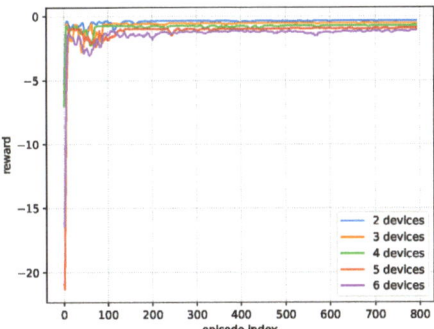

Figure 3. Learning curves under various number of devices.

5.2. Testing Stage

In the testing stage, we verify the performance of the near-optimal policy obtained in the training stage. Existing works have adopted GA [32] and random power allocation policy [33] as the baseline algorithm for power allocation; therefore, we selected these two algorithms for comparison. Here, random power allocation policy and GA are introduced as follows:

- Random policy: Randomly allocate the power of each device m within $[0, P_{m,max}]$ and the sample collection commands is obtained according to Equation (19).
- GA-based power allocation: In each time slot, the BS randomly generates a population vector according to $P_{m,max}$ and a population size B. Each individual element in the population vector stands for the power allocation for all devices. The BS selects the best individuals in the population vector as offspring according to their fitness, i.e., the reward function, of each individual. Then, after evolving for N_{GA} times, for each evolution, the probabilities of crossover and mutation for these offspring become F_c and F_m, respectively, where crossover means that two individuals in the offspring exchange the power allocation of a random device, and mutation means that the power allocation of any device in the offspring is selected within $[0, P_{m,max}]$ randomly. After that, selecting best individuals from the offspring that has experienced crossover and mutation as the input for the next evolution. After all the evolutions, the best individual from the last offspring, which is the near-optimal power allocation derived by GA, is elected. After that, the BS calculates the optimal sample collection based on the near-optimal power allocation derived by GA according to Equation (19), and then executes the near-optimal power allocation derived by GA and the optimal sample collection. In the end, the BS iterates into the next time slot.

Figure 4 presents the AoI of the near-optimal policy derived by DDPG, the random policy and the near-optimal power allocation derived by GA. It can be seen that the AoI of the three policies increases as the number of devices increases. This is because each device will suffer from the interference as the number of devices increases, and thus degrades its transmission delay according to Equation (9), which may further increase the AoI of the system. Meanwhile, the near-optimal policy derived by DDPG and the near-optimal power allocation derived by GA always outperform the random policy, because the near-optimal policy derived by DDPG can adjust the power allocation adaptively according to the observed state, and the near-optimal power allocation derived by GA will find the optimal power allocation according to the fitness in the evolutions to ensure a low AoI, while the random policy just generates power allocation randomly. It also can be seen that the near-optimal policy derived by DDPG outperforms the near-optimal power allocation derived by GA, because the DDPG will consider the influence of the power allocation in each time slot on the subsequent AoI, while GA cannot.

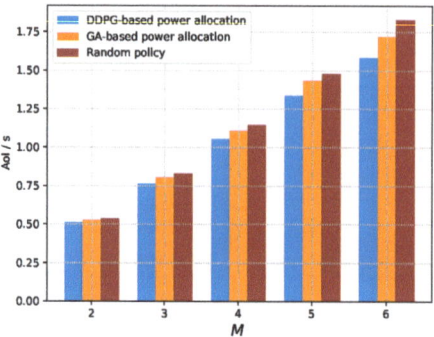

Figure 4. AoI of the system vs. number of devices.

Figure 5 compares the energy consumption of three policies. It can be seen that energy consumption increases as the number of devices increases. This is due to the fact that, according to Equation (4), the increasing number of vehicles increases the interference power, leading to a decrease in SINR. According to Equation (9), the AoI of the system increases as the SINR decreases. Hence, the devices may consume more energy for more frequent sampling and faster transmission to reduce AoI. Moreover, the increasing number of devices contributes to the increasing energy consumption according to Equation (12). Meanwhile, the near-optimal policy derived by DDPG and the near-optimal power allocation derived by GA always outperform the random policy, because DDPG and GA can allocate power adaptively to ensure a low-energy consumption. Moreover, it also can be seen that the near-optimal policy derived by DDPG always outperforms the near-optimal power allocation derived by GA, which is due to the fact that the GA cannot take into account the influence of the power allocation in each time slot on the subsequent energy consumption.

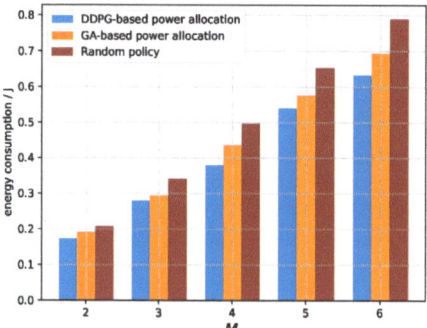

Figure 5. Energy consumption of the system vs. number of devices.

Figure 6 compares the average reward under the three policies, where the reward is obtained by averaging the test results over all slots. We can see that the average reward decreases as the number of devices increases. This is due to the fact that the reward function consists of the AoI and energy consumption of the system according to Equation (21), and both of them increase as the number of devices increases. Moreover, the average reward under the near-optimal policy derived by DDPG and the near-optimal power allocation derived by GA are higher than that of the random policy. This is attributed to the fact that the near-optimal policy allocates power according to the observed state to maximize the long-term discounted reward, and the GA obtains the near-optimal power allocation by maximizing the reward. It can also be seen that the near-optimal policy obtained by the DDPG-based method always outperforms the near-optimal power allocation derived by the

GA. This is due to the fact that the GA aims to find the near-optimal power allocation based on fitness, i.e., the reward in each slot, while ignoring the long-term reward maximization.

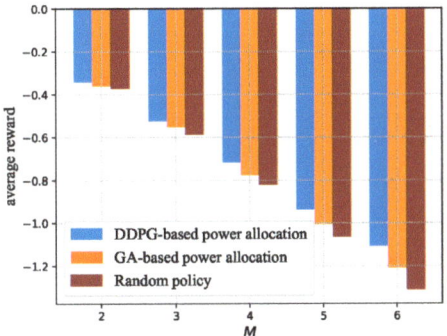

Figure 6. Average reward vs. number of devices.

Figure 7 shows the relationship between the AoI of the system and packet size, i.e., Q, under three policies. It can be seen that the AoI increases as the packet size increases under the three policies. This is due to the fact that, according to Equation (9), the packet size influences the transmission delay. That is, the transmission delay is long when the packet size is large. With regards to Equation (8), the AoI is affected by transmission delay, wherein a smaller transmission delay results in a smaller AoI. In addition, we can see that the AoI of the near-optimal policy obtained by DDPG and the near-optimal power allocation derived by the GA are lower than the AoI under the random policy. This is because the near-optimal policy derived by DDPG can adjust the power allocation based on the observed state, and the GA obtain near-optimal power allocation according to fitness, which can significantly reduce the AoI of the system. The gap between the near-optimal power allocation derived by DDPG and the near-optimal power allocation derived by the GA is caused by the advantage of long-term minimization for DDPG.

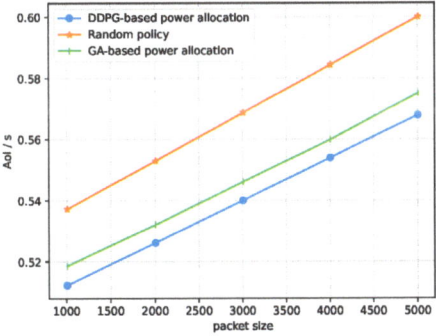

Figure 7. AoI of the system vs. packet size.

Figure 8 shows the relationship between the energy consumption of the system and packet size under three policies. It can be seen that the energy consumption of all three policies increases when the packet size increases. As shown in Figure 7, the transmission delay is long when the packet size is large, thus incurring the increase in energy consumption of the system. We can also see that the energy consumption of the near-optimal policy derived by DDPG and the near-optimal power allocation derived by the GA are lower than that of the random policy. This is due to the fact that the near-optimal policy derived by DDPG can adaptively allocate power and the GA can obtain the near-optimal power allocation according to fitness to ensure a lower energy consumption. However, the near-optimal policy derived by DDPG accounts for the influence of power allocation on the energy consumption of later time slots; thus, the near-optimal policy obtained by DDPG has a lower energy consumption than the near-optimal power allocation derived by the GA.

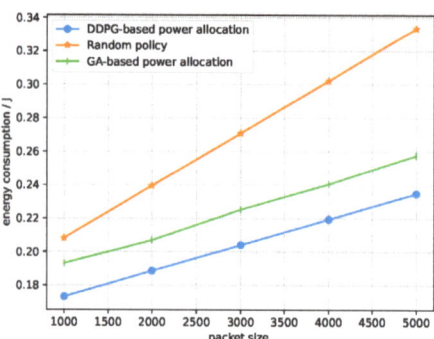

Figure 8. Energy consumption vs. packet size.

6. Conclusions

In this paper, we formulated a problem to minimize the AoI and energy consumption of the MIMO-NOMA IoT system. To solve it, we simplified the formulated problem and proposed the power allocation scheme based on DDPG to maximize the long-term discounted reward. Extensive simulations have demonstrated that the proposed scheme reduces the reward by 6.44% compared to the GA, and by 11.78% compared to the random policy, respectively. According to the theoretical analysis and simulation results, the key findings and contributions of this paper can be summarized as follows: (1) An increase in the number of devices and packet size will increase the AoI of the system. In this case, agents can inform the devices to consume more energy to sample more frequently and transmit faster, thereby reducing the AoI and increasing the energy consumption. (2) The near-optimal policy trained by DDPG outperforms the baseline policy under different numbers of users and packet sizes, which has a good capability to suit the system dynamic variation. We also noted some limitations and future directions for further research in this study: DDPG may face challenges when addressing high-dimensional state and action spaces. In future work, we will consider decomposing the problem into multiple subtasks for independent learning or improving the function approximators to enhance its robustness. In addition, as mentioned in [44], fairness is also a relatively important factor in the NOMA system. Therefore, our future research will focus on achieving a fair resource allocation in MIMO-NOMA systems and evaluating its impact on other performances.

Author Contributions: Conceptualization, Z.Z., H.Z. and Q.W.; Methodology, Z.Z., H.Z. and Q.W.; Software, Z.Z. and H.Z.; Writing—Original Draft Preparation, Z.Z.; Writing—Review and Editing, Q.F., P.F., H.Z. and J.W. All authors have read and agreed to the published version of the manuscript.

Funding: This work was supported in part by the National Natural Science Foundation of China under Grant No. 61701197, in part by the open research fund of State Key Laboratory of Integrated Services Networks under Grant No. ISN23-11, in part by the National Key Research and Development Program of China under Grant No. 2021YFA1000500(4), in part by the 111 project under Grant No. B23008.

Data Availability Statement: Data are contained within the article.

Conflicts of Interest: Author Qiang Fan was employed by the company Qualcomm. The remaining authors declare that the research was conducted in the absence of any commercial or financial relationships that could be construed as a potential conflict of interest.

References

1. Wu, Q.; Wang, X.; Fan, Q.; Fan, P.; Zhang, C.; Li, Z. High stable and accurate vehicle selection scheme based on federated edge learning in vehicular networks. *China Commun.* **2023**, *20*, 1–17. [CrossRef]
2. Wu, Q.; Wang, S.; Ge, H.; Fan, P.; Fan, Q.; Letaief, K.B. Delay-Sensitive Task Offloading in Vehicular Fog Computing-Assisted Platoons. *IEEE Trans. Netw. Serv. Manag.* **2023**, early access. [CrossRef]
3. Wu, Q.; Zhao, Y.; Fan, Q.; Fan, P.; Wang, J.; Zhang, C. Mobility-Aware Cooperative Caching in Vehicular Edge Computing Based on Asynchronous Federated and Deep Reinforcement Learning. *IEEE J. Sel. Top. Signal Process.* **2023**, *17*, 66–81. [CrossRef]
4. Gao, Y.; Xia, B.; Xiao, K.; Chen, Z.; Li, X.; Zhang, S. Theoretical Analysis of the Dynamic Decode Ordering SIC Receiver for Uplink NOMA Systems. *IEEE Commun. Lett.* **2017**, *21*, 2246–2249. [CrossRef]
5. Wu, Q.; Shi, S.; Wan, Z.; Fan, Q.; Fan, P.; Zhang, C. Towards V2I Age-aware Fairness Access: A DQN Based Intelligent Vehicular Node Training and Test Method. *Chin. J. Electron.* **2022**, *32*, 1–15. [CrossRef]
6. Bo, Z.; Saad, W. Joint Status Sampling and Updating for Minimizing Age of Information in the Internet of Things. *IEEE Trans. Commun.* **2019**, *67*, 7468–7482.
7. Zhu, H.; Wu, Q.; Wu, X.J.; Fan, Q.; Fan, P.; Wang, J. Decentralized Power Allocation for MIMO-NOMA Vehicular Edge Computing Based on Deep Reinforcement Learning. *IEEE Internet Things J.* **2022**, *9*, 12770–12782. [CrossRef]
8. Long, D.; Wu, Q.; Fan, Q.; Fan, P.; Li, Z.; Fan, J. A Power Allocation Scheme for MIMO-NOMA and D2D Vehicular Edge Computing Based on Decentralized DRL. *Sensors* **2023**, *23*, 3449. [CrossRef] [PubMed]
9. Volodymyr, M.; Koray, K.; David, S.; Rusu, A.A.; Joel, V.; Bellemare, M.G.; Alex, G.; Martin, R.; Fidjeland, A.K.; Ostrovski, G. Human-level control through deep reinforcement learning. *Nature* **2019**, *518*, 529–533.
10. Zhao, T.; He, L.; Huang, X.; Li, F. DRL-Based Secure Video Offloading in MEC-Enabled IoT Networks. *IEEE Internet Things J.* **2022**, *9*, 18710–18724. [CrossRef]
11. Chen, Y.; Sun, Y.; Yang, B.; Taleb, T. Joint Caching and Computing Service Placement for Edge-Enabled IoT Based on Deep Reinforcement Learning. *IEEE Internet Things J.* **2022**, *9*, 19501–19514. [CrossRef]
12. Grybosi, J.F.; Rebelatto, J.L.; Moritz, G.L. Age of Information of SIC-Aided Massive IoT Networks With Random Access. *IEEE Internet Things J.* **2022**, *9*, 662–670. [CrossRef]
13. Wang, S.; Chen, M.; Yang, Z.; Yin, C.; Saad, W.; Cui, S.; Poor, H.V. Distributed Reinforcement Learning for Age of Information Minimization in Real-Time IoT Systems. *IEEE J. Sel. Top. Signal Process.* **2022**, *16*, 501–515. [CrossRef]
14. Abd-Elmagid, M.A.; Dhillon, H.S.; Pappas, N. AoI-Optimal Joint Sampling and Updating for Wireless Powered Communication Systems. *IEEE Trans. Veh. Technol.* **2020**, *69*, 14110–14115. [CrossRef]
15. Li, C.; Huang, Y.; Li, S.; Chen, Y.; Jalaian, B.A.; Hou, Y.T.; Lou, W.; Reed, J.H.; Kompella, S. Minimizing AoI in a 5G-Based IoT Network Under Varying Channel Conditions. *IEEE Internet Things J.* **2021**, *8*, 14543–14558. [CrossRef]
16. Hatami, M.; Leinonen, M.; Codreanu, M. AoI Minimization in Status Update Control with Energy Harvesting Sensors. *IEEE Trans. Wireless Commun.* **2021**, *69*, 8335–8351. [CrossRef]
17. Sun, M.; Xu, X.; Qin, X.; Zhang, P. AoI-Energy-Aware UAV-assisted Data Collection for IoT Networks: A Deep Reinforcement Learning Method. *IEEE Internet Things J.* **2021**, *8*, 17275–17289. [CrossRef]
18. Hu, H.; Xiong, K.; Qu, G.; Ni, Q.; Fan, P.; Letaief, K.B. AoI-Minimal Trajectory Planning and Data Collection in UAV-Assisted Wireless Powered IoT Networks. *IEEE Internet Things J.* **2021**, *8*, 1211–1223. [CrossRef]
19. Emara, M.; Elsawy, H.; Bauch, G. A Spatiotemporal Model for Peak AoI in Uplink IoT Networks: Time Versus Event-Triggered Traffic. *IEEE Internet Things J.* **2020**, *7*, 6762–6777. [CrossRef]
20. Lyu, L.; Dai, Y.; Cheng, N.; Zhu, S.; Guan, X.; Lin, B.; Shen, X. AoI-Aware Co-Design of Cooperative Transmission and State Estimation for Marine IoT Systems. *IEEE Internet Things J.* **2021**, *8*, 7889–7901. [CrossRef]
21. Wang, X.; Chen, C.; He, J.; Zhu, S.; Guan, X. AoI-Aware Control and Communication Co-Design for Industrial IoT Systems. *IEEE Internet Things J.* **2021**, *8*, 8464–8473. [CrossRef]
22. Hao, X.; Yang, T.; Hu, Y.; Feng, H.; Hu, B. An Adaptive Matching Bridged Resource Allocation Over Correlated Energy Efficiency and AoI in CR-IoT System. *IEEE Trans. Green Commun. Netw.* **2022**, *6*, 583–599. [CrossRef]
23. Yilmaz, S.S.; Özbek, B.; İlgüy, M.; Okyere, B.; Musavian, L.; Gonzalez, J. User Selection for NOMA based MIMO with Physical Layer Network Coding in Internet of Things Applications. *IEEE Internet Things J.* **2021**, *9*, 14998–15006. [CrossRef]
24. Shi, Z.; Wang, H.; Fu, Y.; Yang, G.; Ma, S.; Hou, F.; Tsiftsis, T.A. Zero-Forcing-Based Downlink Virtual MIMO–NOMA Communications in IoT Networks. *IEEE Internet Things J.* **2020**, *7*, 2716–2737. [CrossRef]
25. Wang, Q.; Wu, Z. Beamforming Optimization and Power Allocation for User-Centric MIMO-NOMA IoT Networks. *IEEE Access* **2021**, *9*, 339–348. [CrossRef]
26. Han, L.; Liu, R.; Wang, Z.; Yue, X.; Thompson, J.S. Millimeter-Wave MIMO-NOMA-Based Positioning System for Internet-of-Things Applications. *IEEE Internet Things J.* **2020**, *7*, 11068–11077. [CrossRef]
27. Zhang, S.; Wang, L.; Luo, H.; Ma, X.; Zhou, S. AoI-Delay Tradeoff in Mobile Edge Caching With Freshness-Aware Content Refreshing. *IEEE Trans. Wireless Commun.* **2021**, *20*, 5329–5342. [CrossRef]
28. Chinnadurai, S.; Yoon, D. Energy Efficient MIMO-NOMA HCN with IoT for Wireless Communication Systems. In Proceedings of the 2018 International Conference on Information and Communication Technology Convergence (ICTC), Jeju, Republic of Korea, 17–19 October 2018; pp. 856–859. [CrossRef]

29. Gao, J.; Wang, X.; Shen, R.; Xu, Y. User Clustering and Power Allocation for mmWave MIMO-NOMA with IoT devices. In Proceedings of the 2021 IEEE Wireless Communications and Networking Conference (WCNC), Nanjing, China, 29 March–1 April 2021; pp. 1–6. [CrossRef]
30. Feng, W.; Zhao, N.; Ao, S.; Tang, J.; Zhang, X.; Fu, Y.; So, D.K.C.; Wong, K.K. Joint 3D Trajectory and Power Optimization for UAV-Aided mmWave MIMO-NOMA Networks. *IEEE Trans. Commun.* **2021**, *69*, 2346–2358. [CrossRef]
31. Ding, Z.; Dai, L.; Poor, H.V. MIMO-NOMA Design for Small Packet Transmission in the Internet of Things. *IEEE Access* **2016**, *4*, 1393–1405. [CrossRef]
32. Bulut, I.S.; Ilhan, H. Energy Harvesting Optimization of Uplink-NOMA System for IoT Networks Based on Channel Capacity Analysis Using the Water Cycle Algorithm. *IEEE Trans. Green Commun. Netw.* **2021**, *5*, 291–307. [CrossRef]
33. Ullah, S.A.; Zeb, S.; Mahmood, A.; Hassan, S.A.; Gidlund, M. Deep RL-assisted Energy Harvesting in CR-NOMA Communications for NextG IoT Networks. In Proceedings of the 2022 IEEE Globecom Workshops (GC Wkshps), Rio de Janeiro, Brazil, 4–8 December 2022; pp. 74–79. [CrossRef]
34. Kang, J.M.; Kim, I.M.; Chun, C.J. Deep Learning-Based MIMO-NOMA With Imperfect SIC Decoding. *IEEE Syst. J.* **2020**, *14*, 3414–3417. [CrossRef]
35. He, X.; Huang, Z.; Wang, H.; Song, R. Sum Rate Analysis for Massive MIMO-NOMA Uplink System with Group-Level Successive Interference Cancellation. *IEEE Wirel. Commun. Lett.* **2023**, *12*, 1194–1198. [CrossRef]
36. Wang, S.; Chen, M.; Saad, W.; Yin, C.; Cui, S.; Poor, H.V. Reinforcement Learning for Minimizing Age of Information under Realistic Physical Dynamics. In Proceedings of the GLOBECOM 2020—2020 IEEE Global Communications Conference, Taipei, Taiwan, 7–11 December 2020; pp. 1–6. [CrossRef]
37. Arulkumaran, K.; Deisenroth, M.P.; Brundage, M.; Bharath, A.A. Deep Reinforcement Learning: A Brief Survey. *IEEE Signal Process. Mag.* **2017**, *34*, 26–38. [CrossRef]
38. Qiao, G.; Leng, S.; Maharjan, S.; Zhang, Y.; Ansari, N. Deep Reinforcement Learning for Cooperative Content Caching in Vehicular Edge Computing and Networks. *IEEE Internet Things J.* **2020**, *7*, 247–257. [CrossRef]
39. Lillicrap, T.P.; Hunt, J.J.; Pritzel, A.; Heess, N.; Erez, T.; Tassa, Y.; Silver, D.; Wierstra, D. Continuous control with deep reinforcement learning. *arXiv* **2015**, arXiv:1509.02971.
40. Silver, D.; Lever, G.; Heess, N.; Degris, T.; Wierstra, D.; Riedmiller, M. Deterministic Policy Gradient Algorithms. In Proceedings of the 2014 International Conference on Machine Learning(ICML), Beijing, China, 21–26 June 2014; pp. 387–395.
41. Kingma, D.P.; Ba, J. ADAM: A method for stochastic optimization. *arXiv* **2015**, arXiv:1412.6980.
42. Uhlenbeck, G.E.; Ornstein, L.S. On the Theory of the Brownian Motion. *Rev. Latinoam. Microbiol.* **1973**, *15*, 29–35. [CrossRef]
43. Ngo, H.Q.; Larsson, E.G.; Marzetta, T.L. Energy and Spectral Efficiency of Very Large Multiuser MIMO Systems. *IEEE Trans. Commun.* **2013**, *61*, 1436–1449. [CrossRef]
44. Darsena, D.; Gelli, G.; Iudice, I.; Verde, F. A Hybrid NOMA-OMA Scheme for Inter-plane Intersatellite Communications in Massive LEO Constellations. *arXiv* **2023**, arXiv:2307.08340.

Disclaimer/Publisher's Note: The statements, opinions and data contained in all publications are solely those of the individual author(s) and contributor(s) and not of MDPI and/or the editor(s). MDPI and/or the editor(s) disclaim responsibility for any injury to people or property resulting from any ideas, methods, instructions or products referred to in the content.

Article

Age of Information Analysis for Multi-Priority Queue and Non-Orthoganal Multiple Access (NOMA)-Enabled Cellular Vehicle-to-Everything in Internet of Vehicles

Zheng Zhang [1], Qiong Wu [1,*], Pingyi Fan [2] and Qiang Fan [3]

1 School of Internet of Things Engineering, Jiangnan University, Wuxi 214122, China; zhengzhang@stu.jiangnan.edu.cn
2 Department of Electronic Engineering, Beijing National Research Center for Information Science and Technology, Tsinghua University, Beijing 100084, China; fpy@tsinghua.edu.cn
3 Qualcomm, San Jose, CA 95110, USA; qf9898@gmail.com
* Correspondence: qiongwu@jiangnan.edu.cn; Tel.: +86-0510-8591-0633

Abstract: With the development of Internet of Vehicles (IoV) technology, the need for real-time data processing and communication in vehicles is increasing. Traditional request-based methods face challenges in terms of latency and bandwidth limitations. Mode 4 in cellular vehicle-to-everything (C-V2X), also known as autonomous resource selection, aims to address latency and overhead issues by dynamically selecting communication resources based on real-time conditions. However, semi-persistent scheduling (SPS), which relies on distributed sensing, may lead to a high number of collisions due to the lack of centralized coordination in resource allocation. On the other hand, non-orthogonal multiple access (NOMA) can alleviate the problem of reduced packet reception probability due to collisions. Age of Information (AoI) includes the time a message spends in both local waiting and transmission processes and thus is a comprehensive metric for reliability and latency performance. To address these issues, in C-V2X, the waiting process can be extended to the queuing process, influenced by packet generation rate and resource reservation interval (RRI), while the transmission process is mainly affected by transmission delay and success rate. In fact, a smaller selection window (SW) limits the number of available resources for vehicles, resulting in higher collisions when the number of vehicles is increasing rapidly. SW is generally equal to RRI, which not only affects the AoI part in the queuing process but also the AoI part in the transmission process. Therefore, this paper proposes an AoI estimation method based on multi-priority data type queues and considers the influence of NOMA on the AoI generated in both processes in C-V2X system under different RRI conditions. Our experiments show that using multiple priority queues can reduce the AoI of urgent messages in the queue, thereby providing better service about the urgent message in the whole vehicular network. Additionally, applying NOMA can further reduce the AoI of the messages received by the vehicle.

Keywords: C-V2X; resource reservation interval; non-orthogonal multiple access; age of information

1. Introduction

With advanced IoV technology, the development of intelligent transportation systems has facilitated various applications such as automated navigation, collision warning, and multimedia entertainment [1–3]. Due to the high-speed mobility of vehicles, ensuring timely data access through user requests is crucial [4–6]. In the traditional request-based approach, users connect their base stations, via which they can access the data center to retrieve the requested data [7–12]. However, this method suffers from end-to-end delays and limited backhaul bandwidth [13–16]. To address these issues, C-V2X proposes an interface called PC5 for communication between autonomous vehicles. PC5 offers two resource

allocation methods: Mode 3, where user equipment (UE) requests time and frequency-domain transmission resources from the eNodeB, and Mode 4, where UE autonomously selects resources without involving the cellular infrastructure. Mode 4 not only eliminates the limited coverage drawback but also minimizes interaction between base stations and vehicles, thereby resolving excessive delays and overhead [17].

In Mode 4, vehicles autonomously select communication resources using the SPS protocol based on SPS, allowing vehicles to choose among RRI. However, this increases the collision probability when multiple vehicles occupy the same resources with the same message transmission interval, leading to higher block error rate (BLER) [18]. NOMA is a potential solution for C-V2X communication, promising to enhance spectrum efficiency and handle large-scale vehicle communications, mitigating latency and packet reception probability degradation caused by high vehicle density [19,20]. Successive interference cancellation (SIC) is a well-known multi-user detection technique used to extract overlapping signals, decoding different power levels of received signals occupying the same resource. Thus, the high power signals no longer interfere with other low-power signals after decoding, improving the signal-to-interference-plus-noise ratio (SINR) of low-power signals block error rate (BLER) [21–24].

Some studies have examined the effectiveness of NOMA applied in C-V2X. In [25], a NOMA receiver based on SIC and joint decoding was proposed to reduce BLER compared to traditional orthogonal multiple access (OMA) methods. In [26], TAKESHI et al. introduced SPS-NOMA based on uplink non-orthogonal multiple access to improve the SIC under broadcast scenarios and alleviate channel congestion. In [27], Utpal et al. proposed a model with a large-scale MIMO Jacobi detection algorithm for the PHY layer of C-V2X, enhancing reliability by reducing bit error rate compared to existing PHY layer frameworks. These works demonstrated improvement of NOMA in terms of reliability and transmission delay in the C-V2X system.

It is worth noting that these two metrics are often in trade-off, where an increase in reliability performance may come at the cost of increased delay. Therefore, a new metric is necessary to comprehensively reflect both reliability and latency performance, such as AoI. A lower average AoI indicates lower latency with the same reliability and higher reliability with the same delay Therefore, a new metric like AoI is necessary as it provides a comprehensive reflection of both reliability and latency performance. A lower average AoI indicates either lower latency with the same reliability or higher reliability with the same delay, thereby better measuring the timeliness of information and the overall performance of the system [28–30]. In [31], Peng et al. adopted AoI to evaluate the MAC layer performance of the C-V2X sidelink, proposing a Piggyback-based cooperative method for vehicles to inform each other of potential resource occupation, reducing collisions and exhibiting good AoI performance. In [32], Zoubeir et al. presented a resource allocation problem based on NOMA, optimizing resource allocation to provide minimum AoI and high reliability for vehicle safety information.

However, the aforementioned studies did not consider the AoI of the packet in the queue in the C-V2X system. The packet at the receiver includes the packet generation time, which should be a key factor for the queuing process. In [33], Akar et al. investigated the freshness of information in IoT-based state update systems using the AoI performance metric. They studied discrete-time servers in multi-source IoT systems, assuming Bernoulli arrivals of information packets and universally distributed discrete phase-type service times across all sources. Their analysis of AoI under various queuing disciplines was formulated in matrix-geometric terms. In [34], Zhang et al. considered a dual-server short-block wireless communication system to ensure real-time delivery of newly generated information at a relatively high update rate to its destination. Information is generated at a relatively high update rate, encoded into two short-block queues and delivered in real-time through two parallel paths. The study based on the Markov-chain process investigated the AoI performance of the dual-queue system in the presence of block delivery errors. In [35], Liu et al. propose a hybrid TDMA and NOMA protocol that takes advantage of the two protocols and a clustering-based dynamic adjustment of the shortest path algorithm to the long-term average AoI in an unmanned-aerial-vehicle-assisted wireless-powered communication network.

The AoI of the packet generated in the queue is influenced by packet generation rates and service rates, where the service rate in C-V2X depends on the RRI [36]. However, for the same RRI, while the AoI of transmitter undergoes a short queuing time, the collision probability increases, leading to a potential large AoI at the receiver [37]. This motivates us to consider this work. In this paper, we propose an AoI calculation approach based on a multi-priority data type queue in C-V2X. We first design a queue model with four types of messages of different priorities, in which the AoI of receiver better reflects the ability to observe the status of the transmitter in a timely manner. We then consider the impact of NOMA on AoI in both processes, calculate AoI for different RRIs, and analyze the effect of multi-priority queues and NOMA on the C-V2X communication system. (The source code has been released at: https://github.com/qiongwu86/Analysis-of-the-Impact-of-Multi-Priority-Queue-and-NOMA-on-Age-of-Information-in-C-V2X, accessed on 14 July 2024). The remaining parts of this article are as follows, Section 2 provides a brief introduction to the system model. Section 3 presents a description of the proposed modeling of the approach in detail. In Section 4, we present some simulation results, followed by the conclusion in Section 5.

2. System Model

We consider the system model as shown in Figure 1. In this model, a C-V2X-based communication system is with N half-duplex vehicles. In C-V2X Mode 4, all communication resources are in a resource pool, with the basic unit of the resource block(RB) with a size of 180 kHz. A sub-channel typically consists of 10 RBs. The time axis is set to discrete values, with each time slot being 1ms, noted as the duration of an RB, representing the size of one sub-frame in the resource pool. In C-V2X Mode 4, there are four priority levels of message generating types. When the multi-priority queue is not empty, vehicles reserve communication resources from the resource pool. The time interval between reserved resources is defined as the RRI, commonly taking values of 20 ms, 50 ms, or 100 ms. The initial value of the reserved resources (RC), which refers to the set of resources allocated in advance to ensure efficient operation in a network, is defined as $500/RRI + rand(1000/RRI)$. That is to say, when the RRI is 20 ms, the output range is from 25 to 75. Moreover, due to half-duplex communication, vehicles cannot receive signals while using communication resources.

We use AoI as the performance metric for this system and calculate the AoI generated in the multi-priority queue and the transmission process. Regarding the AoI in the queue, using communication resources will change the AoI and position of messages in the queue, while not using communication resources will keep them unchanged. Regarding the AoI in the transmission process, the AoI at the receiver is updated to the AoI of the received message when communication is completed. Otherwise, it will grow over time. For simplicity, the receiver determines whether it receives the signal by comparing the SINR of the received signal with the SINR threshold as the criteria of successful transmission. When C-V2X employs OMA, it will compute SINR, where interferences mainly come from the collisions of other signals. For NOMA mode, SIC (see Equation (10) for detail) is used to improve the SINR of low-power signals in collisions, where the SINR calculated for received signals are sorted in descending order of their power and only signals with lower power than the current signal are considered as interference. Additionally, in C-V2X, for the same RRI, the AoI in the queue may be very small while the receiver's AoI can be large. To better reflect the improvement in collisions by NOMA, it is necessary to separately calculate the AoI generated in the two processes and analyze the effect of NOMA on the C-V2X communication system.

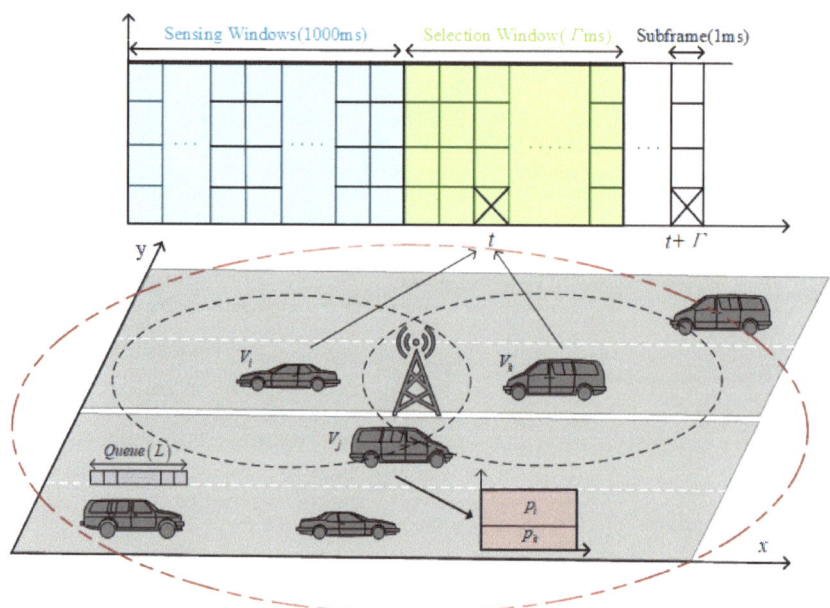

Figure 1. System model.

3. Mathematical Model

In this section, we first establish a computational model to measure the AoI for a C-V2X system with a multi-priority message queue. Subsequently, we propose to employ NOMA based on SIC to improve the AoI of the C-V2X system. Table 1 summarizes the notations of the symbols.

Table 1. The summary for notations.

Notation	Description	Notation	Description
RB	The time slot where vehicle reserves resources.	T_C	The generation interval of CAM messages.
T_D	DENM repeat transmission interval.	T_H	HPD repeat transmission interval.
K_D	Number of repeated transmissions of DENM.	K_H	Number of repeated transmissions of HPD.
$\lambda_{H,D,M}$	The new message arrival rates for the three message types: HPD, DENM, and MHD.	L	The maximum length of the message queue.
q	The queue length.	s	The transmission action.
SPS	The method for vehicles to autonomously choose communication resources.	SW	The value of the reselection counter for vehicles.
RC	The value of the reselection counter for vehicles.	RRI	Resource selection buffer time.
b	The position of the message in queue.	$\varphi_{i,n}^{t,b}$	AoI of messages at position b in queue n of vehicle i.
$\Phi_{i \to j}^t$	the AoI of the receiving end j for vehicle i.	$u_{i \to j}^t$	Indicates that i transmitted the message to j.
Q	The size of the transmitted message.	R_{th}	Transmission rate threshold.
p_i^t	The transmission power of vehicle i.	σ^2	Noise energy.
P_{col}	Resource collision probability.	π	The probability that the vehicle is at the moment when it is ready to select resources.

3.1. AoI in C-V2X

(1) C-V2X Queue Model

The C-V2X Model 4 communication system includes four types of messages: high priority data (HPD), decentralized environmental notification message (DENM), signal awareness message (CAM), and miscellaneous high-density data (MHD), with the following priority order: HPD > DENM > CAM > MHD. CAM-type messages are generated periodically, while the rest of the types are triggered. The generation probability of a new CAM packet is $1/T_c$, where T_c is the fixed packet generation period. The new packet generation probabilities for HPD, DENM, and MHD are expressed by a Poisson distribution:

$$P(\text{arr}_{i,n}^t = k) = \frac{\lambda_n^k}{k!} e^{-\lambda_n} \quad (1)$$

where $n \in \{\text{HPD}, \text{DENM}, \text{MHD}\}$, and λ_n^k represents the number of packet arrivals for each type of message in a certain time period, and k represents the number of data packets arriving, which is typically set to 1. For DENM and MHD, the new packets need to be retransmitted multiple times to ensure successful transmission, and this retransmission process is periodic. Thus, four corresponding first-in-first-out queues are established based on different signal types. Their queue capacity is L, and the queue length is q. When the queues meet $q < L$, new packets can be added and share the same transmission opportunity.

Given the transmission opportunity, vehicles can use their own reserved resources for transmission. At this point, the vehicle transmits messages from different queues based on their priority. When the q of the high-priority message queue is non-zero, the corresponding transmission action s is set to 1, while s of other queues with lower priorities are set to 0. Only when the length q of the high-priority queue is zero, can s of the second highest priority queue be set to 1. Here, $s = 1$ indicates that the messages in that queue can be transmitted. Otherwise, they cannot be transmitted. Therefore, the expression for the transmission action s of the multi-priority queues is

$$\begin{cases} s_{i,H}^t = 1, s_{i,C}^t = 0 & q_{i,H}^t = 1 \\ s_{i,H}^t = 0, s_{i,C}^t = 1 & q_{i,C}^t \left(1 - q_{i,H}^t\right) = 1 \\ s_{i,H}^t = s_{i,C}^t = 0 & \text{otherwise} \end{cases} \quad (2)$$

Since the generation methods of HDP, DENM, and MHD-type messages are the same, and CAM is different from them, the two priority queues with HDP and CAM-type messages can be used to represent the relationship between the representations of s of different queues. Let $s_{i,C}^t$ represent the transmission opportunity of the HPD queue at time t and $s_{i,H}^t$ represent the transmission opportunity of the CAM queue. When the q of all queues are zero, it indicates that no packets can be transmitted, and at this moment, all s values are set to 0.

(2) AoI Model

The cumulative AoI of messages in the queue is impacted by the queuing process, which is defined as the time from packet generation to transmission. It is worth noting that during the interval when the vehicle is using the reserved resource, the AoI will continuously increase. The AoI expression for each message in each queue is given by

$$\varphi_{i,n}^{t+1,b} = \begin{cases} \varphi_{i,n}^{t,b+1} + 1 & s_{i,n}^t = 1 \\ \varphi_{i,n}^{t,b} + 1 & s_{i,n}^t = 0 \end{cases} \quad (3)$$

where n is the index of queues, $b \in [1, q-1]$ represents the position of the message in the queue at time t for vehicle i, and $\varphi_{i,n}^{t,b}$ denotes the message AoI in queue n at position b for vehicle i at time t. When $s_{i,n}^t = 1$, the position of all messages in the queue will be updated except for the first message. When $s_{i,n}^t = 0$, all message positions in the queue remain unchanged. Here, the probability of $s_{i,n}^t = 1$ represents the processing rate, depending on the 1/RRI in C-V2X. Thus, the RRI will impact the size of the AoI. In addition, when a message in the queue is transmitted, the AoI at the last position in the queue will be reset to 0.

The cumulative AoI during the communication process can be considered as the time spent to complete the transmission between the receiver and the transmitter, which is influenced by both the transmission delay and the transmission process. Additionally, since the AoI of the received message reflects the situation at the time of packet generation by the transmitter, the receiver needs to inherit it to indicate whether it communicates with the transmitter in a timely manner. The AoI expression for receiver j regarding transmitter i is given by

$$\Phi_{i \to j}^{t+1} = \begin{cases} \varphi_i^{t,1} + 1 & u_{i \to j}^t = 1 \\ \Phi_{i \to j}^t + 1 & u_{i \to j}^t = 0 \end{cases} \quad (4)$$

where $\Phi_{i \to j}^t$ represents the AoI of vehicle j from vehicle i at time t, and $u_{i \to j}^t$ indicates whether the message is successfully transmitted to j by i. If $u_{i \to j}^t = 1$, $\Phi_{i \to j}^t$ is equal to the $\varphi_{i \to j}^{t,1}$ of the highest-priority message in the queue transmitted by vehicle i, plus the transmission delay. If $u_{i \to j}^t = 0$, $\Phi_{i \to j}^{t+1}$ is equal to $\Phi_{i \to j}^t$ plus the subframe size. According to Equations (3) and (4), it can be observed that transmission failure will cause a greater increase in $\Phi_{i \to j}^{t+1}$. In C-V2X model 4, each transmission failure requires waiting for an RRI interval, owing to the reserved resource. The main reason for transmission failure is a low SINR caused by collisions, which can be addressed by NOMA.

3.2. NOMA in C-V2X

(1) Collisions in C-V2X

In C-V2X model 4, vehicles autonomously allocate resources and transmit data in a broadcast manner. When multiple vehicles reserve resources in the neighboring time slots, they may select the same resources. Moreover, due to the broadcast nature, multiple vehicles can simultaneously communicate with the same receiver. This situation leads to collisions when multiple vehicles use the same resources to communicate with the same receiver. According to [38], the non collision probability in C-V2X model 4 can be expressed as

$$P_{ncol} \approx \left[1 - \left[1 - \prod_{i=0}^{\Gamma-1} \left(1 - \frac{\pi}{1 - \pi i} \right) \right] \frac{1 - P_{rk}}{CSR - N_v + 1} \right]^{N_v - 1} \quad (5)$$

where π represents the probability that a vehicle is preparing to select a resource, where it needs to satisfy the three following conditions: the vehicle queue is not empty, the RC is zero, and a new resource is being rescheduled. P_{rk} represents the probability of a vehicle selecting a new resource when the reselection counter resets to zero. CSR represents the total number of resources in the selection window. N_v represents the total number of vehicles. Γ is the size of SW, which refers to the range of available time from which vehicles can select for transmission, , so CSR is consistent with Γ. However, the range of Γ is small, typically being 20, 50, or 100. Thus, when a specific value of Γ is chosen, $\left[1 - \prod_{i=0}^{\Gamma-1} \left(1 - \frac{\pi}{1 - \pi i} \right) \right] \frac{1 - P_{rk}}{CSR - N_v + 1}$ is a functions of the number of vehicles N_v. Thus, as N_v increases, P_{col} also increases. Furthermore, changing the value of Γ for data transmission may reduce the collision probability, but it can also increase the waiting time of messages in the queue, leading to a larger AoI. Therefore, considering the impact of collisions, we also explore the use of NOMA to mitigate this issue while changing Γ.

(2) SINR Calculation

In C-V2X model 4, because the system uses a dynamic resource allocation approach based on the network conditions and the traffic load, the size of transmission resources is not fixed. Vehicles calculate the required bandwidth B and the transmission rate threshold R_{th} for successful transmission within one subframe based on the size of the message Q, as follows:

$$R_{\text{th}} = B_i \log_2(1 + \text{SINR}_{\text{th}}) \tag{6}$$

where $R_{\text{th}} = Q$ because the time taken to complete data transmission for a message must be less than the length of one subframe, i.e., $Q/R \leq 1$. Next, perform a transformation on Equation (6); the SINR threshold SINR_{th} is calculated based on the bandwidth and the transmission rate threshold, as shown in the following expression:

$$\text{SINR}_{\text{th}} = 2^{Q_i/B_i} - 1 \tag{7}$$

Due to the time-varying distances between vehicles, the corresponding channel condition leads to varying communication rates at the receiver. Additionally, since multiple vehicles may use different communication resources within the same subframe, the receiver also receives signals at different rates within different communication resources. If the receiver receives only one signal within a resource, then the SINR between vehicle i and vehicle j at time t is given by

$$\text{SINR}_{i \to j}^{t,n} = \frac{p_{i \to j}^t \left| h_{i \to j}^{t,n} \right|^2}{\sigma^2} \tag{8}$$

where $p_{i \to j}^t$ is the transmission power of vehicle i, $\left| h_{i \to j}^{t,n} \right|^2$ is the channel gain between vehicles i and j for resource n, and σ^2 is the noise power. However, due to the half-duplex resource selection scheme in C-V2X, vehicles may select the same resource block when choosing resources, leading to collisions when they use the same resource for transmission. In this case, the wireless information transmission rate is defined as

$$\text{SINR}_{i \to r}^{t,n} = \frac{p_{i \to j}^t \left| h_{i \to j}^{t,n} \right|^2}{\sum_{m \in N_m} p_{m \to j}^t \left| h_{m \to j}^{t,n} \right|^2 + \sigma^2} \tag{9}$$

where m represents interfering vehicles in the C-V2X scenario, and their transmission energy is denoted as P_m. It can be observed that when multiple vehicles use the same resource, the transmission rate decreases, and if the transmission rate is too low, it may result in transmission failure.

Therefore, to address this issue, we introduce NOMA based on SIC. The receiver sorts the received signals in descending order of received power, considering the maximum received power signal as the target signal and the rest as interference signals. After decoding and removing the target signal, this process is repeated for all signals to compute their SINR. Let $N_k = \{k \in N \setminus i | p_{i \to j}^t \left| h_{i \to j}^{t,n} \right|^2 > p_{k \to j}^t \left| h_{k \to j}^{t,n} \right|^2 \}$ represent the set of vehicles who has weaker received power than vehicle i. Then, the SINR for vehicle i is given by

$$\text{SINR}_{i \to r}^{t,n} = \frac{p_{i \to j}^t \left| h_{i \to j}^{t,n} \right|^2}{\sum_{k \in N_k} p_{k \to j}^t \left| h_{k \to j}^{t,n} \right|^2 + \sigma^2} \tag{10}$$

It can be seen that using NOMA in the case of a collision can reduce the interference at the target signal and increase SINR, thereby reducing the possibility of SINR to lower than the threshold. It is generally believed that $\left| h_{i \to j}^{t,n} \right|^2 = c_{ij} d_{ij}$, c_{ij} represents the coefficient between vehicle i and vehicle j, d_{ij} denotes the distance between vehicle i and vehicle j, and η is the path loss exponent. Here, the value of d_{ij} depends on the positions of vehicles i and j at time t.

(3) Vehicle Mobility

Consider a two-way road with a length of D and $U/2$ lanes in each direction. We establish a coordinate system where the position of vehicle i at time t is defined as (x_i^t, y_i^u). The origin of the coordinate system is set as the leftmost position of the road, with the x-axis representing the direction of vehicle movement, and the y-axis representing different lanes. Assuming that the vehicle updates its position periodically after a short time, the speed V of vehicle i can be considered as constant:

$$x_i^{t+1} = x_i^t + \delta V \tau, x_i^t \in [0, D] \tag{11}$$

where δ represents the direction of vehicle, and $x_i^0 \in [0, D]$ is the initial position of the vehicle. Furthermore, y_i^g depends on the lane index $g \in [1, \ldots, G]$ of vehicle i and is calculated as

$$y_i^g = g d_y - y_0 \tag{12}$$

where d_y is the width of one lane, and y_0 represents the distance from vehicle i to the edge of the lane. Typically, y_0 is taken as $1/2 d_y$, which is half of the lane width.

4. Numerical Results and Discussion

We have conducted the simulations to demonstrate our work based on the simulation tool, i.e., MATLAB R2021b [39]. In this section, we present the simulation results by comparing AoI in C-V2X under NOMA and OMA, validating the impact of NOMA on AoI in C-V2X.

4.1. Simulation Settings

According to the C-V2X standard, we use a 10 MHz bandwidth with a total of 50 RBs. The message size is set to 500 Bytes, and QPSK modulation is applied for propagation. The T_C has a granularity of 100 ms between 100 ms and 1 s. The arrival rates λ_H, λ_D, and λ_M for different message types, as shown in Equation (1), are set to 0.0001. The T_H and T_D are set to 100 ms and 500 ms, respectively. The K_H and K_D are set to 8 and 5, respectively. The transmission power of all vehicles is uniformly set to 23 dBm, and the speed is set to 120 km/h.

4.2. Performance Evaluation

The average AoI of packets in the queue can be expressed as a function of $\varphi_{i,n}^{t,b}$, as follows:

$$\overline{\varphi} = \frac{1}{N_v} \frac{1}{N} \frac{1}{b} \sum_{i=1}^{N_v} \sum_{n \in N} \sum_{b}^{L} \varphi_{i,n}^{t,b} \tag{13}$$

where Nv represents the total number of vehicles in the scene, where N = 4 indicates four types of messages. The Δ^t is defined as the average of $\Phi_{i \to j}^t$ between each pair of vehicles.

$$\Delta^t = \frac{1}{N_v} \frac{1}{N_v - 1} \sum_{j=1}^{N_v} \sum_{i=1}^{N_v-1} \Phi_{i \to j}^t \tag{14}$$

where $\Phi_{i \to j}^t$ represents the AoI between vehicles i and j at time t, taking into consideration the impact of half-duplex communication. Thus, the summation involves calculating the mean AoI between different vehicle pairs i and j. In general, a higher Δ^t indicates a higher number of transmission failures.

Table 2 presents the communication success rate for different vehicle counts under number of vehicles and Γ, which represents the proportion of successfully received messages compared to all received messages. The vehicle density is set at 60 and 100 vehicles/km, and the length of load is set as 500 m. So it can be observed that the success rate is generally higher for 30 vehicles compared to 50 vehicles. This is because an increase in the number of vehicles enhances the probability of reserving resources at the same time, deteriorating resource contention. Furthermore, the success rate at $\Gamma = 20$ ms is consistently lowest. This is because a smaller CSR will result in a higher probability of each resource being reserved by multiple vehicles. Moreover, it is worth noting that for a given number of vehicles, regardless of the value of Γ, employing NOMA technology consistently yields better transmission success rates compared to C-V2X with OMA.

Table 2. Communication success rate in C-V2X.

N_v		30			50	
RRI	20	50	100	20	50	100
OMA	0.82891	0.83738	0.91560	0.75367	0.80050	0.85184
NOMA	0.89488	0.93332	0.97274	0.87902	0.92636	0.95356

Figure 2 shows the AoI variation trend in various messages when the number of vehicles is set to 50 and Γ is set to 100 ms. The upper part of the graph shows the AoI change in information when a vehicle uses a single queue, while the lower part shows the situation when a vehicle uses a multi-priority queue. At this point, the messages in the queue will pile up, leading to an increasing AoI. It can be seen when there is only one queue, their overall trend in AoI change is similar. In a multi-priority queue, although the probability of generating MHD messages is the lowest, its AoI is the highest due to its lowest priority. So in multi-priority queues, the AoI of high-priority messages decreases to a certain extent, ensuring the timeliness of high-priority messages.

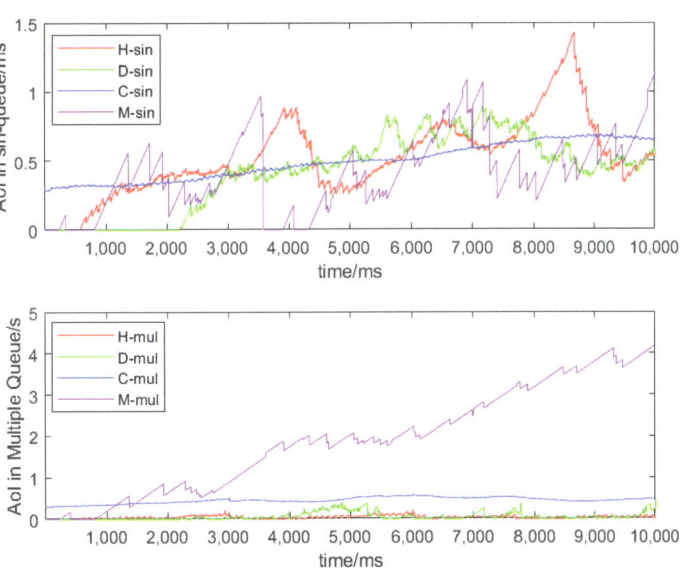

Figure 2. AvgAoI in different queues.

Figure 3 presents the $\overline{\varphi}$ in multi-priority queues with different values of Γ when the number of vehicles is 30 and 50, respectively. When Γ is 20 ms, the inter-arrival interval of various types of packets is close to 100 ms (much larger than Γ), so the queue remains empty for most of the time; the $\varphi_{i,n}^{t,b}$ hardly increases. Similarly, when Γ is 50 ms, the $\overline{\varphi}$ is also relatively small. However, when Γ is 100 ms, which is greater than the packet inter-arrival interval, packets experience more time while waiting in the queue. In addition, it can be observed that when there are more vehicles, the value of $\overline{\varphi}$ increases to a certain extent compared to that of fewer vehicles.

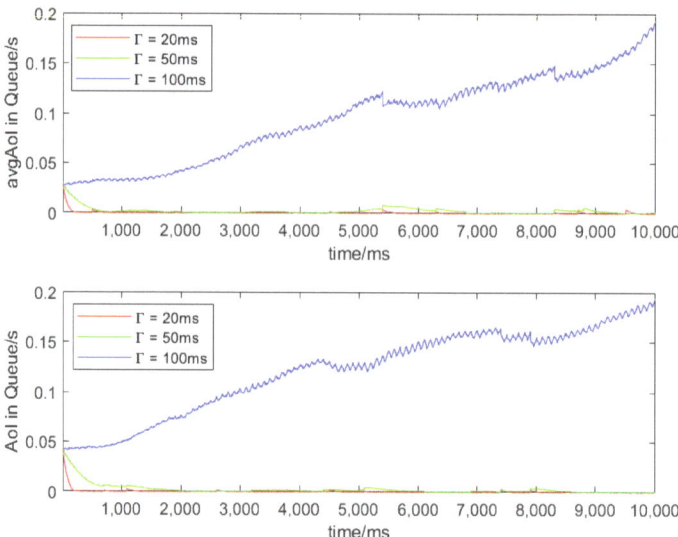

Figure 3. AvgAoI in queues.

Figure 4 shows the Δ^t of 30 vehicles with respect to different Γ. It can be observed that for different Γ, NOMA consistently decreases Δ^t. As shown in Table 1, when Γ is 20 ms, the collision probability is relatively high, resulting in the shorter staying time of messages in the queue, but the AoI of the receiving end is not becoming smaller. By contrast, when Γ is 100 ms, the collision probability is lower. However, due to the accumulation of $\varphi_{i,n}^{t,b}$, it continues to rise and becomes the oldest AoI in the three situations. When Γ is 50 ms, the collision rate is the lowest among the three considered cases, and $\varphi_{i,n}^{t,b}$ does not accumulate, resulting in the lowest Δ^t throughout the entire observation period.

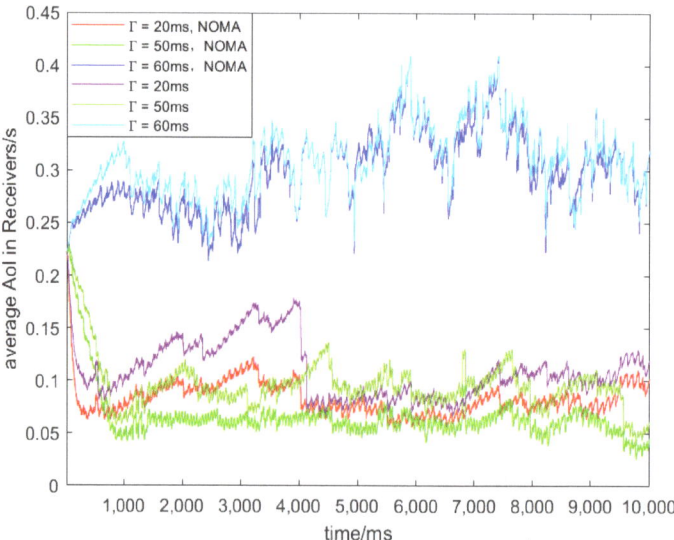

Figure 4. $N_v = 30$.

Figure 5 depicts the Δ^t for 100 vehicles under different Γ. A comparison between Figures 4 and 5 reveals that an increase in the number of vehicles leads to a general increasing trend for all cases, which is attributed to the rising collision probability with an increasing number of vehicles. Additionally, the trends in these two cases are similar to those shown in Figure 4, with the highest AoI at Γ = 100 ms and the lowest AoI at Γ = 50 ms. At 20 ms, communication suffers more interference, leading to higher AoI. At 100 ms, message accumulation in the queue causes delays and increased overload. The 50 ms interval selection can better balance these effects, resulting in a lower AoI. NOMA has positive impacts on AoI for different Γ.

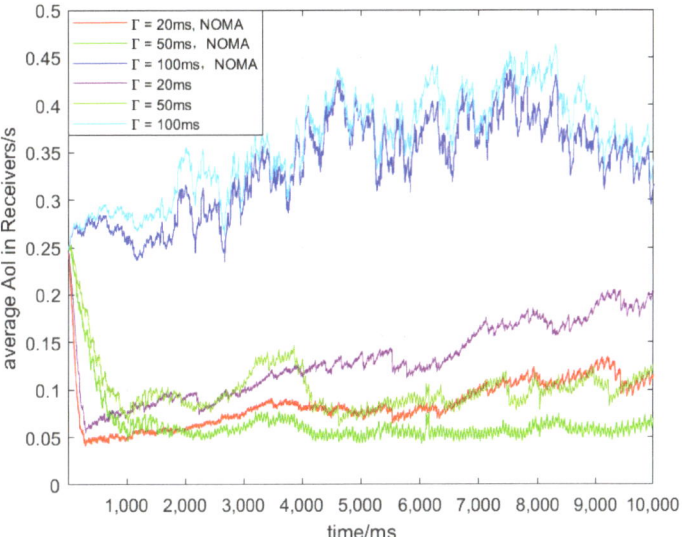

Figure 5. N_v = 50.

5. Conclusions

This paper considers vehicle mobility and analyzes the impact of NOMA on the AoI in different scenarios. Firstly, we propose a novel multi-priority queue AoI calculation model for C-V2X communication. Then, we investigate the improvement in transmission success rate using NOMA based on SIC to observe changes in AoI. The following conclusions are drawn from our analysis:

- Under the same Γ, the collision probability varies with different vehicle counts. Adjusting Γ can reduce collision occurrences, but it must be carefully selected to avoid increased AoI.
- NOMA enhances SINR and reduces the impact of collisions, leading to decreased AoI in various scenarios in the C-V2X system.

Author Contributions: Conceptualization, Z.Z. and Q.W.; methodology, Z.Z. and Q.W.; software, Z.Z.; writing—original draft preparation, Z.Z.; writing—review and editing, P.F. and Q.F. All authors have read and agreed to the published version of the manuscript.

Funding: This work was supported in part by the National Natural Science Foundation of China under Grant No. 61701197, in part by the National Key Research and Development Program of China under Grant No. 2021YFA1000500(4), in part by the 111 project under Grant No. B23008.

Institutional Review Board Statement: Not applicable.

Informed Consent Statement: Not applicable.

Data Availability Statement: The data are contained within the article.

Conflicts of Interest: Author Qiang Fan was employed by the company Qualcomm. The remaining authors declare that the research was conducted in the absence of any commercial or financial relationships that could be construed as a potential conflict of interest.

References

1. Fan, J.; Yin, S.; Wu, Q.; Gao, F. Study on Refined Deployment of Wireless Mesh Sensor Network. In Proceedings of the IEEE International Conference on Wireless Communications, Networking and Mobile Computing (WICOM'10), Chengdu, China, 23–25 September 2010; pp. 370–375.
2. Zhang, C.; Xu, X.; Wu, Q.; Fan, P.; Fan, Q.; Zhu, H.; Wang, J. Anti-Byzantine Attacks Enabled Vehicle Selection for Asynchronous Federated Learning in Vehicular Edge Computing. *China Commun.* **2024**, *73*, 1–17.
3. Ji, M.; Wu, Q.; Cheng, N.; Chen, W.; Wang, J.; Letaief, K.B. Graph Neural Networks and Deep Reinforcement Learning Based Resource Allocation for V2X Communications. *IEEE Internet Things J.* **2024**. [CrossRef]
4. Qi, K.; Wu, Q.; Fan, P.; Cheng, N.; Chen, W.; Letaief, K.B. Reconfigurable Intelligent Surface Aided Vehicular Edge Computing: Joint Phase-Shift Optimization and Multi-User Power Allocation. *IEEE Internet Things J.* **2024**. [CrossRef]
5. Liu, L.; Chen, C.; Pei, Q.; Maharjan, S.; Zhang, Y. Vehicular Edge Computing and Networking: A Survey. *Mob. Networks Appl.* **2021**, *26*, 1145–1168. [CrossRef]
6. Wu, Q.; Wang, W.; Fan, P.; Fan, Q.; Zhu, H.; Letaief, K.B. Cooperative Edge Caching Based on Elastic Federated and Multi-Agent Deep Reinforcement Learning in Next-Generation Networks. *IEEE Trans. Netw. Serv. Manag.* **2024**, *21*, 4179–4196. [CrossRef]
7. Wu, Q.; Wang, X.; Fan, Q.; Fan, P.; Zhang, C.; Li, Z. High Stable and Accurate Vehicle Selection Scheme based on Federated Edge Learning in Vehicular Networks. *China Commun.* **2023**, *20*, 1–17. [CrossRef]
8. Shao, Z.; Wu, Q.; Fan, P.; Cheng, N.; Fan, Q.; Wang, J. Semantic-Aware Resource Allocation Based on Deep Reinforcement Learning for 5G-V2X HetNets. *IEEE Commun. Lett.* **2024**, *28*, 2452–2456. [CrossRef]
9. Qi, K.; Wu, Q.; Fan, P.; Cheng, N.; Chen, W.; Wang, J.; Letaief, K.B. Deep-Reinforcement-Learning-Based AoI-Aware Resource Allocation for RIS-Aided IoV Networks. *IEEE Trans. Veh. Technol.* **2024**. [CrossRef]
10. Zhang, C.; Zhang, W.; Wu, Q.; Fan, P.; Fan, Q.; Wang, J.; Letaief, K.B. Distributed Deep Reinforcement Learning Based Gradient Quantization for Federated Learning Enabled Vehicle Edge Computing. *IEEE Internet Things J.* **2024**. [CrossRef]
11. Gu, X.; Wu, Q.; Fan, P.; Fan, Q.; Cheng, N.; Chen, W.; Letaief, K.B. DRL-Based Resource Allocation for Motion Blur Resistant Federated Self-Supervised Learning in IoV. *IEEE Internet Things J.* **2024**. [CrossRef]
12. Wan, S.; Lu, J.; Fan, P.; Shao, Y.; Peng, C.; Letaief, K.B. Convergence Analysis and System Design for Federated Learning Over Wireless Networks. *IEEE J. Sel. Areas Commun.* **2021**, *39*, 3622–3639. [CrossRef]
13. Dai, Y.; Xu, D.; Maharjan, S.; Qiao, G.; Zhang, Y. Artificial Intelligence Empowered Edge Computing and Caching for Internet of Vehicles. *IEEE Wireless Commun.* **2019**, *26*, 12–18. [CrossRef]
14. Wang, K.; Yu, F.R.; Wang, L.; Li, J.; Zhao, N.; Guan, Q.; Li, B.; Wu, Q. Interference Alignment with Adaptive Power Allocation in Full-Duplex-Enabled Small Cell Networks. *IEEE Trans. Veh. Technol.* **2019**, *68*, 3010–3015. [CrossRef]
15. Wu, Q.; Wang, S.; Ge, H.; Fan, P.; Fan, Q.; Letaief, K.B. Delay-sensitive Task Offloading in Vehicular Fog Computing-Assisted Platoons. *IEEE Trans. Netw. Serv. Manag.* **2024**, *21*, 2012–2026. [CrossRef]
16. Fan, P.; Feng, C.; Wang, Y.; Ge, N. Investigation of the time-offset-based QoS support with optical burst switching in WDM networks. In Proceedings of the 2002 IEEE International Conference on Communications. Conference Proceedings, ICC 2002 (Cat. No.02CH37333), New York, NY, USA, 28 April–2 May 2002; Volume 5, pp. 2682–2686. [CrossRef]
17. Molina-Masegosa, R.; Gozalvez, J. LTE-V for sidelink 5G V2X vehicular communications: A new 5G technology for short-range vehicle to-everything communications. *IEEE Veh. Technol. Mag.* **2017**, *12*, 30–39. [CrossRef]
18. 3G Partnership Project. *Evolved Universal Terrestrial Radio Access (E-UTRA); Medium Access Control (MAC) Protocol Specification (v14.3.0, Release 14)*; Technical Report. 36.321; 3G Partnership Project (3GPP); Sophia-Antipolis: Valbonne, France, 2017.
19. Di, B.; Song, L.; Li, Y.; Li, G.Y. Non-orthogonal multiple access for high-reliable and low-latency V2X communications in 5G systems. *IEEE J. Sel. Areas Commun.* **2017**, *35*, 2383–2397. [CrossRef]
20. Chen, X.; Lu, J.; Fan, P.; Letaief, K.B. Massive MIMO Beamforming With Transmit Diversity for High Mobility Wireless Communications. *IEEE Access* **2017**, *5*, 23032–23045. [CrossRef]
21. Zhang, Y.; Peng, K.; Chen, Z.; Song, J. SIC vs. JD: Uplink NOMA techniques for M2M random access. In Proceedings of the 2017 IEEE International Conference on Communications (ICC), Paris, France, 21–25 May 2017; pp. 1–6.
22. Liu, J.; Xiong, K.; Ng, D.W.K.; Fan, P.; Zhong, Z.; Letaief, K.B. Max-Min Energy Balance in Wireless-Powered Hierarchical Fog-Cloud Computing Networks. *IEEE Trans. Wirel. Commun.* **2020**, *19*, 7064–7080. [CrossRef]
23. Jiang, R.; Xiong, K.; Fan, P.; Zhang, Y.; Zhong, Z. Power Minimization in SWIPT Networks With Coexisting Power-Splitting and Time-Switching Users Under Nonlinear EH Model. *IEEE Internet Things J.* **2019**, *6*, 8853–8869. [CrossRef]
24. Guo, Y.; Xiong, K.; Lu, Y.; Wang, D.; Fan, P.; Letaief, K.B. Achievable Information Rate in Hybrid VLC-RF Networks With Lighting Energy Harvesting. *IEEE Trans. Commun.* **2021**, *69*, 6852–6864. [CrossRef]
25. Situ, Z.; Ho, I.W.-H.; Hou, Y.; Li, P. The Feasibility of NOMA in C-V2X. In Proceedings of the IEEE INFOCOM 2020—IEEE Conference on Computer Communications Workshops (INFOCOM WKSHPS), Toronto, ON, Canada, 6–9 July 2020; pp. 562–567. [CrossRef]
26. Hirai, T.; Murase, T. Performance Evaluation of NOMA for Sidelink Cellular-V2X Mode 4 in Driver Assistance System With Crash Warning. *IEEE Access* **2020**, *8*, 168321–168332. [CrossRef]

27. Dey, U.K.; Akl, R.; Chataut, R. Performance Improvement in Cellular V2X (C-V2X) by Using Massive MIMO Jacobi Detector. In Proceedings of the 2022 IEEE 19th International Conference on Smart Communities: Improving Quality of Life Using ICT, IoT and AI (HONET), Marietta, GA, USA, 19–21 December 2022; pp. 122–127. [CrossRef]
28. Kaul, S.; Yates, R.; Gruteser, M. Real-time status: How often should one update? *Proc. IEEE Infocom. Mar.* **2012**, 2731–2735.
29. Ge, Y.; Xiong, K.; Wang, Q.; Ni, Q.; Fan, P.; Letaief, K.B. AoI-minimal Power Adjustment in RF-EH-powered Industrial IoT Networks: A Soft Actor-Critic-Based Method. *IEEE Trans. Mob. Comput.*. [CrossRef]
30. Zheng, H.; Xiong, K.; Fan, P.; Zhong, Z.; Letaief, K.B. Age of Information-Based Wireless Powered Communication Networks With Selfish Charging Nodes. *IEEE J. Sel. Areas Commun.* **2021**, *39*, 1393–1411. [CrossRef]
31. Peng, F.; Jiang, Z.; Zhang, S.; Xu, S. Age of Information Optimized MAC in V2X Sidelink via Piggyback-Based Collaboration. *IEEE Trans. Wirel. Commun.* **2021**, *20*, 607–622. [CrossRef]
32. Mlika, Z.; Cherkaoui, S. Deep Deterministic Policy Gradient to Minimize the Age of Information in Cellular V2X Communications. *IEEE Trans. Intell. Transp. Syst.* **2022**, *23*, 23597–23612. [CrossRef]
33. Akar, N.; Dogan, O. Discrete-Time Queueing Model of Age of Information With Multiple Information Sources. *IEEE Internet Things J.* **2021**, *8*, 14531–14542. [CrossRef]
34. Zhang, Z.; Zhu, X.; Jiang, Y.; Cao, J.; Liu, Y. Closed-Form AoI Analysis for Dual-Queue Short-Block Transmission with Block Error. In Proceedings of the 2021 IEEE Wireless Communications and Networking Conference (WCNC), Nanjing, China, 29 March–1 April 2021; pp. 1–6. [CrossRef]
35. Liu, X.; Liu, H.; Zheng, K.; Liu, J.; Taleb, T.; Shiratori, N. AoI-minimal Clustering, Transmission and Trajectory Co-design for UAV-assisted WPCNs. *IEEE Trans. Veh. Technol.* **2024**, 1–16. [CrossRef]
36. Wijesiri N.B.A., G.P.; Haapola, J.; Samarasinghe, T. A Markov Perspective on C-V2X Mode 4. In Proceedings of the 2019 IEEE 90th Vehicular Technology Conference (VTC2019-Fall), Honolulu, HI, USA, 22–25 September 2019; pp. 1–6. [CrossRef]
37. Gonzalez-Martín, M.; Sepulcre, M.; Molina-Masegosa, R.; Gozalvez, J. Analytical Models of the Performance of C-V2X Mode 4 Vehicular Communications. *IEEE Trans. Veh. Technol.* **2019**, *68*, 1155–1166. [CrossRef]
38. Haapola, G.P.W.N.B.A.J.; Samarasinghe, T. A Discrete-Time Markov Chain Based Comparison of the MAC Layer Performance of C-V2X Mode 4 and IEEE 802.11p. *IEEE Trans. Commun.* **2021**, *69*, 2505–2517. [CrossRef]
39. Cecchini, G.; Bazzi, A.; Masini, B.M.; Zanella, A. LTEV2Vsim: An LTE-V2V simulator for the investigation of resource allocation for cooperative awareness. In Proceedings of the 2017 5th IEEE International Conference on Models and Technologies for Intelligent Transportation Systems (MT-ITS), Naples, Italy, 26–28 June 2017; pp. 80–85. [CrossRef]

Disclaimer/Publisher's Note: The statements, opinions and data contained in all publications are solely those of the individual author(s) and contributor(s) and not of MDPI and/or the editor(s). MDPI and/or the editor(s) disclaim responsibility for any injury to people or property resulting from any ideas, methods, instructions or products referred to in the content.

Article

A Novel Framed Slotted Aloha Medium Access Control Protocol Based on Capture Effect in Vehicular Ad Hoc Networks

Lianyou Lai, Zhongzhe Song and Weijian Xu *

School of Ocean Information Engineering, Jimei University, Xiamen 361000, China; kaikaixinxinlly@jmu.edu.cn (L.L.); 202111810010@jmu.edu.cn (Z.S.)
* Correspondence: xwjxwj@jmu.edu.cn; Tel.: +86-135-1596-6506

Abstract: The capture effect is a frequently observed phenomenon in vehicular ad hoc networks (VANETs) communication. When conflicts arise during time slot access, failure to access does not necessarily occur; instead, successful access may still be achieved. The capture effect can enhance the likelihood of multiple access and improve communication efficiency. The security of VANETs communication is undoubtedly the primary concern. One crucial approach to enhance security involves the design of an efficient and reliable medium access control (MAC) protocol. Taking into account both aspects, we propose a novel framed slotted Aloha (FSA) MAC protocol model. Firstly, we derive the closed-form expression for the capture probability in the Rician fading channel in this paper. Subsequently, we analyze how the number of vehicles and time slots influence the success probability of vehicle access channels as well as examine the impact of the capture effect on this success probability. Then, under constraints regarding vehicle access channel success probability, we derive optimal values for slot numbers, access times, and transmission power while proposing a comprehensive implementation method to ensure high access channel success probabilities. We verify both theoretical derivations and proposed methods through simulation experiments.

Keywords: FSA; slotted Aloha; capture effect; Rician; vehicular ad hoc networks (VANETs); medium access control (MAC)

Citation: Lai, L.; Song, Z.; Xu, W. A Novel Framed Slotted Aloha Medium Access Control Protocol Based on Capture Effect in Vehicular Ad Hoc Networks. *Sensors* **2024**, *24*, 992. https://doi.org/10.3390/s24030992

Academic Editors: Pingyi Fan and Qiong Wu

Received: 11 December 2023
Revised: 31 January 2024
Accepted: 1 February 2024
Published: 3 February 2024

Copyright: © 2024 by the authors. Licensee MDPI, Basel, Switzerland. This article is an open access article distributed under the terms and conditions of the Creative Commons Attribution (CC BY) license (https://creativecommons.org/licenses/by/4.0/).

1. Introduction

With the rapid development of information and communication technology, VANETs have been widely applied in the field of intelligent transportation. The vehicle network protocol is also designed with a layered approach, and the MAC protocol of the vehicle network protocol is particularly important due to security considerations. The Aloha protocol, a multiple-access random access algorithm, plays a crucial role in wireless communication networks [1]. It encompasses the pure-Aloha protocol, slotted Aloha (SA) protocol, and frame slotted Aloha (FSA) protocol [2–4]. The FSA is further categorized into basic frame slotted Aloha (BFSA) with fixed frame length and dynamic frame slotted Aloha (DFSA) with variable frame length [5,6].

The performance of various FSA protocols is significantly affected by the capture effect, which arises from the adoption of an access slot contention mechanism [7]. In the FSA protocol, a frame consists of N slots, and each user randomly selects one slot within the frame for transmitting data packets. Slot selection follows a method of random selection and free competition. If only one user selects a slot, the access is successful. However, if two or more users select the same slot simultaneously, a collision conflict occurs. In case of collision, access may either fail or succeed depending on the power levels involved. When competing users have similar power levels, access fails; however, when one user's power level significantly exceeds that of the others in competition, it successfully accesses while others fail to do so. This phenomenon is known as the capture effect [8,9]. The prudent utilization of the capture effect can thus augment the probability of user access.

In recent years, the research on FSA protocol has garnered significant attention [10,11]. The utilization of FSA has been widespread in various domains including wireless sensor networks [12], vehicular ad hoc networks [13,14], and radio frequency identification networks [15]. Meanwhile, an increasing number of end-users are adopting the FSA protocol for communication, leading to intensified competition for access time slots. The phenomenon of capture within the FSA protocol has emerged as a pivotal consideration.

In this paper, the capture probability of the Rician fading channel in FSA-based networks is analyzed, and a novel FSA MAC protocol model based on the capture effect is proposed. Additionally, the access probability is considered as an essential constraint. The key contributions of this study are outlined below.

- The closed-form expression for the acquisition probability is derived in the context of a Rician fading channel.
- The paper presents an FSA MAC protocol model based on the capture effect and conducts an analysis of the impact of vehicle quantity and slot allocation on the success rate of vehicle access channels.
- The impact of the capture effect on the success rate of the vehicle access channel is analyzed, and the numerical results validate the accuracy of the theoretical analysis.
- The relationship between the number of vehicles and time slot allocation under maximum conflict probability is derived, providing a foundation for the rational distribution of time slots.
- Under the constraint of a vehicle access channel success probability, this paper derives the requirements for slot number, access times, and transmission power. Taking into account comprehensive factors, an implementation method is proposed to comprehensively enhance the success rate of vehicle access channels from these three aspects.

The remaining sections of this article are organized as follows: Section 2 presents an overview of the research work related to this paper. In Section 3, we derive a closed-form expression for the capture probability in a Rician fading channel. Section 4 proposes an FSA MAC protocol model based on the capture effect and analyzes the impact of the number of vehicles and slots on the success rate of vehicle access channels. We present numerical results in Section 5. Finally, we summarize this article in Section 6.

2. Related Work

The medium access control (MAC) protocol plays a pivotal role in wireless networks, exerting a direct impact on their performance. Numerous approaches have been proposed from diverse perspectives.

The paper proposed an FSA protocol with variable frame length in ref. [4], taking into account the channel capture effect. By deriving the expression of the optimal frame length from maximum channel utilization, this study focuses on effectively utilizing the channel. In [16], a novel approach is presented for estimating the population of active nodes using an additive scheme, and this estimate is utilized to optimize throughput by determining frame size and participation probability. In [7], a comprehensive analysis of the correlation between the normalized offered load and the attainable performance is provided in terms of packet loss rate. In [17], the authors focus on the case of messaging from mobile vessels to stationary control centers on land through advanced channel random access schemes that exploit time diversity. This paper extends an existing theoretical framework to evaluate the probability of packet loss in FSA systems by examining both the probability distribution of colliding users within a frame and their replicas in time slots. In [18], the authors investigate the congestion behavior of a IEEE802.11p/1609.4-based MAC protocol by varying the vehicle density under urban and highway conditions. In ref. [19], an adaptive and on-demand TDMA-based MAC protocol is proposed to address the challenges of high-transmission collisions and channel resource wastage in the unbalanced traffic conditions of VANETs. In ref. [20], an analytical model based on the Markov chain model is presented to evaluate the performance of IEEE 802.11 MAC for VANETs in the presence of the capture effect. The study investigates various parameters that can potentially impact the performance and establishes their relationship with performance metrics, including the probability of successful transmission, probability of frame capture, and throughput expressions. In ref. [21], the proposed novel scheme can be seamlessly integrated with IEEE 1609.4 to operate on top of IEEE 802.11p infrastructure. The core concept in the proposed solution revolves around distributing access time initiation across the control channel (CCH) period. This technique, an enhanced iteration of the S-Aloha protocol, effectively mitigates collision risks. The MAC protocol based on framed slotted Aloha for these networks is proposed in ref. [22]. The paper investigates the probable packet sizes, energy consumption, battery lifetime, and success rate of our protocol. In ref. [23], the proposed Coordinated multichannel MAC (C-MAC) protocol incorporates the contention-free broadcasting of safety messages. In our C-MAC, vehicles utilize dynamic framed slotted ALOHA (DFSA) to transmit their channel reservation requests during dedicated channel reservation slots. The paper proposes a low-energy low-latency MAC (LL-MAC) protocol, referred to as the low-energy consumption and low-delay MAC protocol in ref. [24], which is based on receiver-initiated and captured effects. This protocol enables rapid matching between the senders and receivers when the sending nodes have data to transmit. Additionally, an enhanced greedy algorithm is introduced for power allocation among nodes, while an efficient collision response mechanism is employed for effective data transmission. In ref. [25], an MAC algorithm is proposed that utilizes the average waiting time as a common control reference, facilitating nodes to achieve equitable channel access by modifying one of the parameters of the IEEE 802.11-enhanced distributed channel access: contention window, arbitration inter-frame space, or transmission opportunity.

In the aforementioned literature, none of them considered the success rate of access probability as a significant security constraint. The closed-form expression for the acquisition probability in the Rician fading channel is derived in this paper. We establish an FSA system model where N vehicles compete freely for L slots. We analyze the impact of the number of vehicles and slots, as well as the capture effect, on the success probability of vehicle access. From a security perspective, we propose that meeting the constraint condition of successful vehicle access to roadside units is essential. Under this condition, we derive requirements for the number of slots, access times, and vehicle transmission power.

3. Capture Probability of Rician Fading Channel

The PDF of the received signal power of Rice channel fading is

$$f_p(p) = \frac{1}{\sigma^2} e^{-\frac{p}{\sigma^2}} e^{-K} I_0 \left(\sqrt{\frac{4Kp}{\sigma^2}} \right) \tag{1}$$

where p represents the instantaneous power of the received signal, σ^2 represents the power of the received signal's random reflection component, and K is the Rician factor. As shown in Equation (1), the power of the Rician fading channel follows a non-centrosymmetric Chi square distribution. When the Rician factor $K = 0$ corresponds to a Rayleigh fading channel.

Under normal circumstances, when multiple signals are simultaneously connected to a channel, a collision effect occurs, resulting in the failure of collectively accessing these signals. However, if the power of one dominant signal is significantly higher—meaning that the ratio between its power and the sum of all other signals' powers exceeds a certain threshold—and all other signals can be considered as negligible noise, then the successful connection of the dominant signal can still be achieved. This phenomenon is known as the capture effect.

Consider a scenario where n signals are simultaneously connected to a channel, and one of these signals is designated as the main signal with the maximum power value denoted by p_t. When the ratio of p_t to the sum of powers from the remaining $n-1$ signals exceeds a threshold value of z, it is considered significantly larger. In such cases, an acquisition effect occurs, resulting in successful access for p_t while causing access failure for the other $n - 1$ signals. The mathematical representation is given by Equation (2).

$$\frac{p_t}{\sum_{j=1}^{n-1} p_j} > z. \qquad (2)$$

The probability of capture effect is shown in Equation (3)

$$\begin{aligned} P_{cap}(n) &= n \int_0^\infty f_{p_t}(p_t) \cdot \Pr\left(p_t \Big/ \sum_{j=1}^{n-1} p_j > z \right) dp_t \\ &= n \int_0^\infty f_{p_t}(p_t) \cdot \left[\int_0^{p_t/z} f_{p_{n-1}}(p_{n-1}) dp_{n-1} \right] dp_t. \end{aligned} \qquad (3)$$

According to the reference [26], the probability density function (PDF) of the capture effect in a Rician fading channel is presented in Equation (4), while the probability of capture in the Rice fading channel is given by Equation (5).

$$f_{p_n}(p_n) = \left(\frac{1+K}{\bar{p}_n}\right)^{(n+1)/2} e^{-nK} \left(\frac{p_n}{nK}\right)^{(n-1)/2} \exp\left[-\frac{p_n(1+K)}{\bar{p}_n}\right] \cdot I_{n-1}\left(2\sqrt{\frac{(1+K)nKp_n}{\bar{p}_n}}\right) \qquad (4)$$

$$p_{cap}(n,z,K) = n \left\{ \begin{array}{c} 1 - e^{-nK} \cdot \sum_{i=0}^{\infty} \frac{[(n-1)K]^i}{i!} \cdot \sum_{k=0}^{n-2+i} \frac{\left(\frac{1}{z+1}\right)^k}{k!} \cdot \\ \sum_{j=0}^{\infty} \frac{K^j(k+j)!}{(j!)^2} \cdot \left(\frac{z}{z+1}\right)^{j+1} \end{array} \right\} \qquad (5)$$

The function p_{cap} in the Rice fading channel is dependent on the variables n, z, and K. In the case where the Rice factor $k = 0$, the Rice fading channel transforms into a Rayleigh fading channel.

4. System Model and Access Probability

4.1. Access Model Description

The system model depicting the vehicle's access to the roadside unit is illustrated in Figure 1. A total of N vehicles engage in a competitive allocation of L time slots among the available roadside units. Each individual vehicle has the flexibility to select any L idle time slots, with an equal probability $1/L$ assigned to each selection.

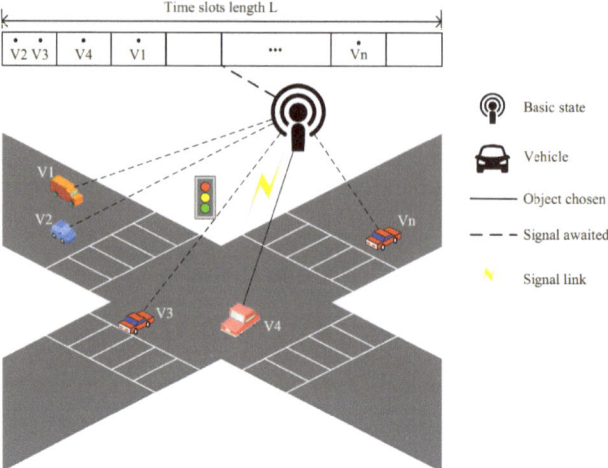

Figure 1. System model (n vehicles compete for L slots).

In Figure 1, L represents the count of available idle time slots for access, and N denotes the total number of vehicles. The regulations governing vehicle access to the roadside unit are as follows. If no vehicle selects a time slot to access, the time slot remains idle. In the case where only one vehicle chooses to access a time slot, there is no competition conflict, resulting in the successful occupation of the slot. However, if two or more vehicles attempt to access the same time slot simultaneously, a competition conflict arises which may lead to either successful or failed access. The probability of success or failure depends on the capture probability in this scenario. Notably, as the number of vehicles competing for a particular time slot increases, the capture probability decreases and vice versa.

4.2. Access Success and Conflict Probability without Considering Capture Effect

N vehicles engage in a competitive selection process for the L slots of roadside units, with each vehicle exercising its discretion. The total number of possible combinations is denoted as Equation (6)

$$N_{cmb} = L^N, L \geq 1, N \geq 1. \tag{6}$$

If only one vehicle accesses a specific slot, there is no competition conflict. The number of successful access combinations is denoted by

$$N_{col}^0 = \binom{N-1}{0} L(L-1)^{N-1} = L(L-1)^{N-1} \tag{7}$$

where N_{col}^0 represents the number of combinations without conflicts. In this context, it signifies that a vehicle is not combined with other $N-1$ vehicles, resulting in $\binom{N-1}{0}$ possible combinations. Subsequently, there are L available slots for making a free choice, providing L options. The remaining $N-1$ vehicles have the freedom to choose from the available $L-1$ slots, resulting in a total of $(L-1)^{N-1}$ possible choices. Thus, there are $\binom{N-1}{0} L(L-1)^{N-1}$ combinations.

In a given time slot, n vehicles simultaneously attempt to gain access, resulting in competition conflicts. The number of combinations in such scenarios is calculated as Equation (8)

$$N_{col}^n = \binom{N-1}{n-1} L(L-1)^{N-n}, n \leq N \tag{8}$$

where N_{col}^n represents the count of combinations involving conflicts among n vehicles, wherein all n vehicles attempt to select a slot simultaneously. The meaning of this expression is that a certain vehicle is combined with any $n - 1$ vehicles among the set of $N - 1$ vehicles. The total number of possible combinations is determined by the value of $\binom{N-1}{n-1}$. Subsequently, exercise your autonomy by selecting freely among L available options. There exist L alternatives to choose from. The remaining $N - n$ vehicles have the freedom to choose from the remaining $L - 1$ slots, resulting in a total of $(L-1)^{N-n}$ possible choices. Overall, there are $\binom{N-1}{n-1} L(L-1)^{N-n}$ combinations available. In the absence of multiple vehicles vying for a slot, there is no occurrence of competition conflict. The probability of successfully accessing the slot remains

$$p_{acc} = p_{col}^0 = \frac{N_{col}^0}{N_{cmb}} = \left(\frac{L-1}{L}\right)^{N-1}. \tag{9}$$

When multiple vehicles concurrently attempt to access a slot, the likelihood of encountering competition conflicts is

$$p_{col}^n = \frac{N_{col}^n}{N_{cmb}} = \frac{\binom{N-1}{n-1}(L-1)^{N-n}}{L^{N-1}}. \tag{10}$$

A slot is selected by multiple vehicles, and the likelihood of conflicts is

$$p_{col} = p(i \geq 2) = \sum_{i=2}^{N} p_{col}^i = \sum_{i=2}^{N} \frac{\binom{N-1}{i-1}(L-1)^{N-i}}{L^{N-1}}. \tag{11}$$

According to Equation (9), we can obtain

$$\frac{\partial}{\partial L} \ln p_{acc} = \frac{\partial}{\partial L} \ln \left(\frac{L-1}{L}\right)^{(N-1)} = \frac{N-1}{L(L-1)} > 0. \tag{12}$$

Therefore, p_{acc} monotonically increases with L. The access probability p_{acc} also increases as the value of L increases.

According to Equation (10), let $\frac{\partial}{\partial L} \ln p_{col}^n = 0$, then obtain the Equation (13)

$$L = \frac{(n-1)!(N-1)}{[(n-1)! - 1]N - [(n-1)! - n]}. \tag{13}$$

The maximum value of p_{col}^n occurs at this point in time. It represents the probability that n vehicles select the same slot for collision, which is maximized. When p_{col}^2, p_{col}^3, and p_{col}^4 are maximum, the relationship between L and N is shown in Equations (14)–(16)

$$L = N - 1 \quad \text{s.t.} \quad \max p_{col}^2 \tag{14}$$

$$L = \frac{2(N-1)}{N+1} \quad \text{s.t.} \quad \max p_{col}^3 \tag{15}$$

$$L = \frac{6(N-1)}{5N-2} \quad \text{s.t.} \quad \max p_{col}^4. \tag{16}$$

4.3. Access Success and Failure Probability When Considering Capture Effect

When considering the capture effect, successful access may occur when multiple vehicles simultaneously contend for a slot. The probability of achieving successful access under such circumstances is enhanced due to the capture effect. The conditional probability in cases of competition conflict can be represented by Equation (17).

$$\Delta p_{suc} = p_{cap} p_{col} = \sum_{i=2}^{N} p_{col}(i) p_{cap}(i). \tag{17}$$

When considering the capture effect, the probability of successful slot access can be expressed as Equation (18), while the probability of slot access failure is represented by Equation (19).

$$p_{suc} = p_{acc} + \Delta p_{suc} = p_{acc} + \sum_{i=2}^{N} p(i) p_{cap}(i) \tag{18}$$

$$p_{fail} = p_{col} - \Delta p_{suc} = p_{col}(1 - p_{cap}) = \sum_{i=2}^{N} p_{col}^{i}\left(1 - p_{cap}^{i}\right) \tag{19}$$

4.4. Access Methods When Access Probability Constraints

Safety is the paramount concern in VANETs, with numerous factors influencing it. Among these, ensuring the timely connectivity of vehicles to VANETs through roadside units stands out as the most crucial aspect. Access delay and access success rate are two key dimensions of this issue for vehicles. A low delay implies a high access success rate and vice versa. Therefore, enforcing a constraint on the vehicle's access success rate to roadside units becomes imperative for guaranteeing VANET safety. The threshold value for this success rate may vary based on different road conditions; however, addressing this problem is essential. This paper proposes three approaches to enhance the access success rate.

4.4.1. Increase the Number of Access Slots to Improve the Success Rate of Vehicle Access

Revised sentence: "In this scenario, the probability of multiple vehicles selecting the same time slot simultaneously is reduced, thereby enhancing the success rate of accessing the slot in a single attempt through the comprehensive configuration of time slot resources".
Assuming that the threshold of the access success rate is p_{th}, then

$$p_{suc} = p_{acc} + p_{cap} p_{col} > p_{acc} > p_{th}. \tag{20}$$

According to Equations (9) and (15), Equation (21) can be inferred

$$L > \left\lfloor \frac{1}{1 - e^{\frac{\ln(p_{th})}{N-1}}} \right\rfloor. \tag{21}$$

where $\lfloor \cdot \rfloor$ denotes the floor function. When Equation (21) is satisfied, the one-time access success probability of the vehicle can be guaranteed to exceed p_{th}.

4.4.2. Improving the Success Rate of Vehicle Access in VANETs by Increasing the Access Times

Theoretically, for a given vehicle V1, the initial attempt to access fails, and the probability of successful access within n attempts is described by Equation (22)

$$p_{in}(n) = (1 - p_{suc})^{n-1} p_{suc}. \tag{22}$$

The time interval is Δt. The method of accessing the slot is implemented using an equal time interval cycle mode, without considering the access back time strategy. The time required for the n-th successful access is represented as Equation (23)

$$t(n) = \frac{(n-1)\Delta t}{p_{in}(n)}. \tag{23}$$

In the practical application scenario, it is imperative for the vehicle to access the roadside unit slot at a specific time $t(n)$ as the Equation (24) in order to satisfy the conditional real-time requirements

$$t(n) < t_{\max}. \tag{24}$$

Therefore, the maximum value of access times, denoted as $n_{\max}$, can be calculated when the access slot interval Δt is constant. In normal circumstances, the maximum duration $t_{\max}$ is determined by vehicle speed and safe distance under road conditions. However, in practice, the TDMA slot cycle time is very short and vehicle travel distance is limited. Thus, it is reasonable for vehicles that fail to access roadside units to attempt multiple accesses in order to improve their success rate.

In the initial scenario, assuming that a total of N_1 vehicles randomly access L_1 available slots, the probability of the first vehicle to successfully gain access can be determined using either Equation (9) or (13), denoted by p_1. The number of vehicles successfully accessing slots after the first random access is $\lfloor N_1 p_1 \rfloor$, and these vehicles occupy the same number of slots. The number of vehicles that have not successfully established a connection can be expressed as $N_2 = N_1 - N_1 p_1$, while the remaining slots can be represented by $L_2 = N_1 - N_1 p_1$. According to Equation (9) or (13), the probability of the second successful vehicle gaining access is denoted by p_2. The number of vehicles successfully accessing slots after the second random access is $\lfloor N_2 p_2 \rfloor$, and they occupy $\lfloor N_2 p_2 \rfloor$ time slots. The number of vehicles that have not been successfully accessed is given by $N_3 = N_2 - N_2 p_2$, where N_2 represents the total number of vehicles and p_2 denotes the probability of successful access. Similarly, the remaining time slots can be calculated as $L_3 = N_2 - N_2 p_2$. Furthermore, Equation (9) or (13) provides an expression for the probability (p_3) of achieving a third successful access to vehicles. The process is $(\; N_1 \quad L_1 \quad p_1 \;) \rightarrow (\; N_2 \quad L_2 \quad p_2 \;) \rightarrow (\; N_3 \quad L_3 \quad p_3 \;)$.

Equations (25) and (26) present the probabilities of total access failure and total access success, respectively, after three random accesses

$$p_{fail} = (1 - p_1)(1 - p_2)(1 - p_3) \tag{25}$$

$$p_{suc} = 1 - p_{fail}. \tag{26}$$

4.4.3. Revising the Vehicle Transmission Power to Enhance the Capture Probability and Thereby Improve the Success Rate of Vehicle Access

In accordance with the fundamental principles of the capture effect, augmenting the transmission power of a vehicle can enhance its likelihood of being allocated slots when competing with other vehicles, thereby improving the success rate of vehicle access to slots. When the transmission power p_t exceeds $z \sum_{j=1}^{n-1} p_j$, it means that the capture effect occurs in the region below $z \sum_{j=1}^{n-1} p_j$.

$$\Pr\left[p_t < z \sum_{j=1}^{n-1} p_j\right] = \Pr\left[p_t < z \cdot E\left[\sum_{j=1}^{n-1} p_j\right]\right] = \Pr[p_t < z \cdot (n-1)E[p_j]] \tag{27}$$

where p_j obeys a Chi-square distribution with one degree of freedom.

The condition $N < L$ is typically necessary to enhance the access success rate and reduce the access delay, particularly for security information frames. This implies that the likelihood of multiple vehicles contending for the same time slot simultaneously is exceedingly low. Consequently, the number of conflicting vehicles can be limited to a maximum of 4, disregarding scenarios where it exceeds this threshold. When $z = 2$, we can derive

$$\Pr\left[p_t < z \sum_{j=1}^{n-1} p_j\right] = \Pr[p_t < z \cdot (n-1)E[p_j]] = \Pr[p_t < 6] = 0.9857. \quad (28)$$

Equation (28) demonstrates that a 7.8 dB increase in vehicle power guarantees an access success rate exceeding 98%. Moreover, when the vehicle power is amplified by 9 dB, it ensures an access success rate surpassing 99.7%.

The access success probability of a vehicle can be improved by increasing the transmission power, while keeping other conditions unchanged. However, this approach is not an overall optimal solution but rather a priority due to the resulting failure of competitive time slots for other vehicles and decrease in access success rate. Nevertheless, it remains feasible and necessary within the context of security information framework.

4.4.4. Access Method Subject to Constraints on Access Probability

After considering the aforementioned cases comprehensively, it becomes evident that adopting a single method has its limitations. Hence, we propose a comprehensive access approach. The specific methodology is as follows: the time slot length is set to $L = 2N$, with access times generally being three-fold, twice in the high-speed section and four times in the low-speed section. In situations requiring safety information transmission during emergencies, an increase of 7 dB to 9 dB in transmission power is implemented.

5. Simulation

In this section, we present the simulation results of the algorithm performance. Most simulations were conducted using both numerical and Monte Carlo methods. The Monte Carlo method was employed to validate the accuracy of the numerical simulation results, with an average value obtained from 10,000 independent random experiments. All simulations were performed using MATLAB 2022b software, with random numbers generated by internal MATLAB function. In the provided simulation figures, the legend for the numerical simulation method is labeled as 'Ana', while the legend for the Monte Carlo simulation method is labeled as 'Sim'.

5.1. Simulation of Capture Effect

The relationship between the acquisition probability and the number of vehicle conflict n, the Rician factor K, and the power ratio threshold z of capture effect is shown in Figure 2 based on Equation (5).

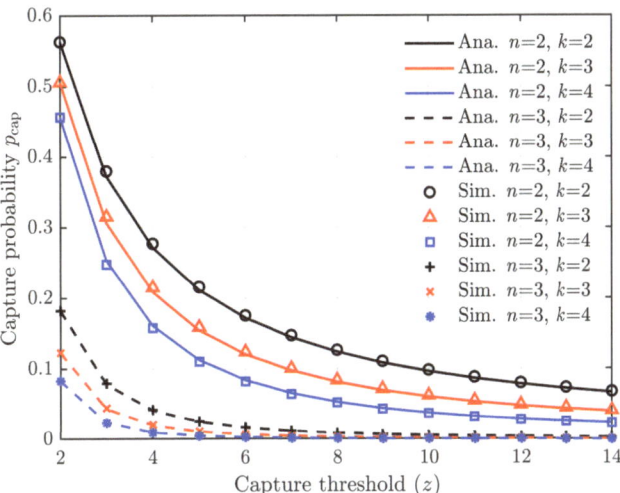

Figure 2. Capture probability vs. power ratio threshold.

In Figure 2, the relationship between the capture probability p_{cap} and the power ratio threshold z is depicted for six combinations of vehicle conflict number n and Rician factor K. The simulation curves are divided into two groups, with three curves corresponding to $n = 2$ forming one group, and three curves corresponding to $n = 3$ forming the other group. This observation highlights that the impact of n on the capture probability p_{cap} outweighs that of K. Moreover, when both K and n remain constant, p_{cap} exhibits a rapid decrease as z increases. These findings indicate that p_{cap}, particularly when considering values of $n \geq 3$, tends to be very small. Additionally, for values of $z \geq 10$, p_{cap} also remains negligible. These empirical laws serve as a foundation for approximating the calculation of capture probability.

5.2. Simulation of the Model

According to Equations (9) and (10), Table 1 presents the probabilities of successful vehicle access to the roadside unit slot and conflicts arising from 2, 3, or more vehicles simultaneously selecting a slot. The capture effect is not considered in this analysis.

Table 1. Illustrative calculations of the probabilities of successful and unsuccessful access.

	L	N	p_{col}^0	p_{col}^2	p_{col}^3	p_{col}^4	p_{col}^5
	1	1	1	0	0	0	0
	1	2	0	1	0	0	0
	1	3	0	0	1	0	0
Extreme conditions	2	1	1	0	0	0	0
	2	2	0.50	0.50	0	0	0
	2	3	0.25	0.50	0.25	0	0
	10	15	0.22877	0.35586	0.25701	0.11423	0.04413
More vehicles with fewer slots	10	20	0.13509	0.28518	0.28518	0.17956	0.11500
	10	25	0.07977	0.21271	0.27180	0.22146	0.21426
	30	15	0.62212	0.30033	0.06732	0.009285	0.00095
Fewer vehicles with more slots	30	20	0.52512	0.34404	0.10677	0.020864	0.003201
	30	25	0.44324	0.36682	0.14546	0.036784	0.007688
	50	40	0.45480	0.36198	0.14036	0.035329	0.007534
General conditions	50	50	0.37160	0.37160	0.18201	0.058193	0.016594
	50	60	0.30363	0.36559	0.21637	0.083898	0.030516

In Table 1, the probabilities of successful access for vehicle 0 under various conditions are presented, including "extreme conditions", "more vehicles with fewer slots", "fewer vehicles with more slots", and "general conditions", as well as the probability of conflicts arising from multiple vehicles selecting a slot simultaneously. These probabilities can be analyzed quantitatively to assess both successful access and conflict occurrences. Furthermore, they serve as a means to validate the accuracy of the algorithm.

According to Equations (9) and (10), Figure 3 depicts the relationship between the probability of successful vehicle access to the roadside unit slot, as well as the probabilities of 2, 3, and 4 vehicles simultaneously selecting a slot leading to conflicts, in relation to the number of vehicles. In this analysis, we do not consider the capture effect and assume a constant time slot length.

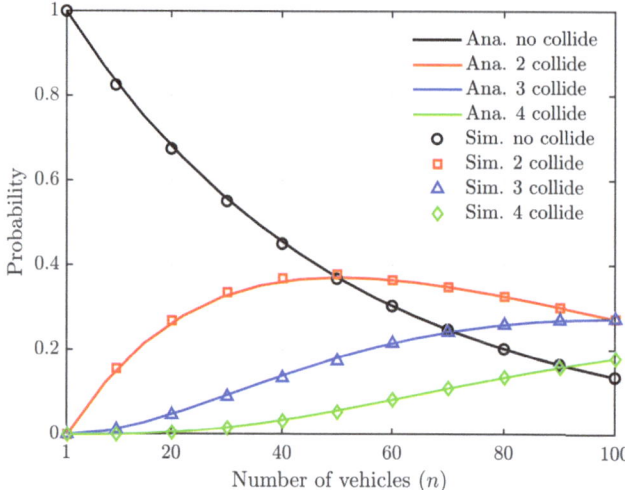

Figure 3. Relationship between access probability, collision probability of n vehicles, and the number of vehicles (the slot number is constant, $L = 50$).

In Figure 3, with a fixed number of slots $L = 50$, the range of vehicles N varies from 1 to 100. The term "no collide" refers to the scenario where only one vehicle successfully selects and connects to a slot without any conflicts. On the other hand, "2 collide" indicates that two vehicles simultaneously select the same slot, resulting in a conflict. Similar patterns can be observed for other scenarios as well. Figure 3 demonstrates that, when the number of slots remains constant, the probability of successful access without slot selection conflicts decreases as the number of vehicles increases. When N is twice as large as L, this probability reduces to less than 10%. In cases where there are slot selection conflicts, it is predominantly observed between two or three cars, while conflicts involving four or more cars are relatively rare. Additionally, there is a high likelihood of the capture effect when two or three cars experience slot selection conflicts; however, this effect becomes negligible when four or more cars encounter such conflicts. Therefore, our focus on studying capture effects should primarily revolve around scenarios involving two or three cars slot selection conflicts while disregarding those with four or more cars selections.

According to Equations (9) and (10), in the absence of considering the capture effect and with a constant number of vehicles, Figure 4 illustrates the relationship between the probability of successful vehicle access to the roadside unit slot, as well as the probabilities of two, three, and four vehicles simultaneously selecting a slot causing conflicts.

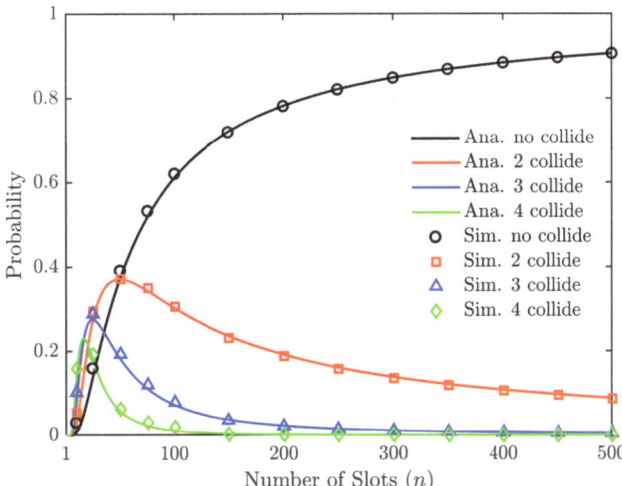

Figure 4. Relationship between access probability, collision probability of N vehicles, and slot number (the number of vehicles is constant, $N = 50$).

In Figure 4, for a fixed number of vehicles $N = 50$, the range of slots L is from 1 to 500. The term "no collide" refers to a scenario where only one vehicle selects a slot and successfully accesses it without any conflict, while "2 collide" indicates that two vehicles select the same slot simultaneously, resulting in a conflict. The remaining cases follow similar patterns. Figure 4 illustrates that, as the number of slots increases, there is a rapid escalation in the probability of slot selection conflicts, with successively increasing probabilities observed for scenarios involving four, three, and two vehicles. Subsequently, this probability gradually decreases but at a slower rate for conflicts between two vehicles. When the number of slots exceeds 100, the likelihood of conflict between car 4 and car 3 becomes negligible and can be disregarded. These findings demonstrate that, when dealing with large numbers of slots and relatively fewer vehicles, the capturing effect primarily considers conflicts between pairs of vehicles while other situations have minimal impact.

When considering the capture effect, successful access can occur when multiple vehicles simultaneously contend for a slot. According to Formula (13), the probability of successful access under this condition represents an increment in the probability due to the capture effect. This corresponds to the conditional probability in cases of competition conflict, as illustrated in Figures 5 and 6.

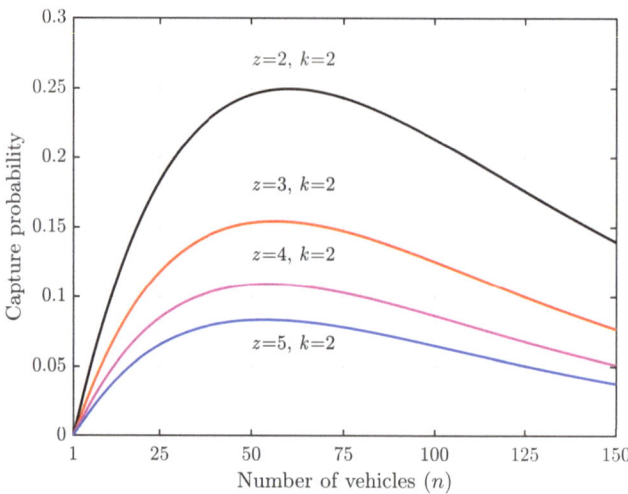

Figure 5. Increment in successful access probabilities due to capture effect vs. number of vehicles.

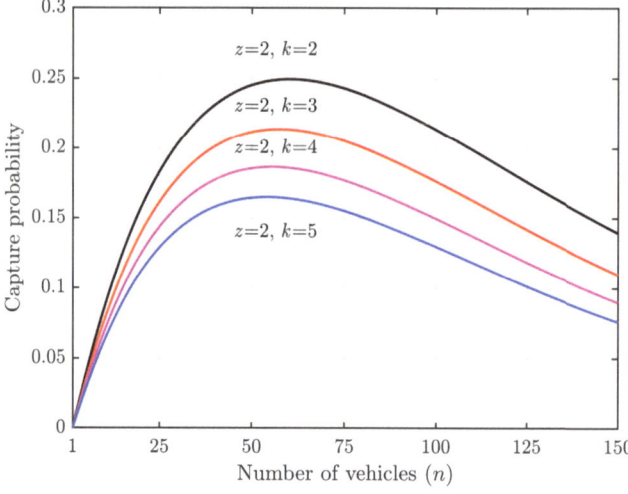

Figure 6. Increment in successful access probabilities due to capture effect vs. number of vehicles.

According to Equation (13), the factors affecting the acquisition probability are L, N, and K. In Figures 5 and 6, we set $L = 50$ and varied the number of vehicles, N, from 1 to 150. For Figure 5, we fixed $K = 2$ and varied spreading factor, z, from 2 to 5. For Figure 6, we fixed $z = 2$ and varied K from 2 to 5. Our results show that (1) when there are as many vehicles as slots ($N = L$), the capture probability is the highest with a significant capture effect; (2) when there are a few vehicles relative to slots ($N \ll L$), acquisition probability is very low due to low collision probabilities; (3) when there are many more vehicles than slots ($N \gg L$), the capture probability decreases due to high collision probabilities but also because competition for access reduces the likelihood of successful captures; (4) comparing Figures 5 and 6, it can be seen that spreading factor has a greater impact on capture probability than does value of K.

The relationship between the probability of a vehicle successfully accessing a roadside unit slot and the number of vehicles is depicted in Figure 7, based on Equations (10) and (14).

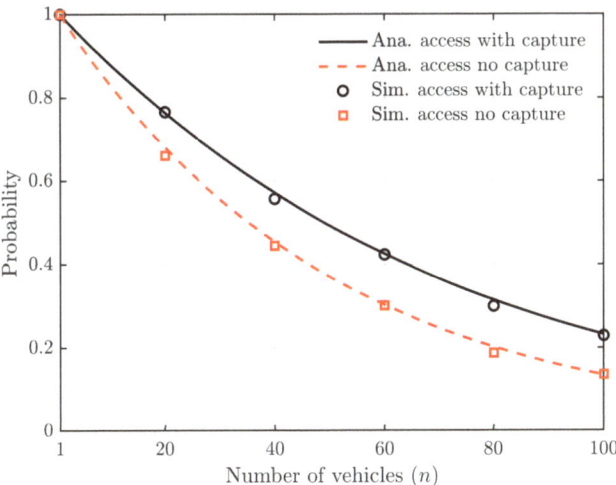

Figure 7. Relationship between the probability of successful access to slots and the number of vehicles (the number of slots is constant).

The simulation parameters in Figure 7 are $L = 50$, $z = 3$, $K = 3$, and the number of vehicles, denoted as N, ranges from 1 to 100. As the number of vehicles increases while keeping the number of slots constant, there is an elevated likelihood for multiple vehicles to select the same slot simultaneously, leading to a continuous decline in successful access probability. Figure 7 demonstrates that considering the capture effect significantly enhances the probability of successful vehicle-to-roadside unit access under similar conditions.

According to Equations (10) and (14), Figure 8 illustrates the correlation between the probability of successful vehicle access to a roadside unit slot and the number of slots.

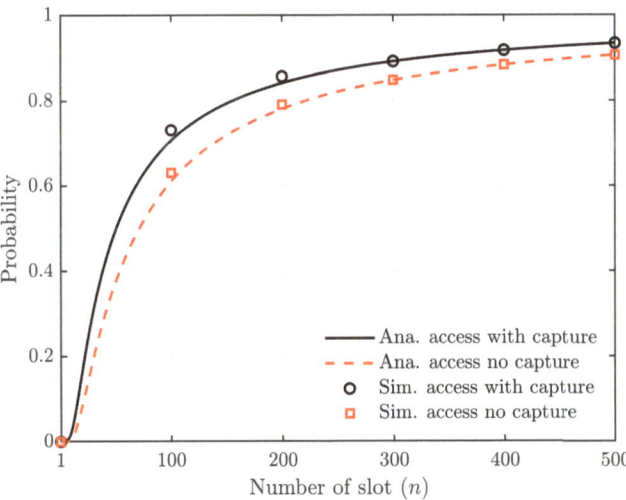

Figure 8. Relationship between the probability of successful access to slots and the number of slots (the number of vehicles is constant).

The simulation parameters in Figure 8 are $N = 50$, $z = 3$, $K = 3$, and the number of time slots L ranges from 1 to 500. As the number of time slots increases while keeping the number of vehicles constant, the likelihood of multiple vehicles selecting the same time slot simultaneously decreases, leading to an increased probability of successful time slot access. Figure 8 demonstrates a significant enhancement in vehicle's successful access to roadside units when considering the capture effect under similar conditions.

5.3. Simulation of Performance

5.3.1. Time Responsiveness

The probability of successful access for vehicle V1 until the n-th attempt is as follows. Figure 9 illustrates a rapid decline in the probability of success for each subsequent access attempt, with the fourth and subsequent attempts having a very low probability of success. Consequently, if an access failure occurs, there is a high likelihood of achieving access success after three consecutive attempts.

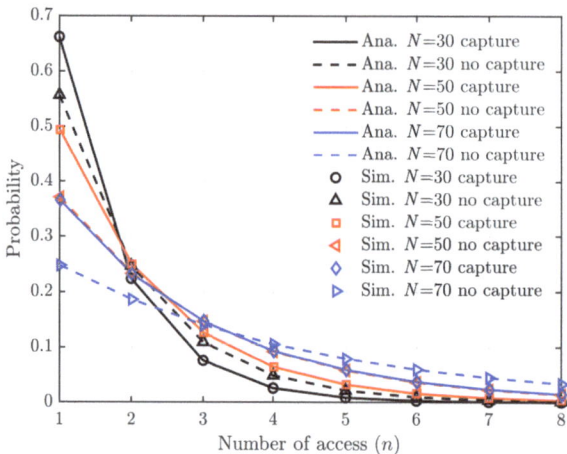

Figure 9. Probability of successful access for the first time.

We considered three cases in which the number of vehicles and the number of slots are, respectively, 100/200, 50/100, and 100/100 as Figure 10.

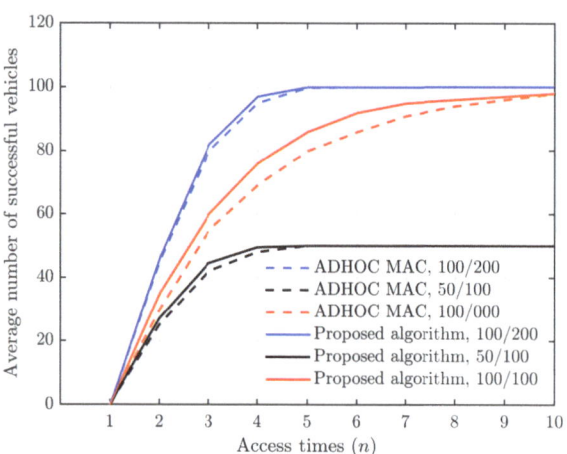

Figure 10. Average number of successful vehicles vs. access time (n).

In the 100/200 and 50/100 cases, all the vehicles access successfully within 5 times. In the case 100/100, due to intense competition among vehicles, they need to go through multiple rounds of competition in order to successfully connect with the RSU. The competition for vehicle access time slots is more intense, requiring more attempts to establish connections. Figure 10 meanwhile demonstrates that the proposed method in this paper outperforms ADHOC MAC, particularly when there are many vehicles vying for limited time slots and the competition is fierce. The ADHOC MAC simulation data are sourced from ref. [2].

We consider a highway equipped with a roadside unit, where vehicles enter and exit the coverage area of the RSU. Assuming a constant speed of 100 km/h for vehicles, the inter-vehicle distance follows a uniform distribution. The RSU coverage ranges are 300 m, 400 m, and 500 m, respectively. Figure 11 illustrates the direct correlation between vehicle density and the successful connection of vehicles to roadside units.

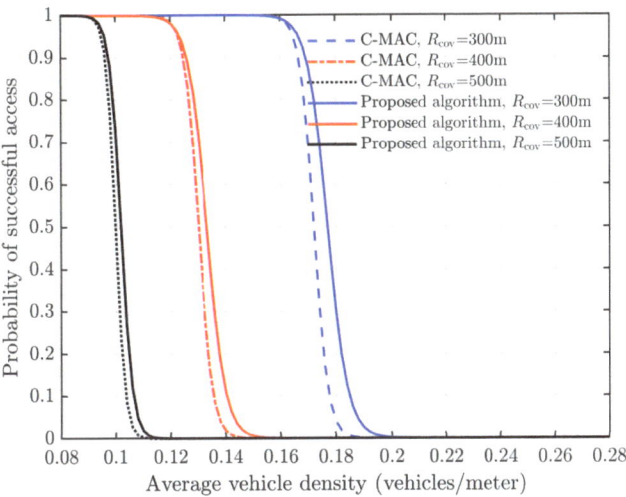

Figure 11. Probability of successful access vs. vehicle density.

The graph in Figure 11 demonstrates that, at low vehicle densities, the probability of successful access is 1. However, as the vehicle density surpasses a certain threshold, there is a significant decrease in the likelihood of successful access. Specifically, when the RSU coverage range is set to 300 m, the probability of successful access starts declining once the vehicle density exceeds 0.16.

The algorithm demonstrates superiority over the C-MAC algorithm, as depicted in Figure 11. This advantage becomes more pronounced when the RSU coverage range is 300 m. With an increase in vehicle density, the probability of successful vehicle access rapidly decreases. However, compared to the C-MAC method, this decrease occurs at a slower rate due to the capture effect. Consequently, this characteristic renders this algorithm more suitable for urban traffic scenarios characterized by significant fluctuations in vehicle density. The C-MAC simulation data are sourced from ref. [23].

5.3.2. Selection of Parameters under the Constraint of Successfully Accessing Probability

For ensuring security, it is imperative that newly arrived vehicles can access the VANETs in a timely manner through the roadside unit with a high probability of success. In the FSA network, there exist three primary approaches to enhance the vehicle's access success rate: (1) augmenting the configuration of access slot numbers; (2) mitigating failure probability via multiple accesses; and (3) amplifying the transmission power of vehicles to improve the capture probability.

(1) Increase the number of access slots to improve the success rate of vehicle access.

The objective of this approach is to prioritize access delay, allocate resources comprehensively, and enhance the success rate of one-time access. The simulation results in Figure 12 demonstrate the impact of different values (0.6, 0.7, 0.8, and 0.9) for p_{th} as per Equation (21).

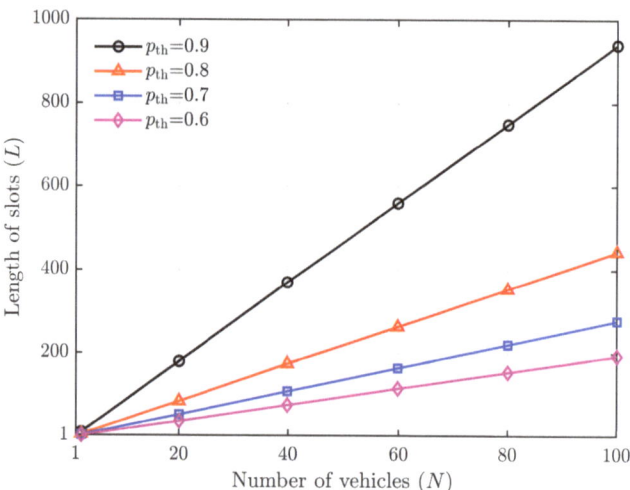

Figure 12. The number of slots vs. the number of the vehicle under the condition of access probability ($p_{th} = 0.6, 0.7, 0.8, 0.9$).

It can be observed from Figure 12 that a significantly high access success rate can be achieved when the number of slots is abundant. Specifically, with ten times more slots than vehicles, the one-time success rate of vehicle access to the roadside unit exceeds 90%.

(2) Increase access times to improve vehicle access success rate.

The objective of this approach is to comprehensively consider both the delay and the availability of slot resources in order to enhance the access success rate through multiple access. This method requires fewer slot resources, albeit resulting in a higher vehicle access delay to the roadside unit, making it particularly suitable for low-speed scenarios. Based on Equations (25) and (26), the simulation results are presented in Table 2.

Table 2. Success probability after three accesses (the number of slots is twice the number of vehicles).

N_1	L_1	p_1	N_2	L_2	p_2	N_3	L_3	p_3	p_{fail}	p_{suc}
20	40	0.60269	8	28	0.74756	3	23	0.87515	1.25220	98.7478
50	100	0.60501	20	70	0.74993	6	56	0.89753	1.01220	98.9878
100	200	0.60577	40	140	0.75071	10	110	0.91272	0.85776	99.1422
200	400	0.60615	79	279	0.75302	20	220	0.91291	0.84713	99.1529

The results presented in Table 2 demonstrate that, when the number of slots is twice the number of vehicles, the final access success rate exceeds 98.7% after three consecutive attempts. In comparison to method (1), this approach significantly reduces the number of provided slots while maintaining a remarkably high access success rate.

(3) Enhance the transmission power to elevate the success rate of vehicle access by augmenting the capture probability.

The capture probability can be enhanced by increasing the transmitting power of the vehicle. Figure 13 illustrates the relationship between the access success probability and the number of slots, considering $N = 50$, as derived from Equations (9), (10) and (27). In this scenario, we account for the capture probability by augmenting the transmission power of vehicles.

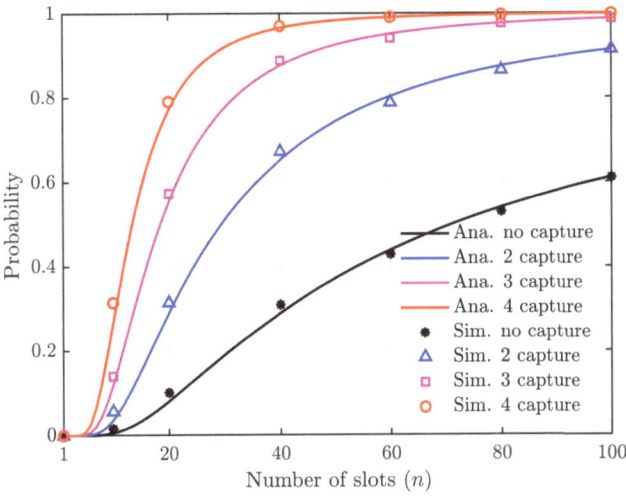

Figure 13. Access success probability vs. the number of slots (*N*=50).

The access success probability exhibits a slow increase with the number of slots when the transmission power is small and the capture probability is neglected, as depicted in Figure 13. However, as the transmitting power of vehicles increases, a collision capture effect between two vehicles emerges. Further increasing the transmitting power leads to collision capture effects involving three or four vehicles. Under these circumstances, the access success probability experiences a rapid growth with an increasing number of slots. Remarkably enhanced access success probabilities are observed under identical conditions.

It is worth noting that increasing the transmission power for a specific vehicle can enhance the access success probability. However, this enhancement comes at the expense of reduced success rates for other vehicles' access attempts. Consequently, while not an overall optimal approach, it becomes indispensable in emergency situations for individual vehicles.

6. Conclusions

In this paper, we propose a novel FSA MAC protocol based on the capture effect and derive the closed-form expression of access probability in the Rician fading channel. We establish a system model where N vehicles compete for L slots, and analyze the impact of vehicle and slot numbers on the success probability of vehicle access. From a security perspective, we suggest that vehicle access to roadside units must adhere to constraints regarding access success rate. Under these conditions, we derive requirements for the slot number, access times, and vehicle transmission power. Finally, through simulation verification, we validate our proposed model and algorithm.

Author Contributions: Conceptualization, L.L. and W.X.; methodology, L.L. and W.X.; software, Z.S.; validation, W.X.; formal analysis, Z.S. and W.X.; investigation, L.L. and W.X.; resources, L.L.; data curation, Z.S.; writing—original draft preparation, L.L.; writing—review and editing, Z.S.; visualization, L.L. All authors have read and agreed to the published version of the manuscript.

Funding: This work was supported by the Fujian Provincial Natural Science Foundation (Grant number 2023J01810), the Guidance Projects of FuJian Science and Technology Agency: 2022H0022, the Project of the Xiamen Science and Technology Bureau: 2022CXY0315, the Project of the Xiamen Science and Technology Bureau: 2022CXY0317.

Institutional Review Board Statement: Not applicable.

Informed Consent Statement: Not applicable.

Data Availability Statement: The data that support the findings of this study are available from the corresponding authors upon reasonable request.

Conflicts of Interest: The authors declare no conflicts of interest.

References

1. Baccelli, F.; Blaszczyszyn, B.; Muhlethaler, P. An Aloha protocol for multihop mobile wireless networks. *IEEE Trans. Inf. Theory* **2006**, *52*, 421–436. [CrossRef]
2. Borgonovo, F.; Capone, A.; Cesana, M.; Fratta, L. ADHOC MAC: New MAC Architecture for Ad Hoc Networks Providing Efficient and Reliable Point-to-Point and Broadcast Services. *Wirel. Netw.* **2004**, *10*, 359–366. [CrossRef]
3. Han, C.; Dianati, M.; Tafazolli, R.; Liu, X.; Shen, X. A Novel Distributed Asynchronous Multichannel MAC Scheme for Large-Scale Vehicular Ad Hoc Networks. *IEEE Trans. Veh. Technol.* **2012**, *61*, 3125–3138. [CrossRef]
4. Wang, Y.; Shi, J.; Chen, L.; Lu, B.; Yang, Q. A Novel Capture-Aware TDMA-Based MAC Protocol for Safety Messages Broadcast in Vehicular Ad Hoc Networks. *IEEE Access* **2019**, *7*, 116542–116554. [CrossRef]
5. Choi, S.S.; Kim, S. A dynamic framed slotted aloha algorithm using collision factor for RFID identification. *IEICE Trans. Commun.* **2009**, *92*, 1023–1026. [CrossRef]
6. Ferreira, H.P.A.; Assis, F.M.D.; Serres, A.R. A Novel RFID Method for Faster Convergence of Tag Estimation on Dynamic Frame Size ALOHA Algorithms. *IET Commun.* **2019**, *13*, 1218–1224. [CrossRef]
7. Cassará, P.; Cola, T.D.; Gotta, A. A Statistical Framework for Performance Analysis of Diversity Framed Slotted Aloha With Interference Cancellation. *IEEE Trans. Aerosp. Electron. Syst.* **2020**, *56*, 4327–4337. [CrossRef]
8. Wang, Y.; Shi, J.; Chen, L. Capture Effect in the FSA-Based Networks under Rayleigh, Rician and Nakagami-m Fading Channels. *Appl. Sci.* **2018**, *8*, 414. [CrossRef]
9. Hadzi-Velkov, Z.; Spasenovski, B. Capture effect in IEEE 802.11 basic service area under influence of Rayleigh fading and near/far effect. In Proceedings of the 13th IEEE International Symposium on Personal, Indoor and Mobile Radio Communications, Lisbon, Portugal, 18 September 2002; Volume 1, pp. 172–176.
10. Tian, N.; Cai, X.; Cheng, J.; Yue, W.; Luo, M. Short-Packet Transmission in Irregular Repetition Slotted ALOHA System over the Rayleigh Fading Channel. *Int. J. Pattern Recognit. Artif. Intell.* **2022**, *36*, 2259016. [CrossRef]
11. Pejoski, S.; Hadzi-Velkov, Z. Slotted ALOHA Wireless Networks with RF Energy Harvesting in Nakagami-m Fading. *Ad Hoc Netw.* **2020**, *107*, 102235. [CrossRef]
12. Yue, Z.; Yang, H.H.; Zhang, M.; Pappas, N. Age of Information Under Frame Slotted ALOHA-Based Status Updating Protocol. *IEEE J. Sel. Areas Commun.* **2023**, *41*, 2071–2089. [CrossRef]
13. Sun, M.; Guo, Y.; Zhang, D.; Jiang, M.M. Anonymous Authentication and Key Agreement Scheme Combining the Group Key for Vehicular Ad Hoc Networks. *Complexity* **2021**, *2021*, 5526412. [CrossRef]
14. Qiong, W.; Shuai, S.; Ziyang, W.; Qiang, F.; Pingyi, F.; Cui, Z. Towards V2I Age-aware Fairness Access: A DQN Based Intelligent Vehicular Node Training and Test Method. *Chin. J. Electron.* **2023**, *32*, 1230–1244. [CrossRef]
15. Akyıldız, T.; Ku, R.; Harder, N.; Ebrahimi, N.; Mahdavifar, H. ML-Aided Collision Recovery for UHF-RFID Systems. In Proceedings of the 2022 IEEE International Conference on RFID (RFID), Las Vegas, NV, USA, 17–19 May 2022; pp. 41–46.
16. Yu, J.; Zhang, P.; Chen, L.; Liu, J.; An, J. Stabilizing Frame Slotted Aloha Based IoT Systems: A Geometric Ergodicity Perspective. *IEEE J. Sel. Areas Commun.* **2020**, *39*, 714–725. [CrossRef]
17. Bacco, M.; Cassara, P.; Gotta, A.; Cola, T.D. Diversity Framed Slotted Aloha with Interference Cancellation for Maritime Satellite Communications. In Proceedings of the ICC 2019–2019 IEEE International Conference on Communications (ICC), Shanghai, China, 20–24 May 2019; pp. 1–6.
18. Rajeswar, R.G.; Ramanathan, R. An Empirical study on MAC layer in IEEE 802.11p/WAVE based Vehicular Ad hoc Networks. *Procedia Comput. Sci.* **2018**, *143*, 720–727.
19. Liu, J.X.; Ding, S.B.; Zhang, L.; Xie, X.P. Event-driven intermittent control for vehicle platooning over vehicular ad hoc networks. *Int. J. Robust Nonlinear Control* **2023**, *33*, 1214–1230. [CrossRef]
20. Shah, A.F.M.S.; Ilhan, H.; Tureli, U. Modeling and Performance Analysis of the IEEE 802.11 MAC for VANETs under Capture Effect. In Proceedings of the 2019 IEEE 20th Wireless and Microwave Technology Conference (WAMICON), Cocoa Beach, FL, USA, 8–9 April 2019; pp. 1–5.
21. Menouar, H.; Yagoubi, M.B.; Ouladdjedid, L.K. CSSA MAC: Carrier sense with slotted-Aloha multiple access MAC in vehicular network. *Int. J. Veh. Inf. Commun. Syst.* **2018**, *3*, 336–354.
22. Chowdhury, M.S.; Ullah, N.; Al Ameen, M.; Kwak, K.S. Framed slotted aloha based MAC protocol for low energy critical infrastructure monitoring networks. *Int. J. Commun. Syst.* **2015**, *27*, 1783–1797. [CrossRef]
23. Kim, Y.; Lee, M.; Lee, T.J. Coordinated Multichannel MAC Protocol for Vehicular Ad Hoc Networks. *IEEE Trans. Veh. Technol.* **2016**, *65*, 6508–6517. [CrossRef]

24. Qi, H.M.; Chen, J.Q.; Yuan, Z.Y.; Fan, L.L. A Low Energy Consumption and Low Delay MAC Protocol Based on Receiver Initiation and Capture Effect in 5G IoT. In Proceedings of the Algorithms and Architectures for Parallel Processing: 21st International Conference, Virtual Event, 3–5 December 2022; pp. 709–723.
25. Jeong, J.; Choi, S.; Yoo, J.; Lee, S.; Kim, C.K. Physical layer capture aware MAC for WLANs. *Wirel. Netw.* **2013**, *19*, 533–546. [CrossRef]
26. Sanchez-Garcia, J.; Smith, D.R. Capture probability in Rician fading channels with power control in the transmitters. *Commun. IEEE Trans.* **2002**, *50*, 1889–1891. [CrossRef]

Disclaimer/Publisher's Note: The statements, opinions and data contained in all publications are solely those of the individual author(s) and contributor(s) and not of MDPI and/or the editor(s). MDPI and/or the editor(s) disclaim responsibility for any injury to people or property resulting from any ideas, methods, instructions or products referred to in the content.

Communication

Integrated Sensing and Secure Communication with XL-MIMO

Ping Sun [1], Haibo Dai [2] and Baoyun Wang [1,*]

[1] School of Communication and Information Engineering, Nanjing University of Posts and Telecommunications, Nanjing 210023, China; 2018010214@njupt.edu.cn

[2] School of Internet of Things, Nanjing University of Posts and Telecommunications, Nanjing 210046, China; hbdai@njupt.edu.cn

* Correspondence: bywang@njupt.edu.cn

Abstract: This paper studies extremely large-scale multiple-input multiple-output (XL-MIMO)-empowered integrated sensing and secure communication systems, where both the radar targets and the communication user are located within the near-field region of the transmitter. The radar targets, being untrusted entities, have the potential to intercept the confidential messages intended for the communication user. In this context, we investigate the near-field beam-focusing design, aiming to maximize the achievable secrecy rate for the communication user while satisfying the transmit beampattern gain requirements for the radar targets. We address the corresponding globally optimal non-convex optimization problem by employing a semidefinite relaxation-based two-stage procedure. Additionally, we provide a sub-optimal solution to reduce complexity. Numerical results demonstrate that beam focusing enables the attainment of a positive secrecy rate, even when the radar targets and communication user align along the same angle direction.

Keywords: near field; integrated sensing and communication; beam focusing; vehicular ad hoc networks (VANETs)

1. Introduction

Integrated sensing and communication (ISAC) has emerged as a significant advancement in upcoming wireless systems [1,2]. These systems enable the dual use of hardware platforms and limited spectrum/power resources for both communication and sensing tasks [3]. One of the most promising areas of ISAC application is in *vehicular ad hoc networks* (VANETs), which are at the forefront of the evolution of connected and autonomous vehicles [4]. In VANETs, vehicles communicate with each other and possibly with infrastructure components, such as traffic signals or roadside units. These communications can be used for sharing traffic conditions, alerting nearby vehicles of emergency braking, or even for cooperative driving in the future [5]. With the increasing complexity and density of modern traffic systems, vehicles also need advanced sensing capabilities for safety and navigation purposes. By integrating ISAC into VANETs, vehicles can dynamically optimize resources, streamline hardware, and enhance safety through real-time communication of sensed data, ensuring efficient and adaptive network performance [6,7].

Currently, ISAC systems allow the reuse of information-bearing signals for radar sensing [2]. While this dual operation improves spectrum efficiency and reduces the cost of systems, it introduces potential risks of information leakage. This becomes particularly concerning when sensing targets, being untrusted entities, might intercept these confidential signals [8]. To address this issue, physical layer security techniques have been integrated into ISAC systems to ensure communication security [9–11]. In [9], the authors examine an ISAC scenario consisting of one communication user and one untrusted target, potentially an eavesdropper, and they design the transmit covariance matrices to maximize the achievable secrecy rate. In [10], the authors jointly optimize the transmit information and sensing beamforming, aiming to minimize the beampattern matching error for sensing while ensuring the minimum secrecy rate requirement. Meanwhile, in [11], optimization-based beamforming designs are presented, focusing on the security of information transmissions within ISAC systems.

It should be noted that all the aforementioned papers on secrecy ISAC address the far-field scenario, where both radar targets and communication users reside within the far-field region of the transmitter. Motivated by the great success of multiple-input multiple-output (MIMO) technology in 4G and 5G [12,13], in future 6G networks, extremely large-scale (XL)-MIMO is envisioned to be utilized. XL-MIMO, deploying hundreds, or even thousands, of antenna elements at base stations or reflecting elements at reconfigurable intelligent surfaces, can significantly improve performance, such as spectral efficiency [14]. The envisioned deployment of XL-MIMO in future 6G networks will likely lead to wireless operations in the radiating near-field (Fresnel) region, in contrast to conventional wireless systems, which typically operate in the far-field region [15]. Thus, there will be a shift in 6G: radar targets and communication users may predominantly operate within the near-field region of the transmitter [16]. While far-field propagation assumes a planar wavefront, the near-field region is characterized by a spherical wavefront. This unique near-field waveform propagation characteristic paves the way for the beam focusing technique, as detailed in [17,18], offering a notable boost in secrecy communication performance by mitigating information leakage. Nevertheless, this promising avenue has garnered limited attention in the current secrecy ISAC literature.

Motivated by the preceding discussion, in this paper, we explore the near-field beam-focusing design for secrecy ISAC systems. Here, one transmitter equipped with an extremely large-scale antenna array senses multiple radar targets and communicates with a single communication user concurrently. Both the radar targets and the communication users are situated within the near-field region of the transmitter. These radar targets, being untrusted, might eavesdrop on the confidential messages intended for the communication user. In this context, we investigate the near-field beam-focusing design to characterize the trade-off between radar sensing and secure communication. The main contributions of this paper are summarized below:

- To the best of the authors' knowledge, this work represents the first study of the near-field secrecy ISAC scenario, with a focus on revealing the benefits of near-field operation for secure communication. To achieve this, we formulate a near-field secrecy beam-focusing problem to maximize the secrecy rate for the communication user while meeting the transmit beampattern gain requirement for each radar target. The considered problem is new and has not been previously studied in either the conventional far-field or near-field contexts.
- While the formulated problem is non-convex, we achieve a global optimum solution by employing a two-stage procedure based on semidefinite relaxation (SDR). Additionally, we propose a novel low-complexity sub-optimal beam-focusing design tailored for this new problem.
- Finally, we provide numerical results to validate the effectiveness of our proposed secrecy beam-focusing designs. Notably, our proposed beam-focusing designs enable a positive secrecy rate even when the radar targets and communication user align along the same angle direction, which is unattainable with far-field beam steering. Additionally, we illustrate the performance trade-off between radar sensing and secrecy communication in near-field ISAC systems.

The remainder of this paper is organized as follows: Section 2 introduces the system model and problem formulation for near-field secrecy ISAC systems. Section 3 presents the development of an optimal solution and a sub-optimal, lower-complexity alternative. In Section 4, we provide extensive simulation results to validate our designs and illustrate the trade-offs involved. Finally, Section 5 concludes the paper with a summary of our findings and potential future research directions.

Let boldface lower-case and upper-case letters denote vectors and matrices, respectively. Use $\|\cdot\|$ for the ℓ_2 norm, $(\cdot)^T$ for the transpose, $(\cdot)^H$ for the Hermitian transpose, $\text{Null}(\cdot)$ for the null space of a matrix, and $\mathbb{C}$ to represent complex numbers.

2. System Model and Problem Formulation

Consider a near-field secrecy ISAC system, as illustrated in Figure 1. Here, one transmitter equipped with an extremely large-scale uniform linear array (ULA) of N antenna elements serves a single-antenna communication user and simultaneously senses L potential targets. Both the radar targets and the communication user are situated within the near-field region of the transmitter. These radar targets, being untrusted, might eavesdrop on the confidential messages intended for the communication user. With an inter-element spacing of d, the total antenna array aperture is $D = (N-1)d$. The boundary separating the near-field and far-field regions, known as either the Fraunhofer or Rayleigh distance, is defined as $d_F = \frac{2D^2}{\lambda}$ [19]. As the antenna count N increases, d_F can extend over several hundred meters, encompassing numerous communication users and sensing targets. Given these considerations, research on near-field ISAC systems becomes increasingly crucial.

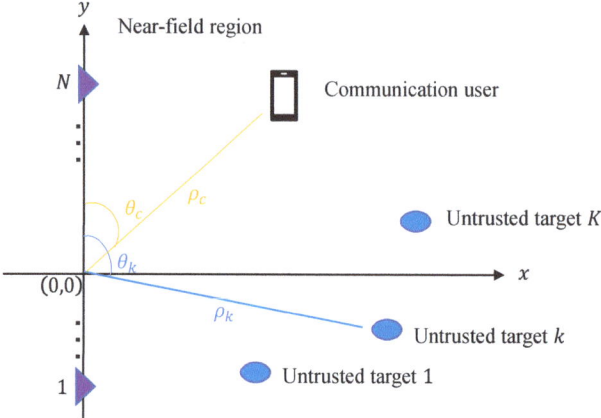

Figure 1. Illustration of near-field secrecy ISAC systems, consisting of one extremely large-scale antenna array that senses potential radar targets and serves communication users simultaneously.

2.1. Near-Field Channel Model

The ULA of the transmitter is positioned along the y-axis. The Cartesian coordinate of the nth antenna element is represented as $p_n = \left(0, \left(n - \frac{N+1}{2}\right)d\right)$, with $n = 1, 2, \ldots, N$. The communication user is located at $p_c = (\rho \sin\theta, \rho \cos\theta)$, with ρ and θ denoting the distance and angle between the communication user and the array center, respectively. Let $d_n \triangleq \sqrt{(\rho \sin\theta)^2 + \left(\rho \cos\theta - \left(n - \frac{N+1}{2}\right)d\right)^2}$ denote the distance between the nth antenna element of the transmitter and the communication user. Denote c and f_c as the light speed and the carrier, respectively. Then, the near-field channel between the transmitter and the communication user is given by

$$\mathbf{h}(\rho, \theta) = g e^{-j2\pi f_c \frac{\rho}{c}} \mathbf{a}(\rho, \theta), \tag{1}$$

where $g \triangleq \sqrt{\left(\frac{c}{4\pi f_c d_1}\right)^2}$, and $\mathbf{a}(\rho, \theta)$ denotes the steering vector, given by

$$\mathbf{a}(\rho, \theta) = \left[e^{-j2\pi f_c \frac{(d_1 - \rho)}{c}}, \ldots, e^{-j2\pi f_c \frac{(d_N - \rho)}{c}}\right]^T.$$

Let ρ_l and θ_l denote the distance and angle between the lth untrusted radar target and the array center, respectively. The corresponding near-field channel between the transmitter and the lth untrusted radar target is given by

$$\mathbf{h}(\rho_l, \theta_l) = g_l e^{-j2\pi f_c \frac{\rho_l}{c}} \mathbf{a}(\rho_l, \theta_l), \quad 1 \leq l \leq L, \tag{2}$$

where $g_l \triangleq \sqrt{\left(\frac{c}{4\pi f_c d_1^l}\right)^2}$, and the steering vector $\mathbf{a}(\rho_l, \theta_l)$, given by

$$\mathbf{a}(\rho_l, \theta_l) = \left[e^{-j2\pi f_c \frac{(d_1^l - \rho_l)}{c}}, \cdots, e^{-j2\pi f_c \frac{(d_N^l - \rho_l)}{c}} \right]^T.$$

2.2. Signal Model

The transmitter uses beam focusing to transmit a combined signal for both communication and radar-sensing tasks. Represented by $x \in \mathbb{C}^{N \times 1}$, the baseband signal at the transmitter is given by

$$x = \mathbf{w}s + t_0, \tag{3}$$

where $\mathbf{w} \in \mathbb{C}^{N \times 1}$ is the beam-focusing vector for the communication user, with s being the information-bearing signal. On the other hand, t_0 is the dedicated radar signal with zero mean, and its covariance matrix is denoted as $\mathbf{R} = \mathbb{E}[t_0 t_0^H] \succeq \mathbf{0}$.

According to (1) and (3), the received signal at the communication user can be written as $y_c = \mathbf{h}^H(\rho, \theta)(\mathbf{w}s + t_0) + n_c$, where n_c denotes the additive noise at the communication user, with zero mean and variance of δ_c^2. Thus, the signal-to-interference-plus-noise ratio (SINR) at the communication user is expressed as

$$\Gamma_c(\mathbf{w}, \mathbf{R}) = \frac{\left| \mathbf{h}(\rho, \theta)^H \mathbf{w} \right|^2}{\mathbf{h}^H(\rho, \theta) \mathbf{R} \mathbf{h}(\rho, \theta) + \sigma_c^2}. \tag{4}$$

In the near-field ISAC scenarios, both communication and radar signals can contribute to radar-sensing tasks. Consequently, the transmit beampattern gain at the lth radar target is defined as [20]

$$\begin{aligned} \mathcal{B}_l(\mathbf{w}, \mathbf{R}) &= \mathbb{E}\left[\left| a^H(\rho_l, \theta_l)(\mathbf{w}s + t_0) \right|^2 \right] \\ &= a^H(\rho_l, \theta_l)\left(\mathbf{w}\mathbf{w}^H + \mathbf{R} \right) a(\rho_l, \theta_l). \end{aligned} \tag{5}$$

The radar targets are untrusted nodes, which potentially intercept the confidential messages intended for the communication user. The SINR at the lth target is given by

$$\Gamma_l(\mathbf{w}, \mathbf{R}) = \frac{\left| \mathbf{h}(\rho_l, \theta_l)^H \mathbf{w} \right|^2}{\mathbf{h}^H(\rho_l, \theta_l) \mathbf{R} \mathbf{h}(\rho_l, \theta_l) + \sigma_l^2}, \quad 1 \leq l \leq L. \tag{6}$$

Based on (4) and (6), the achievable secrecy rate of the communication user can be expressed as

$$r(\mathbf{w}, \mathbf{R}) = \left[\log_2(1 + \Gamma_c(\mathbf{w}, \mathbf{R})) - \max_{1 \leq l \leq L} \log_2(1 + \Gamma_l(\mathbf{w}, \mathbf{R})) \right]^+, \tag{7}$$

where $[x]^+ \triangleq \max(0, x)$.

2.3. Problem Formulation

We are interested in characterizing the performance trade-off between radar sensing and secrecy communication. Specifically, we aim to maximize the secrecy rate of the communication user while satisfying the transmit beampattern gains for radar targets, by jointly optimizing the beam-focusing vector $\mathbf{w}$ and the radar covariance matrix $\mathbf{R}$. Mathematically, the problem of interest is formulated as

$$\max_{\mathbf{w},\mathbf{R}} \ r(\mathbf{w},\mathbf{R})$$
$$\text{s.t.} \ \mathcal{B}_l(\mathbf{w},\mathbf{R}) \geq b_l, \ 1 \leq l \leq L, \tag{8}$$
$$\|\mathbf{w}\|^2 + \text{Tr}(\mathbf{R}) \leq P_t, \ \mathbf{R} \succeq 0,$$

where b_l denotes the transmit beampattern gain threshold for the lth radar target, and P_t is the transmit power at the transmitter.

3. Proposed Solution

In this section, we first propose a two-stage optimization approach to solve the non-convex problem (8) globally optimally in Section 3.1. Then, in Section 3.2, we develop a low-complexity sub-optimal solution.

3.1. Optimal Solution

In this subsection, we present an SDR-based two-stage procedure to achieve the optimal solution for (8). This is accomplished by decomposing (8) into two sub-problems. First, by introducing a variable γ, we reformulate (8) as

$$\max_{\mathbf{w},\mathbf{R}} \ \frac{|\mathbf{h}(\rho,\theta)^H\mathbf{w}|^2}{\mathbf{h}^H(\rho,\theta)\mathbf{R}\mathbf{h}(\rho,\theta) + \sigma_c^2}$$
$$\text{s.t.} \ \frac{|\mathbf{h}(\rho_l,\theta_l)^H\mathbf{w}|^2}{\mathbf{h}^H(\rho_l,\theta_l)\mathbf{R}\mathbf{h}(\rho,\theta) + \sigma_l^2} \leq \gamma, \ 1 \leq l \leq L, \tag{9}$$
$$\mathbf{a}^H(\rho_l,\theta_l)\left(\mathbf{w}\mathbf{w}^H + \mathbf{R}\right)\mathbf{a}(\rho_l,\theta_l) \geq b_l, \ 1 \leq l \leq L,$$
$$\|\mathbf{w}\|^2 + \text{Tr}(\mathbf{R}) \leq P_t, \ \mathbf{R} \succeq 0.$$

Similar to [21], it can be proven that there always exists a γ for which problem (9) shares the same optimal solution as problem (8). To be specific, let $f(\gamma)$ denote the optimal value of (9) for a given $\gamma > 0$. Consequently, the optimal value of (8) is equivalent to the following problem:

$$\max_{\gamma > 0} \ \log_2\left(\frac{1+f(\gamma)}{1+\gamma}\right) \tag{10}$$

Based on the above discussion, we conclude that problem (8) can be solved through a two-step process. Firstly, we solve (9) to obtain the optimal value $f(\gamma)$ for any given γ. Secondly, we solve (10) to find the optimal γ^* by conducting a one-dimensional search over the parameter space. Consequently, we next focus on solving (9).

Problem (9) is still non-convex. Next, we apply the SDR technique to solve (9) globally optimally. To this end, by defining $\mathbf{W} = \mathbf{w}\mathbf{w}^H$ and ignoring the non-convex rank-one constraint on $\mathbf{W}$, we recast (9) as

$$\max_{\mathbf{W},\mathbf{R}} \ \frac{\text{Tr}[\mathbf{H}(\rho,\theta)\mathbf{W}]}{\text{Tr}[\mathbf{H}(\rho,\theta)\mathbf{R}] + \sigma_c^2}$$
$$\text{s.t.} \ \frac{\text{Tr}[\mathbf{H}(\rho_l,\theta_l)\mathbf{W}]}{\text{Tr}[\mathbf{H}(\rho_l,\theta_l)\mathbf{R}] + \sigma_l^2} \leq \gamma, \ 1 \leq l \leq L,$$
$$\text{Tr}[\mathbf{A}(\rho_l,\theta_l)(\mathbf{W}+\mathbf{R})] \geq b_l, \ 1 \leq l \leq L, \tag{11}$$
$$\text{Tr}(\mathbf{W}) + \text{Tr}(\mathbf{R}) \leq P_t,$$
$$\mathbf{W} \succeq 0, \ \mathbf{R} \succeq 0,$$

where $\mathbf{H}(\rho,\theta) \triangleq \mathbf{h}(\rho,\theta)\mathbf{h}(\rho,\theta)^H$, $\mathbf{H}(\rho_l,\theta_l) \triangleq \mathbf{h}(\rho_l,\theta_l)\mathbf{h}(\rho_l,\theta_l)^H$, and $\mathbf{A}(\rho_l,\theta_l) \triangleq \mathbf{a}(\rho_l,\theta_l)\mathbf{a}(\rho_l,\theta_l)^H$.

Problem (11) is non-convex due to the non-concave nature of its objective function. Nevertheless, we can effectively address this by employing the Charnes–Cooper transformation [22], which allows us to equivalently reformulate (11) as

$$\begin{aligned}
\max_{\bar{\mathbf{W}},\bar{\mathbf{R}},\tau} \quad & \operatorname{Tr}[\mathbf{H}(\rho,\theta)\bar{\mathbf{W}}] \\
\text{s.t.} \quad & \operatorname{Tr}[\mathbf{H}(\rho,\theta)\bar{\mathbf{R}}] + \tau\sigma_c^2 = 1, \\
& \operatorname{Tr}[\mathbf{H}(\rho_l,\theta_l)(\bar{\mathbf{W}} - \gamma\bar{\mathbf{R}})] \leq \tau\sigma_l^2 \gamma, 1 \leq l \leq L, \\
& \operatorname{Tr}[\mathbf{A}(\rho_l,\theta_l)(\bar{\mathbf{W}} + \bar{\mathbf{R}})] \geq \tau b_l, \ 1 \leq l \leq L, \\
& \operatorname{Tr}(\bar{\mathbf{W}}) + \operatorname{Tr}(\bar{\mathbf{R}}) \leq \tau P_t, \\
& \bar{\mathbf{W}} \succeq 0, \ \bar{\mathbf{R}} \succeq 0, \tau > 0.
\end{aligned} \quad (12)$$

Problem (12) is convex, and thus, it can be solved directly using existing solvers such as CVX [23].

Let $\{\bar{\mathbf{W}}^*, \bar{\mathbf{R}}^*, \tau^*\}$ represent the optimal solution to (12). It can be readily demonstrated that $\mathbf{W}^* = \frac{\bar{\mathbf{W}}^*}{\tau^*}$ and $\mathbf{R}^* = \frac{\bar{\mathbf{R}}^*}{\tau^*}$ are the optimal solution to (11). If $\operatorname{Rank}(\mathbf{W}^*) = 1$, the relaxation of the rank-one constraint in (11) does not impact its optimality. We now present a theorem affirming that we can always construct an optimal solution to (12) with a rank-one matrix $\bar{\mathbf{W}}$.

Theorem 1. *Suppose that $\{\bar{\mathbf{W}}^*, \bar{\mathbf{R}}^*, \tau^*\}$ represents the optimal solution to problem (12) with $\operatorname{Rank}(\bar{\mathbf{W}}^*) > 1$. Then, we can construct an alternative feasible solution to (12), denoted as $\{\tilde{\mathbf{W}}^*, \tilde{\mathbf{R}}^*, \tilde{\tau}^*\}$. This alternative solution not only attains an equivalent objective value as $\{\bar{\mathbf{W}}^*, \bar{\mathbf{R}}^*, \tau^*\}$, but also satisfies $\operatorname{Rank}(\tilde{\mathbf{W}}^*) = 1$.*

Proof of Theorem 1. The Lagrangian of problem (12) is given by

$$\mathcal{L} = \operatorname{Tr}((\mathbf{A} + \mathbf{Z})\bar{\mathbf{W}}) + \operatorname{Tr}(\mathbf{B}\bar{\mathbf{R}}) + \rho\tau + \mu, \quad (13)$$

where $\mathbf{Z} \succeq 0$ is a matrix Lagrange multiplier associated with the constraint $\bar{\mathbf{W}} \succeq 0$; $\mathbf{A}$, $\mathbf{B}$, and ρ are, respectively, defined as

$$\mathbf{A} = \mathbf{H}(\rho,\theta) - \sum_{l=1}^{L} \beta_l \mathbf{H}(\rho_l,\theta_l) + \sum_{l=1}^{L} \eta_l \mathbf{A}(\rho_l,\theta_l) - \nu \mathbf{I}, \quad (14)$$

$$\mathbf{B} = -\mu \mathbf{H}(\rho,\theta) + \sum_{l=1}^{L} \beta_l \gamma \mathbf{H}(\rho_l,\theta_l) + \sum_{l=1}^{L} \eta_l \mathbf{A}(\rho_l,\theta_l) - \nu \mathbf{I}, \quad (15)$$

$$\rho = -\mu + \sum_{l=1}^{L} \beta_l \sigma_l^2 - \sum_{l=1}^{L} \eta_l b_l + \nu P_t, \quad (16)$$

where $\mu \geq 0$, $\{\beta_l \geq 0\}$, $\{\eta_l \geq 0\}$, and $\nu \geq 0$ represent the Lagrange multipliers corresponding to the first through fourth constraints in (12), respectively.

Given (13), the relevant Karush–Kuhn–Tucker (KKT) conditions for the proof are formulated as

$$\frac{\partial \mathcal{L}}{\partial \bar{\mathbf{W}}} = \mathbf{A}^* + \mathbf{Z}^* = \mathbf{0}, \quad (17a)$$

$$\mathbf{Z}^* \bar{\mathbf{W}}^* = \mathbf{0}, \quad (17b)$$

$$\mu^*, \beta_l^*, \eta_l^*, \nu^* \geq 0, \quad (17c)$$

where $\mathbf{Z}^*$, μ^*, β_l^*, η_l^*, and ν^* denote the optimal Lagrange multipliers for the dual problem of (12), and the resulting $\mathbf{A}$ and $\mathbf{B}$ are $\mathbf{A}^*$ and $\mathbf{B}^*$. Here, (17b) is obtained from the complementary slackness condition.

According to $\mathbf{Z}^* \succeq 0$, (17a) and (17b), we have

$$\mathbf{A}^* \preceq 0 \tag{18a}$$
$$\mathbf{A}^* \bar{\mathbf{W}}^* = 0. \tag{18b}$$

Define $\mathbf{C}^* = -\sum_{l=1}^{L} \beta_l^* \mathbf{H}(\rho_l, \theta_l) + \sum_{l=1}^{L} \eta_l^* \mathbf{A}(\rho_l, \theta_l) - \nu^* \mathbf{I}$. Then, we have

$$\mathbf{A}^* = \mathbf{C}^* + \mathbf{H}(\rho, \theta). \tag{19}$$

Based on (18b), the columns of $\bar{\mathbf{W}}^*$ must lie in the null space of $\mathbf{A}^*$ when $\bar{\mathbf{W}}^* \neq \mathbf{0}$. Define $M \triangleq \text{Rank}(\mathbf{C}^*)$. If $M = N$, we can conclude that $\text{Rank}(\mathbf{A}^*) \geq \text{Rank}(\mathbf{C}^*) - \text{Rank}(\mathbf{H}) = N - 1$. In this case, we have $\text{Rank}(\bar{\mathbf{W}}^*) = 1$ when $\bar{\mathbf{W}}^* \neq \mathbf{0}$.

We next discuss the case of $M < N$. In this case, let $\mathbf{X} \in \mathbb{C}^{N \times (N-M)}$ denote the orthonormal basis of the null space of $\mathbf{C}^*$, i.e., $\mathbf{C}^* \mathbf{X} = \mathbf{0}$. Let $\mathbf{x}_m, 1 \leq m \leq (N-M)$, denote the mth column of $\mathbf{X}$. For each $\mathbf{x}_m$, it follows

$$\mathbf{x}_m^H \mathbf{A}^* \mathbf{x}_m = \mathbf{x}_m^H \mathbf{C}^* \mathbf{x}_m + \mathbf{x}_m^H \mathbf{H} \mathbf{x}_m = \mathbf{x}_m^H \mathbf{H} \mathbf{x}_m \preceq 0. \tag{20}$$

From (20), we conclude that $\mathbf{x}_m^H \mathbf{H} \mathbf{x}_m = 0$. Thus, we have

$$N - M + 1 \geq \text{Rank}(\text{Null}(\mathbf{A}^*)) \geq N - M. \tag{21}$$

If $\text{Rank}(\text{Null}(\mathbf{A}^*)) = N - M$, we have $\text{Null}(\mathbf{A}^*) = \mathbf{X}$. This means $\bar{\mathbf{W}}^*$ can be expressed as $\bar{\mathbf{W}}^* = \sum_{m=1}^{M} a_m \mathbf{x}_m \mathbf{x}_m^H$. However, since $\mathbf{x}_m^H \mathbf{H} \mathbf{x}_m = 0$, in this case, the achievable secrecy rate is equal to zero. Thus, we obtain $\text{Rank}(\text{Null}(\mathbf{A}^*)) = N - M + 1$. As a result, the optimal $\bar{\mathbf{W}}^*$ can be expressed as

$$\bar{\mathbf{W}}^* = \sum_{m=1}^{M} a_m \mathbf{x}_m \mathbf{x}_m^H + d \mathbf{u} \mathbf{u}^H, \tag{22}$$

where $a_m \geq 0, \forall m$ and $d > 0$. $\mathbf{u}$ is an extra dimension of orthonormal basis, which lies in the null space of $\mathbf{A}^*$ and is orthogonal to the span of $\mathbf{X}$, i.e., $\mathbf{A}^* \mathbf{u} = \mathbf{0}$ and $\mathbf{X}\mathbf{u} = \mathbf{0}$.

Based on (22), for the case of $\text{Rank}(\bar{\mathbf{W}}^*) > 1$, we can construct another feasible solution to problem (12), given by

$$\tilde{\mathbf{W}}^* = \bar{\mathbf{W}}^* - \sum_{m=1}^{M} a_m \mathbf{x}_m \mathbf{x}_m^H = d \mathbf{u} \mathbf{u}^H, \tag{23}$$

$$\tilde{\mathbf{R}}^* = \bar{\mathbf{R}}^* + \sum_{m=1}^{M} a_m \mathbf{x}_m \mathbf{x}_m^H, \tag{24}$$

$$\tilde{\tau}^* = \tau^*. \tag{25}$$

It can be easily verified that $\{\tilde{\mathbf{W}}^*, \tilde{\mathbf{R}}^*, \tilde{\tau}^*\}$ is a feasible solution to problem (12), which can not only achieve the same objective value as $\{\bar{\mathbf{W}}^*, \bar{\mathbf{R}}^*, \tau^*\}$, but also satisfies $\text{Rank}(\tilde{\mathbf{W}}^*) = 1$. The proof of the optimality of $\{\tilde{\mathbf{W}}^*, \tilde{\mathbf{R}}^*, \tilde{\tau}^*\}$ in problem (12) is thus completed. □

Theorem 1 indicates that the optimal solution $\mathbf{w}^*$ to (9) can be accurately extracted by the eigenvalue decomposition of $\mathbf{W}^*$, with $\mathbf{W}^* = \frac{\tilde{\mathbf{W}}^*}{\tilde{\tau}^*}$. Consequently, $\{\mathbf{w}^*, \tilde{\mathbf{R}}^*\}$ are the globally optimal solution to problem (9).

3.2. ZF-Based Sub-Optimal Solution

In this subsection, we propose a sub-optimal solution based on zero-forcing (ZF) for problem (8), which exhibits a significantly lower computational complexity compared to the optimal solution. This sub-optimal approach involves aligning the information beam $\mathbf{w}$ with the null space of the radar target's channel to prevent any information leakage to potential eavesdroppers, as expressed by $\mathbf{h}(\rho_l, \theta_l)^H \mathbf{w} = 0, 1 \leq l \leq L$. Simultaneously, the radar signal is constrained to lie within the null space of the communication user, specified as $\mathbf{h}^H(\rho, \theta)\mathbf{R}\mathbf{h}(\rho, \theta) = 0$, ensuring that it does not interfere with the communication user.

Define $\mathbf{H} \triangleq [\mathbf{h}(\rho_1, \theta_1), \cdots, \mathbf{h}(\rho_L, \theta_L)]^H$. Denote its singular value decomposition as $\mathbf{H} = \mathbf{U}\boldsymbol{\Sigma}\mathbf{V}^H = \mathbf{U}\boldsymbol{\Sigma}[\mathbf{V}_1, \mathbf{V}_2]^H$, where $\mathbf{U} \in \mathbb{C}^{L \times L}$ and $\mathbf{V} \in \mathbb{C}^{N \times N}$ are unitary matrices, $\boldsymbol{\Sigma}$ is a $L \times N$ diagonal matrix, $\mathbf{V}_1 \in \mathbb{C}^{N \times L}$ and $\mathbf{V}_2 \in \mathbb{C}^{N \times (N-L)}$ are the first L and the last $N - L$ right singular vectors of $\mathbf{H}$. In order to guarantee $\mathbf{h}(\rho_l, \theta_l)^H \mathbf{w} = 0, 1 \leq l \leq L$, the ZF-based $\mathbf{w}$ has the following form:

$$\mathbf{w} = \sqrt{p_c}\mathbf{V}_2\tilde{\mathbf{w}}, \tag{26}$$

where p_c denotes the transmit power of the information beam, and $\tilde{\mathbf{w}} \in \mathbb{C}^{(N-L) \times 1}$ is an arbitrary complex vector of unit norm. To maximize the secrecy rate, $\tilde{\mathbf{w}}$ should match the equivalent channel $\mathbf{h}(\rho, \theta)^H \mathbf{V}_2$, and thus $\tilde{\mathbf{w}} = \frac{\mathbf{V}_2^H \mathbf{h}(\rho, \theta)}{\|\mathbf{V}_2^H \mathbf{h}(\rho, \theta)\|}$.

According to (26), the achievable secrecy rate becomes

$$r_{\text{ZF}} = \log_2\left(1 + \frac{p_c \|\mathbf{h}(\rho, \theta)^H \mathbf{V}_2\|^2}{\sigma_c^2}\right). \tag{27}$$

To maximize the secrecy rate in (27), we need to minimize the power consumption required to ensure the radar-sensing task. Consequently, we can obtain the optimal radar-sensing covariance matrix $\mathbf{R}$ by solving the following problem:

$$\begin{aligned}
\min_{\mathbf{R}} \quad & \text{Tr}(\mathbf{R}) \\
\text{s.t.} \quad & \mathbf{a}^H(\rho_l, \theta_l)\mathbf{R}\mathbf{a}(\rho_l, \theta_l) \geq b_l, \ 1 \leq l \leq L, \\
& \mathbf{h}^H(\rho, \theta)\mathbf{R}\mathbf{h}(\rho, \theta) = 0.
\end{aligned} \tag{28}$$

Problem (28) is convex and can be efficiently solved using CVX [23]. Let $\mathbf{R}^*$ represent the optimal solution to (28). Consequently, the transmit power p_c in (26) or (27) can be expressed as $p_c = P_t - \text{Tr}(\mathbf{R}^*)$.

4. Simulation Results and Discussion

In this section, we showcase numerical results to validate our secure beam-focusing design in the context of near-field ISAC systems. Our system setup involves a transmitter with a ULA of $N = 64$ antennas, operating at 2.4 GHz. The Rayleigh distance for this setup is $d_F = 248$ m. We position the communication user at $(32, 0)$ meters on the x-axis. All receivers are assumed to have an identical noise power of $\sigma_c^2 = \sigma_1^2 = \cdots = \sigma_L^2 = -95$ dBm.

We first show the capabilities of beam focusing for ensuring secure communication. For our demonstration, we consider that a single radar target, which can potentially act as an eavesdropper, is positioned at $(16, 0)$ meters. This location coincides with the angular direction of the communication user. Figure 2a,b show the normalized signal power of the communication signal and radar signal at each point of the near-field xy-plane, respectively, given the transmit power $P_t = 1$ dB and the transmit beampattern gain threshold $b_l = 40, \forall l$. Both the communication signal beam $\mathbf{w}$ and radar signal covariance matrix $\mathbf{R}$ are obtained by solving problem (8) based on the given user locations and the predefined threshold values. From Figure 2a, we observe that most of the energy from the communication beam $\mathbf{w}$ is concentrated around the communication user. This results in minimal information leakage to the radar target. In contrast, Figure 2b shows that the energy from the radar covariance matrix $\mathbf{R}$ predominantly focuses on the radar target, ensuring limited interference to the communication user. Analyzing both Figure 2a,b, we conclude that our proposed secure beam focusing is adept at facilitating secure communication even when the radar targets and communication user occupy the same angular position—a feat not feasible with traditional far-field beam steering.

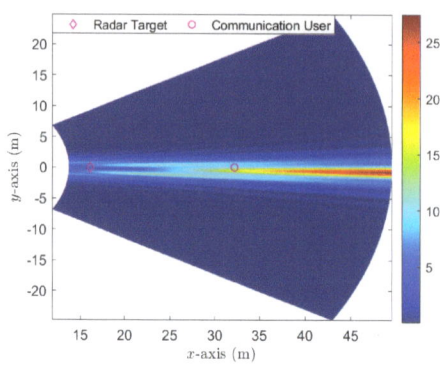

(a)

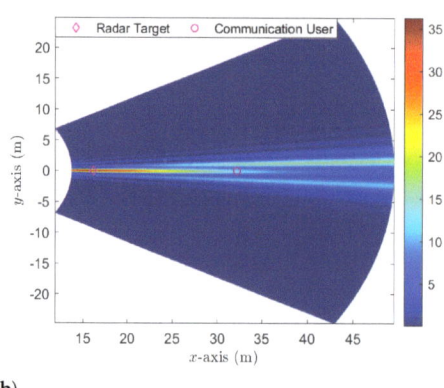

(b)

Figure 2. The normalized signal power of communication and radar signals. (**a**) Communication signal; (**b**) radar signal.

Figure 3 depicts the trade-off between the achievable secrecy rate and the transmit beampattern gain, as evaluated using both the optimal and sub-optimal solutions. The simulation settings are consistent with those used in Figure 2. It is observed that, as the target beampattern gain requirement rises, the achievable secrecy rate declines sharply. This decline occurs because with a fixed and limited total transmit power, dedicating more power to radar sensing reduces the power available for secure communication. This observation indicates that careful design of the secure beam-focusing scheme is crucial for balancing these two performances. Additionally, the optimal solution consistently surpasses the sub-optimal solution regarding the trade-off between secrecy rate and target beampattern gain. This superior performance stems from the more comprehensive and precise optimization techniques used in deriving the optimal solution, albeit at the cost of increased computational complexity. As anticipated, increasing the transmit budget enhances either the secrecy rate or the target beampattern gain.

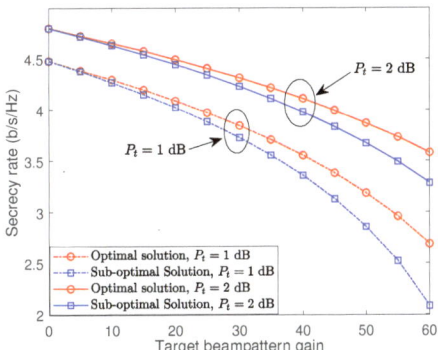

Figure 3. Trade-off between the secrecy rate and the transmit beampattern gain.

Figure 4 effectively illustrates the relationship between the achievable secrecy rate and transmit power, evaluated using both the optimal and sub-optimal near-field solutions, along with the far-field beam steering solution. The far-field beam steering solution is derived by substituting the precise near-field channel model with a standard far-field channel model, thereby neglecting the effects of the near-field. The simulation settings are consistent with those used in Figure 2. Figure 4 shows that the secrecy rates achievable by both the optimal and sub-optimal near-field solutions increase as the transmit power increases. In contrast, the achievable rate of the far-field beam steering remains consistently at zero. This is due to the untrusted radar target, potentially an eavesdropper, sharing an identical angular direction with the communication user, coupled with a relatively smaller distance from the transmitter. Consequently, this potential eavesdropper experiences superior channel conditions compared to the legitimate communication user. Under these circumstances, far-field beam steering fails to differentiate between the untrusted radar target and legitimate communication users, resulting in a zero secrecy rate. Conversely, near-field solutions yield positive secrecy rates, thanks to the beam-focusing capability in the near-field, which proficiently distinguishes users in both angular and distance dimensions.

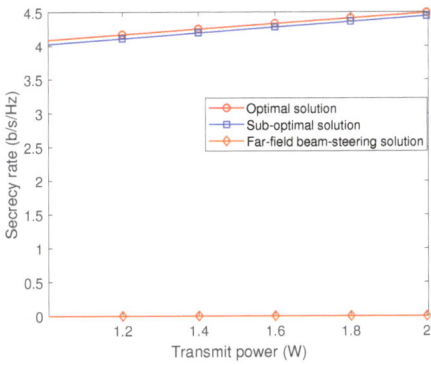

Figure 4. Secrecy rate versus the transmit power.

Finally, we study a general scenario with varying numbers of radar targets randomly positioned on the near-field xy-plane. Figure 5 shows a plot of the achievable secrecy rate against the number of radar targets L, with a consistent radar target threshold $b_l = 20, \forall l$. We successively add radar targets to the near-field xy-plane. As seen in Figure 5, the achievable secrecy rate for both optimal and sub-optimal solutions consistently decreases as the number of radar targets grows. This is because, as the quantity of radar targets rises, so does the likelihood of intercepted communications, consequently leading to a reduced achievable secrecy rate. Additionally, a higher transmit power can improve the secrecy rate for a given radar target count.

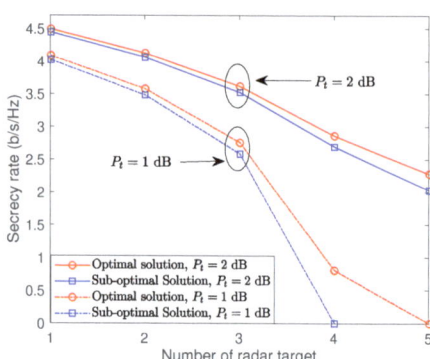

Figure 5. Secrecy rate versus the number of radar targets.

5. Conclusions

This paper addresses beam-focusing design in near-field ISAC systems using XL-MIMO, focusing on maximizing the secrecy rate while meeting radar target requirements. It introduces both globally optimal and practical low-complexity sub-optimal solutions, catering to different computational capacities. Our numerical results demonstrate the efficacy of these solutions, notably achieving positive secrecy rates in challenging scenarios where the radar targets and communication user are positioned at identical angles. This study thus marks a significant advancement in secure communication within near-field ISAC environments, showcasing the potential of XL-MIMO in overcoming traditional security challenges.

One interesting extension of this paper is to use the Cramér–Rao bound (CRB) to evaluate the estimation accuracy. In such a case, the beam-focusing vectors and the radar covariance matrix should be jointly optimized to minimize the CRB. Furthermore, the beam-focusing design can also be extended to the robust beam-focusing case, where perfect channel state information is not necessary. Additionally, while our proposed algorithms have been effectively validated through simulations, we recognize the need for their verification through practical measurements in future research. This step is crucial for a comprehensive assessment of their real-world applicability and performance.

Author Contributions: Conceptualization, P.S. and H.D.; methodology, P.S.; software, P.S.; validation, P.S., H.D. and B.W.; formal analysis, P.S.; investigation, P.S.; resources, P.S.; data curation, P.S.; writing—original draft preparation, P.S.; writing—review and editing, H.D. and B.W.; visualization, P.S.; supervision, H.D. and B.W.; project administration, B.W.; funding acquisition, B.W. All authors have read and agreed to the published version of the manuscript.

Funding: This work was supported by the Open Foundation of National Railway Intelligence Transportation System Engineering Technology Research Center under grant No. RITS2021KF02 and the National Natural Science Foundation of China under grant No. 61971238.

Institutional Review Board Statement: Not applicable.

Informed Consent Statement: Not applicable.

Data Availability Statement: Data is contained within the article.

Conflicts of Interest: The authors declare no conflict of interest.

References

1. Liu, X.; Huang, T.; Shlezinger, N.; Liu, Y.; Zhou, J.; Eldar, Y.C. Joint transmit beamforming for multiuser MIMO communications and MIMO radar. *IEEE Trans. Signal Process.* **2020**, *68*, 3929–3944. [CrossRef]
2. Liu, F.; Cui, Y.; Masouros, C.; Xu, J.; Han, T.X.; Eldar, Y.C.; Buzzi, S. Integrated sensing and communications: Toward dual-functional wireless networks for 6G and beyond. *IEEE J. Sel. Areas Commun.* **2022**, *40*, 1728–1767. [CrossRef]
3. Liu, A.; Huang, Z.; Li, M.; Wan, Y.; Li, W.; Han, T.; Liu, C.; Du, R.; Tan, D.; Lu, J., and others. A survey on fundamental limits of integrated sensing and communication. *IEEE Commun. Surv. Tutor.* **2022**, *24*, 994–1034. [CrossRef]
4. Sheikh, M.S.; Liang, J.; Wang, W. A survey of security services, attacks, and applications for vehicular ad hoc networks (VANETs). *Sensors* **2019**, *19*, 3589. [CrossRef]
5. Hartenstein, H.; Laberteaux, L.P. A tutorial survey on vehicular ad hoc networks. *IEEE Commun. Mag.* **2008**, *46*, 164–171. [CrossRef]
6. Wu, Q.; Wang, S.; Ge, H.; Fan, P.; Fan, Q.; Letaief, K.B. Delay-sensitive task offloading in vehicular fog computing-assisted platoons. *arXiv* **2023**, arXiv:2309.10234.
7. Wu, Q.; Zhao, Y.; Fan, Q.; Fan, P.; Wang, J.; Zhang, C. Mobility-aware cooperative caching in vehicular edge computing based on asynchronous federated and deep reinforcement learning. *IEEE J. Sel. Top. Signal Process.* **2022**, *17*, 66–81. [CrossRef]
8. Wei, Z.; Liu, F.; Masouros, C.; Su, N.; Petropulu, A.P. Toward multi-functional 6G wireless networks: Integrating sensing, communication, and security. *IEEE Commun. Mag.* **2022**, *60*, 65–71. [CrossRef]
9. Deligiannis, A.; Daniyan, A.; Lambotharan, S.; Chambers, J.A. Secrecy rate optimizations for MIMO communication radar. *IEEE Trans. Aerosp. Electron. Syst.* **2018**, *54*, 2481–2492. [CrossRef]
10. Ren, Z.; Qiu, L.; Xu, J. Optimal transmit beamforming for secrecy integrated sensing and communication. In Proceedings of the ICC 2022-IEEE International Conference on Communications, Seoul, Republic of Korea, 16–20 May 2022; pp. 5555–5560.
11. Su, N.; Liu, F.; Masouros, C. Secure Radar-Communication Systems With Malicious Targets: Integrating Radar, Communications and Jamming Functionalities. *IEEE Trans. Wirel. Commun.* **2021**, *20*, 83–95. [CrossRef]
12. Naser, M.; Alsabah, M.; Mahmmod, B.; Noordin, N.; Abdulhussain, S.; Baker, T. Downlink training design for FDD massive MIMO systems in the presence of colored noise. *Electronics* **2020**, *9*, 2155. [CrossRef]
13. Alsabah, M.; Naser, M.; Mahmmod, B.; Noordin, N.; Abdulhussain, S. Sum rate maximization versus MSE minimization in FDD massive MIMO systems with short coherence time. *IEEE Access* **2021**, *9*, 108793–108808. [CrossRef]
14. Zhang, X.; Wang, Z.; Zhang, H.; Yang, L. Near-field channel estimation for extremely large-scale array communications: A model-based deep learning approach. *IEEE Commun. Lett.* **2023**, *27*, 1155–1159. [CrossRef]
15. Zhang, X.; Zhang, H.; Zhang, J.; Li, C.; Huang, Y.; Yang, L. Codebook Design for Extremely Large-Scale MIMO Systems: Near-field and Far-field. *arXiv* **2021**, arXiv:2109.10143.
16. Wang, Z.; Mu, X.; Liu, Y. Near-field integrated sensing and communications. *IEEE Commun. Lett.* **2023**, *27*, 2048–2052. [CrossRef]
17. Zhang, H.; Shlezinger, N.; Guidi, F.; Dardari, D.; Eldar, Y.C. 6G Wireless Communications: From Far-Field Beam Steering to Near-Field Beam Focusing. *IEEE Commun. Mag.* **2023**, *61*, 72–77. [CrossRef]
18. Zhang, H.; Shlezinger, N.; Guidi, F.; Dardari, D.; Imani, M.F.; Eldar, Y.C. Beam focusing for near-field multiuser MIMO communications. *IEEE Trans. Wirel. Commun.* **2022**, *21*, 7476–7490. [CrossRef]
19. Guerra, A.; Guidi, F.; Dardari, D.; Djurić, P.M. Near-field tracking with large antenna arrays: Fundamental limits and practical algorithms. *IEEE Trans. Signal Process.* **2021**, *69*, 5723–5738. [CrossRef]
20. Hua, H.; Xu, J.; Han, T.X. Optimal transmit beamforming for integrated sensing and communication. *IEEE Trans. Veh. Technol.* **2023**, *72*, 10588–10603. [CrossRef]
21. Zhang, H.; Huang, Y.; Li, C.; Yang, L. Secure beamforming design for SWIPT in MISO broadcast channel with confidential messages and external eavesdroppers. *IEEE Trans. Wirel. Commun.* **2016**, *15*, 7807–7819. [CrossRef]
22. Chang, T.-H.; Hsin, C.-W.; Ma, W.-K.; Chi, C.-Y. A linear fractional semidefinite relaxation approach to maximum-likelihood detection of higher-order QAM OSTBC in unknown channels. *IEEE Trans. Signal Process.* **2009**, *58*, 2315–2326. [CrossRef]
23. Grant, M.; Boyd, S. CVX: Matlab Software for Disciplined Convex Programming, Version 2.1. 2014. Available online: http://cvxr.com/cvx/ (accessed on 25 December 2023).

Disclaimer/Publisher's Note: The statements, opinions and data contained in all publications are solely those of the individual author(s) and contributor(s) and not of MDPI and/or the editor(s). MDPI and/or the editor(s) disclaim responsibility for any injury to people or property resulting from any ideas, methods, instructions or products referred to in the content.

Article

A Novel Analytical Model for the IEEE 802.11p/bd Medium Access Control, with Consideration of the Capture Effect in the Internet of Vehicles

Yang Wang [1], Jianghong Shi [1,*], Zhiyuan Fang [2] and Lingyu Chen [1]

[1] School of Informatics, Xiamen University, Xiamen 361005, China; yangwang@stu.xmu.edu.cn (Y.W.); chenly@xmu.edu.cn (L.C.)
[2] China Mobile Communications Group Shanxi Co., Ltd., Taiyuan 030032, China; fangzhiyuan@sx.chinamobile.com
* Correspondence: shijh@xmu.edu.cn; Tel.: +86-139-0601-3456

Abstract: The traditional vehicular ad hoc network (VANET), which is evolving into the internet of vehicles (IoV), has drawn great attention for its enormous potential in road safety improvement, traffic management, infotainment service support, and even autonomous driving. IEEE 802.11p, as the vital standard for wireless access in vehicular environments, has been released for more than one decade and its evolution, IEEE 802.11bd, has also been released for a few months. Since the analytical models for the IEEE 802.11p/bd medium access control (MAC) play important roles in terms of performance evaluation and MAC protocol optimization, a lot of analytical models have been proposed. However, the existing analytical models are still not accurate as a result of ignoring some important factors of the MAC itself and real communication scenarios. Motivated by this, a novel analytical model is proposed, based on a novel two-dimensional (2-D) Markov chain model. In contrast to the existing studies, all the important factors are considered in this proposed model, such as the backoff freezing mechanism, retry limit, post-backoff states, differentiated packet arrival probabilities for empty buffer queue, and queue model of packets in the buffer. In addition, the influence of the capture effect under a Nakagami-*m* fading channel has also been considered. Then, the expressions of successful transmission, collided transmission, normalized unsaturated throughput, and average packet delay are all meticulously derived, respectively. At last, the accuracy of the proposed analytical model is verified via the simulation results, which show that it is more accurate than the existing analytical models.

Keywords: vehicular ad hoc network; internet of vehicles; IEEE 802.11p/bd; medium access control; capture effect; Nakagami-*m* fading

1. Introduction

The vehicular ad hoc network (VANET) has been a widespread concern of academia and industry for its enormous potential in improving road safety, promoting traffic efficiency, providing infotainment services, and even supporting autonomous driving [1,2]. Benefiting from the rapid development of information and communication technology [3–11], the traditional VANET is evolving into the internet of vehicles (IoV) [12]. It can support heterogeneous vehicular communication modes, including vehicle-to-vehicle (V2V), vehicle-to-pedestrian (V2P), vehicle-to-infrastructure (V2I) and vehicle-to-network (V2N), as shown in Figure 1, for satisfying the requirements of different safety or non-safety applications [13,14]. However, the key to accomplish differentiated applications depends on whether the vehicles effectively access the wireless channel. As an important channel access standard for IoV, IEEE 802.11p outlines the specifications of the physical (PHY) layer and medium access control (MAC) layer, where the latter includes the distributed coordination function (DCF) and enhanced distributed channel access (EDCA). According to the EDCA, four access categories (ACs) are defined. In fact, each AC queue is called an enhanced distributed channel access function (EDCAF), which is an enhanced variant of the DCF. It contends for the transmission opportunity (TXOP) by using a set of EDCA parameters [15]. Actually, IEEE 802.11bd, as an evolving version of IEEE 802.11p, also adopts DCF and EDCA protocols [16]. Due to the fact that the DCF protocol is the essential channel access protocol of IEEE 802.11p/bd, it is necessary to propose an effective analytical model for evaluating the precise performance of IEEE 802.11p/bd DCF (or EDCA) in IoV.

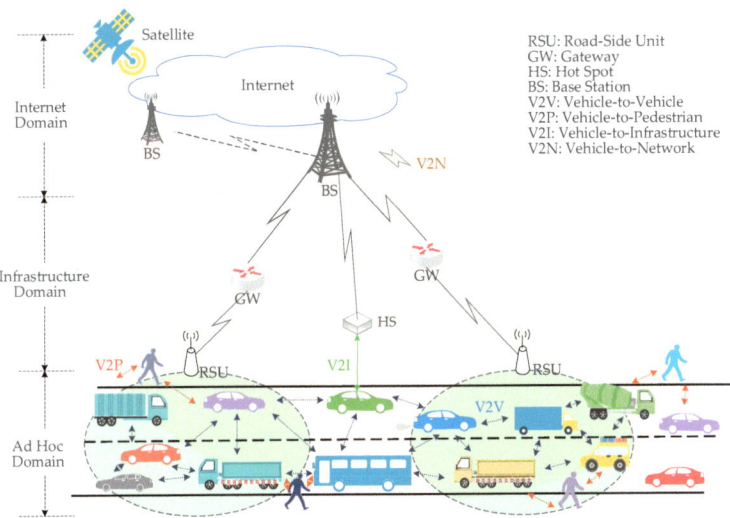

Figure 1. Heterogeneous vehicular communications structure.

Since the DCF is the basis of the IEEE 802.11 series of standards, a lot of analytical models based on Bianchi's pioneering work in [17] have been proposed, under different assumptions, in different communication scenarios [15,18–43]. However, none of them can completely show all the characteristics of the DCF and analyze the performance of the DCF precisely in IoV scenarios, especially for those ignoring the capture effect [17–35]. In fact, when the receiver receives the signal power from one transmitter that is higher than that of the other transmitters, the capture effect, which is a common phenomenon in wireless communication, may occur [43–45]. According to [45], the system performance of IEEE 802.11 networks can be improved by the capture effect. Nevertheless, there are only a few analytical models in the existing literature considering the capture effect under frequently used fading channels in vehicular communication [36–43]. According to [46,47], the Nakagami-m channel model represents small-scale fading in vehicular communication and reflects a realistic driving environment. Therefore, analyzing IEEE 802.11p/bd DCF under this fading channel model is necessary to show its real performance in IoV. Though the capture effect under the Nakagami-m fading channel is considered in [42], the authors only show the non-closed-form formulation for the normalized throughput. In addition, the capture effect under the Nakagami-m fading channel is also considered in our previous work in [43] under the saturated condition, which is a special case that all vehicles in the network always have packets to be transmitted. In fact, the vehicles are often under an unsaturated condition, which means that the buffer queues of vehicles do not always have packets waiting to be transmitted [26,48,49]. Motivated by this, we make the performance analysis of IEEE 802.11p/bd DCF more accurate by proposing a novel two-dimensional (2-D) Markov chain model, where all important characteristics of the DCF itself are included, and the capture effect under the Nakagami-m fading channel is considered, to make the analytical procedure more reasonable. To the best of our knowledge, it is the first analysis of the unsaturated performance of IEEE 802.11p/bd DCF with consideration of the capture effect under the Nakagami-*m* fading channel by proposing a novel 2-D Markov model different from the existing ones. The contributions of this paper are threefold.

1. A novel 2-D Markov chain model is proposed, which is different from the existing ones. In the proposed 2-D Markov chain model, all the key characteristics of the DCF are considered, i.e., backoff freezing mechanism, immediate access mechanism, finite retry limit, post-backoff procedure, different packet arrival probabilities under different channel states for the empty buffer and queuing model of the buffer queue.
2. The capture effect under a Nakagami-*m* fading channel is considered. Then, the closed-form expressions of successful transmission, collided transmission, normalized unsaturated throughput, and average packet delay are all meticulously derived, respectively.
3. To verify the accuracy of the proposed model, simulation results are given. In addition, it is also compared with the existing analytical models. As expected, the proposed model is more accurate than the existing models in terms of normalized unsaturated throughput and average packet delay.

The rest of this paper is organized as follows. Section 2 surveys the related research. Section 3 presents an overview of the DCF and develops a novel analytical model for the DCF. Section 4 validates the accuracy of the proposed analytical model for the DCF by comparing it with the existing models. Finally, Section 5 concludes the paper. In addition, Table 1 presents a list of abbreviations used in the paper.

Table 1. List of abbreviations.

Abbreviation	Definition
2-D	Two-dimensional
3-D	Three-dimensional
AC	Access category
ACK	Acknowledgement
BEB	Binary exponential backoff
BER	Bit error ratio
BS	Base station
CSMA/CA	Carrier sense multiple access with collision avoidance
DCF	Distributed coordination function
DIFS	Distributed inter frame space
EDCA	Enhanced distributed channel access
EDCAF	Enhanced distributed channel access function
FIFO	Fist-in-first-out
GW	Gateway
HS	Hot spot
IoV	Internet of vehicles
MAC	Medium access control
NAV	Network allocation vector
PHY	Physical
RSU	Road-side unit
RTS/CTS	Request-to-send/clear-to-send
SIFS	Short inter-frame space
TXOP	Transmission opportunity
VANET	Vehicular ad hoc network
V2I	Vehicle-to-infrastructure
V2N	Vehicle-to-network
V2P	Vehicle-to-pedestrian
V2V	Vehicle-to-vehicle

2. Related Work

Due to the wide usage of the IEEE 802.11 series of standards in wireless local area networks, a lot of analytical models for the adopted MAC protocols (i.e., DCF or EDCA) under different network scenarios have been proposed for evaluating its performance and then designing a MAC protocol meeting the requirements of different scenarios by researchers around the world.

As is known, Bianchi, G. first proposed a 2-D Markov chain model (which is called Bianchi's model [17]) to analyze the DCF protocol under ideal channel conditions and saturated conditions, without considering the backoff freezing mechanism and retry limit. Based on this pioneering work, a lot of research work has been conducted by worldwide researchers. For example, Duffy, K. et al. extended Bianchi's model to the nonsaturated condition [18], while Madhavi, T. et al. modeled collision-alleviating DCF with a finite retry limit [19]. Though the backoff freezing mechanism and unsaturated condition are considered in [20], the finite retry limit and post-backoff procedure are missed. In addition, the finite retry limit is included in the model in [21], but the post-backoff procedure is still missed. In [22], the performances of saturated throughput and delay for the DCF based on [17], with consideration of the finite retry limit and backoff freezing mechanism, are analyzed. In addition, IEEE 802.11p DCF is analyzed and optimized under saturation conditions in [23], which is just based on the 2-D Markov chain model in [22]. Though the performances of saturated throughput and delay for IEEE 802.11p EDCA are analyzed in [24], the difference between the 2-D Markov chain models in [17,24] is the consideration of retry limit. Moreover, in [15,25], IEEE 802.11p EDCA is analyzed under unsaturation conditions with consideration of the retry limit and backoff freezing mechanism, while the latter considers the queuing model and ignores the post-backoff procedure. Moreover, Cao et al. analyzed the EDCA with consideration of the backoff freezing mechanism, finite retry limit and idle state for four ACs [26], which is more accurate than the model proposed in [15] for considering the queuing model. However, these studies are all on the basis of ideal channel conditions, which is not in line with the reality that the channel does have an effect on the DCF or EDCA protocol in IoV.

Hence, Zhang, Y. et al. analyzed the DCF based on [17] under different channel conditions [27]. In addition, Peng, H. et al. presented a probabilistic analysis of the DCF in a multiplatooning scenario, while a constant probability is used for the transmission error of a packet [28]. However, the backoff freezing mechanism and post-backoff procedure are both ignored. Therefore, Almohammedi, A.A. et al. considered the backoff freezing mechanism and unsaturated condition in the 2-D Markov chain model and analyzed the throughput of the DCF under a varying bit-error ratio (BER) [29]. In addition, Peng, J. et al. also investigated the impact of channel transmission error with a constant probability for a packet in [30]. Moreover, Alshanyour, A. et al. evaluated IEEE 802.11 DCF based on a three-dimensional (3-D) Markov chain model under saturated conditions, and just a constant BER was considered [31]. Meanwhile in [32], an hierarchical 3-D Markov model was proposed for analyzing the non-saturated IEEE 802.11 DCF-based networks under error-prone channel conditions (accomplished by varying the constant block error probability). In [33], Wang, N. et al. evaluated the IEEE 802.11p EDCA based on a 3-D Markov chain model under both saturated and unsaturated conditions, while the impact of channel fading and modulation was modeled with a constant BER for simplicity. In addition, Harkat, Y. et al. analyzed the saturation throughput and average access delay with a constant BER too [34]. In [35], a 3-D Markov chain model is also used to analyze the throughput and average access delay for EDCA under different values of BER.

Unfortunately, the above-mentioned analytical models all ignore the influence from the capture effect. Since the capture effect exists in wireless communication systems, it is necessary to consider it to make the analytical results more accurate [36,44,45,50]. Therefore, Shah, A.F.M.S. et al. analyzed the saturation throughput of DCF by considering the capture effect in a Rayleigh fading environment based on Bianchi's model [36]. In [37], Lei, L. et al. analyzed the saturation throughput of the DCF with the consideration of the capture effect under the free-space propagation model based on a 3-D Markov chain model. However, the backoff freezing mechanism and post-backoff procedure are both disregarded. Meanwhile in [38,39], Daneshgaran, F. et al. analyzed the saturation throughput and unsaturated throughput with consideration of the capture effect under the Rayleigh fading channel, respectively. However, they ignored the post-backoff procedure and retry limit in their 2-D Markov models. In [40], Han, H. et al. also gave the saturation throughput of the DCF with consideration of the capture effect in a Rayleigh fading channel and the retry limit. Again, in [41], Sutton G.J. et al. modeled the DCF with the capture effect under a Rayleigh fading channel based on a 3-D Markov model, but the backoff freezing mechanism was ignored. Moreover, Leonardo, E. J. et al. analyzed throughput of the DCF with consideration of the capture effect under Hoyt, Rice, and Nakagami-m fading channels in [42], while ignoring the backoff freezing mechanism and the retry limit. In addition, the closed-form expressions of capture effect and throughput are missed. Though the capture effect under the Nakagami-m fading channel is included in our previous work [43], the saturation condition is assumed for convenience, which makes this model not very in line with reality.

Therefore, a novel analytical model considering all the important factors (i.e., backoff freezing mechanism, immediate access mechanism, finite retry limit, post-backoff procedure, different packet arrival probabilities under different channel states for the empty buffer, queuing model of buffer queue, and the capture effect under a Nakagami-m fading channel) is proposed for the performance analysis of IEEE 802.11p/bd DCF in real vehicular communication scenarios. Then, we carefully derive the closed-form expressions of successful transmission, collided transmission, unsaturated throughput and average packet delay, respectively. In fact, the proposed analytical model can be easily extended to the performance analysis of the EDCA. Similar extension methods can be referenced in [15,25,26].

3. The Proposed Analytical Model

In this section, a novel analytical model is proposed to evaluate the performance of the IEEE 802.11p/bd DCF protocol. Different from the existing work, we develop a novel 2-D Markov model to derive the closed-form expressions of normalized unsaturated throughput and average packet delay, which are the two main commonly used evaluation indicators. For convenience, the significant notations and variables used in the analysis procedure are summarized in Table 2.

Table 2. Notions used in the proposed analytical model.

Notion	Definition
W_j	Contention window of backoff stage j
CW_{min}	Minimum contention window
CW_{max}	Maximum contention window
M	Maximum backoff stage
f	Retransmission times in the maximum backoff stage
σ	Duration of a backoff slot
λ_{pkt}	Packet arrival rate of upper layer
λ_{eff}	Effective packet arrival rate
q	The probability that the cache queue is not empty
a_b	The probability of packet arrival when the channel is busy
a_i	The probability of a packet arriving when the channel is idle
p_i	The probability that the channel is idle during one backoff slot
τ_{tra}	Transmission probability under unsaturated condition
τ_{tra}^{sat}	Transmission probability under saturated condition
p_{tra}	The probability that at least one vehicle transmits
p_s	The probability of successful transmission under unsaturated condition
p_c	The probability of collided transmission under unsaturated condition
p_c^{sat}	The probability of collided transmission under saturated condition
p_s'	The probability of successful transmission when at least one vehicle transmits
z_{th}	Capture threshold
m	The parameter of Nakagami fading
p_{cap}	The probability of capture effect
σ_{ave}	Duration of the virtual slot under unsaturated condition
T_b	Duration of busy channel

Table 2. *Cont.*

Notion	Definition
T_s	Average time for successful transmission
T_c	Average time for collided transmission
μ_{eff}	Effective service rate of packets
μ_{suc}	Service rate of successfully transmitted packets
μ_{dis}	Overflow rate of packets
ρ	Service intensity
p_k	Steady-state probability when the queue length is k
L_p	The length of payload
$T_s^{L_p}$	The time required to successfully transmit the payload
R_t	The data rate
R_b	Basic transmission rate
p_{of}	Overflow probability of cache queue
D_{que}	Queue delay
D_{MAC}	Delay of MAC layer
L_{ave}	Average number of packets in the cache queue
D_{ave}	Average packet delay
n	Number of vehicles

3.1. Brief Description of DCF

According to the DCF protocol, the vehicles in the network contend for the wireless channel by the carrier sense multiple access with collision avoidance (CSMA/CA) mechanism, which is based on the slotted binary exponential backoff (BEB) scheme. In fact, each vehicle with a packet to be transmitted needs to sense the channel before transmission. If the channel is idle for a duration exceeding the distributed interframe space (DIFS), the vehicle transmits the packet. This is a so-called immediate access mechanism. Otherwise, the backoff procedure is invoked to defer the transmission to avoid collision. According to the BEB scheme, the random backoff time is uniformly chosen in the range $[0, CW - 1]$, where CW is the contention window with the minimum $CW_{min} = W_0$. The backoff counter is decremented by one at the end of each idle slot and the vehicle transmits immediately when the backoff counter reaches zero. However, if the channel is busy, the backoff counter will be frozen. When the channel is idle again for more than one DIFS, the backoff counter will be resumed. The transmission for a data packet (DATA) from the source vehicle is successful if an acknowledgement (ACK) from the destination vehicle can be received by the source vehicle after a period of short interframe space (SIFS). Otherwise, this transmission has failed and a retransmission is scheduled by starting another backoff period with CW doubled. If the maximum of contention window ($CW_{max} = W_M = 2^M W_0$) is reached and then CW can be set to W_M at most for f times before discarding this packet. Hence, the value of CW is reset to CW_{min} after a successful transmission or being discarded due to reaching the retry limit ($M + f$). A backoff procedure shall be performed immediately after the end of every transmission, even if no additional transmissions are currently queued. This is the so-called post-backoff mechanism.

The basic access mode and the request-to-send/clear-to-send (RTS/CTS) access mode are two access techniques supported by the DCF protocol. In fact, the basic access mode is a two-way handshaking mechanism using DATA/ACK packets, while the RTS/CTS access mode is a four-way handshaking mechanism using RTS/CTS packets to reserve the channel resource before transmission. In fact, the latter follows the same backoff rules as the former and reduces the risk of large packet collision by short RTS/CTS packets. Since the duration of the ongoing transmission is included in the above-mentioned control packet, each vehicle updates its network allocation vector (NAV) by the RTS or CTS and then defers transmission for a specified duration to avoid collision.

3.2. A Novel 2-D Markov Chain Model

In Figure 2, a novel 2-D Markov chain model is proposed for modeling the behavior of the DCF protocol in IoV. In this Markov chain model, the unsaturated condition is considered, i.e., the buffer of each vehicle will be empty with probability $(1-q)$, where q denotes the probability that there exist packets in the buffer after a successful transmission or dropping a packet due to reaching the retry limit. The states for vehicles at time t are represented as $\{s(t), b(t)\}$, where $s(t)$ with values from $\{0, 1, \ldots, M+f\}$ is defined as the random backoff stage and $b(t)$ with values from $\{0, 1, \ldots, W_i - 1\}$ is defined as the value of the backoff counter at time t. Moreover, the states $\{s(t)_e, b(t)\}$ refer to the states with empty buffer, which means that the buffer queue of one vehicle is empty after a successful transmission or a failure. These random variables are dependent because the maximum value of the backoff counter depends on the backoff stage

$$W_i = \begin{cases} 2^i W_0, & 0 \le i \le M \\ W_M, & M < i \le M+f \end{cases} \tag{1}$$

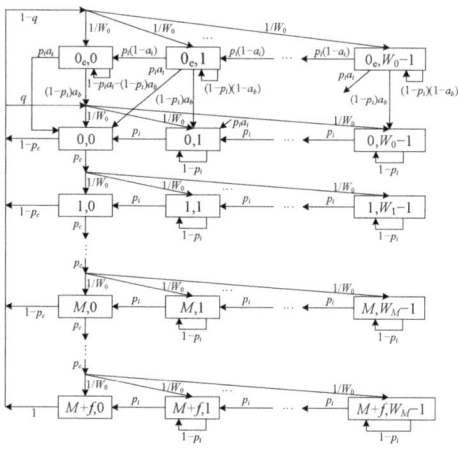

Figure 2. The proposed 2-D Markov chain model.

Let $b(j,k)$ be the stationary distribution of the 2-D Markov chain model in Figure 2. Then, the one-step state transition probabilities can be expressed as

$$\begin{cases} P\{(0_e, k-1)|(0_e, k)\} = p_i(1-a_i), \ 1 \le k \le W_0 - 1 \\ P\{(0, k-1)|(0_e, k)\} = p_i a_i, \ 1 \le k \le W_0 - 1 \\ P\{(0, k)|(0_e, k)\} = (1-p_i)a_b, \ 1 \le k \le W_0 - 1 \\ P\{(0_e, k)|(0_e, k)\} = (1-p_i)(1-a_b), \ 1 \le k \le W_0 - 1 \\ P\{(0, k)|(0_e, 0)\} = (1-p_i)a_b/W_0, \ 1 \le k \le W_0 - 1 \\ P\{(0_e, 0)|(0_e, 0)\} = 1 - p_i a_i - (1-p_i)a_b \\ P\{(0, 0)|(0_e, 0)\} = p_i a_i + (1-p_i)a_b/W_0 \\ P\{(j, k-1)|(j, k)\} = p_i, \ 0 \le j \le M+f, 1 \le k \le W_j - 1 \\ P\{(j, k)|(j, k)\} = 1 - p_i, \ 0 \le j \le M+f, 1 \le k \le W_0 - 1 \\ P\{(0, k)|(j, 0)\} = (1-p_c)q/W_0, \ 0 \le j < M+f, 0 \le k \le W_0 - 1 \\ P\{(0_e, k)|(j, 0)\} = (1-p_c)(1-q)/W_0, \ 0 \le j < M+f, 0 \le k \le W_0 - 1 \\ P\{(j+1, k)|(j, 0)\} = p_c/W_{\min(j+1, M)}, \ 0 \le j < M+f, 0 \le k \le W_0 - 1 \\ P\{(0, k)|(M+f, 0)\} = q/W_0, \ 0 \le k \le W_0 - 1 \\ P\{(0_e, k)|(M+f, 0)\} = (1-q)/W_0, \ 0 \le k \le W_0 - 1 \end{cases} \tag{2}$$

Therefore, based on Figure 2 and Equation (2), we can further obtain the following steady-state probabilities, i.e.,

$$b(j,0) = (p_c)^j b(0,0), \quad 1 \leq j \leq M+f \tag{3}$$

$$b(j,k) = \frac{(W_j - k)(p_c)^j}{W_j p_i} b(0,0) \tag{4}$$

$$b(0,0) = (1 - p_c) \sum_{j=0}^{M+f-1} b(j,0) + b(M+f,0) \tag{5}$$

$$b(0_e, 0) = b(0,0) \frac{1-q}{W_0(a_b + p_i a_i - p_i a_b)} \left(1 + \sum_{k=1}^{W_0-1} \left(\frac{p_i - p_i a_i}{a_b + p_i - p_i a_b}\right)^k\right) \tag{6}$$

The detail derivation processes of the above expressions are omitted to save space, and interested readers are encouraged to refer to [15,17]. By using the above expressions, we can easily obtain

$$\sum_{j=1}^{M+f} \sum_{k=0}^{W_j-1} b(j,k) = b(0,0) \left\{ \frac{p_c - (p_c)^{M+f+1}}{1 - p_c} + \frac{1}{2p_i} \left[\frac{2W_0 p_c(1 - (2p_c)^M)}{1 - 2p_c} + \frac{W_0 2^M (p_c)^{M+1}(1 - (p_c)^f) + (p_c)^{M+f+1} - p_c}{1 - p_c} \right] \right\} \tag{7}$$

$$\sum_{k=0}^{W_0-1} b(0_e, k) + \sum_{k=0}^{W_0-1} b(0,k) = \left[1 + \frac{W_0 - 1}{2p_i}\right] b(0,0) + \left[1 + \frac{(W_0 - 1)(1 - p_i)a_b}{2p_i}\right] b(0_e, 0) \tag{8}$$

Then, according to the normalization condition for stationary distribution, we have

$$\sum_{k=0}^{W_0-1} b(0_e, k) + \sum_{j=0}^{M+f} \sum_{k=0}^{W_j-1} b(j,k) = 1 \tag{9}$$

After substituting Equations (6)–(8) into Equation (9), we can obtain

$$\frac{1}{b(0,0)} = 1 + \frac{W_0 - 1}{2p_i} + \frac{p_c - (p_c)^{M+f+1}}{1 - p_c} + \frac{1-q}{W_0(a_b + p_i a_i - p_i a_b)} \cdot \left[1 + \frac{(W_0-1)(1-p_i)a_b}{2p_i}\right] \left[1 + \sum_{k=1}^{W_0-1} \left(\frac{p_i - p_i a_i}{a_b + p_i - p_i a_b}\right)^k\right] \\ + \frac{1}{2p_i} \left[\frac{2W_0 p_c(1-(2p_c)^M)}{1-2p_c} + \frac{W_0 2^M (p_c)^{M+1}(1-(p_c)^f) + (p_c)^{M+f+1} - p_c}{1-p_c}\right] \tag{10}$$

Therefore, the probability that a concerned vehicle transmits in a randomly chosen slot can be expressed as

$$\tau_{tra} = \sum_{j=0}^{M+f} b(j,0) = \frac{1 - (p_c)^{M+f+1}}{1 - p_c} b(0,0) \tag{11}$$

Then, substituting (10) into (11), we can obtain

$$\tau_{tra} = \frac{1-(p_c)^{M+f+1}}{1-p_c} \left\{ 1 + \frac{W_0-1}{2p_i} + \frac{p_c-(p_c)^{M+f+1}}{1-p_c} + \frac{1-q}{W_0(a_b+p_i a_i - p_i a_b)} \left[1 + \frac{(W_0-1)(1-p_i)a_b}{2p_i}\right] \\ \cdot \left[1 + \sum_{k=1}^{W_0-1} \left(\frac{p_i - p_i a_i}{a_b + p_i - p_i a_b}\right)^k\right] + \frac{1}{2p_i} \left[\frac{2W_0 p_c(1-(2p_c)^M)}{1-2p_c} + \frac{W_0 2^M (p_c)^{M+1}(1-(p_c)^f) + (p_c)^{M+f+1} - p_c}{1-p_c}\right] \right\}^{-1} \tag{12}$$

where a_i and a_b are the probabilities of packet arrivals during an idle slot and a busy slot, respectively. If the arrival of a packet obeys Possion distribution, these two values are calculated as

$$\begin{cases} a_i = \sum_{k=1}^{\infty} \frac{(\lambda\sigma)^k}{k!} e^{-\lambda\sigma} = 1 - \frac{(\lambda\sigma)^0}{0!} e^{-\lambda\sigma} = 1 - e^{-\lambda\sigma} \\ a_b = \sum_{k=1}^{\infty} \frac{(\lambda T_b)^k}{k!} e^{-\lambda T_b} = 1 - \frac{(\lambda T_b)^0}{0!} e^{-\lambda T_b} = 1 - e^{-\lambda T_b} \end{cases} \quad (13)$$

where σ is the duration of an idle slot and T_b is the duration of a busy slot. Here, the durations of a successful slot and a collided slot are assumed to be the same for simplicity. In addition, the probability that the channel is idle for the vehicle concerned is calculated as

$$p_i = (1 - \tau_{tra})^{n-1} \quad (14)$$

where n is the number of vehicles and τ_{tra} is the transmission probability calculated by Equation (12). Because of the consideration of the capture effect, the probability of a collided transmission in a given slot can be calculated by

$$p_c = \sum_{k=1}^{n-1} \left[1 - \frac{p_{cap}(k+1, z_0)}{k+1}\right] C_{n-1}^k (\tau_{tra})^k (1 - \tau_{tra})^{n-k-1} \quad (15)$$

where $C_{n-1}^k = (n-1)!/[k!(n-k-1)!]$ and $p_{cap}(\cdot,\cdot)$ is the occurrence probability of capture effect. According to [45,51], the capture effect occurs at the targeted vehicle if the received signal power from some vehicles is larger than the sum of the others'. For an inference-limited system, the capture condition is $\gamma_t / \sum_{k=1,k\neq t}^n \gamma_k > z_0$, where γ_t, γ_k and z_0 are the signal power from one vehicle, the interference signal power from the other vehicles, and the capture threshold, respectively. Under the hypothesis of perfect power control, the capture probability conditioned on $n-1$ interferers ($n \geq 2$) can be calculated by [51]

$$\begin{aligned} p_{cap}(n, z_0) &= n \int_0^{\infty} f_{\gamma_t}(\gamma_t) \cdot \Pr\left[\gamma_t / \sum_{k=1,k\neq t}^n \gamma_k > z_0\right] d\gamma_t \\ &= n \int_0^{\infty} f_{\gamma_t}(\gamma_t) \left[\int_0^{\gamma_t/z_0} f_{\gamma_{n-1}}(\gamma_{n-1}) d\gamma_{n-1}\right] d\gamma_t \end{aligned} \quad (16)$$

where $f_{\gamma_t}(\gamma_t)$ is the instantaneous received power and $f_{\gamma_{n-1}}(\gamma_{n-1})$ is the $(n-1)$-fold convolution of $f_{\gamma_t}(\gamma_t)$. That the Nakagami-m fading is more suitable to the IoV scenario leads to its wide adoption in the research of VANETs [52]. Therefore, the Nakagami-m fading channel is considered here. Then, $f_{\gamma_t}(\gamma_t)$ and $f_{\gamma_{n-1}}(\gamma_{n-1})$ of (16) can be given by

$$f_{\gamma_t}(\gamma_t) = \frac{m^m \gamma_t^{m-1}}{\overline{\gamma}^m \Gamma(m)} e^{-\frac{m\gamma_t}{\overline{\gamma}}}, \gamma \geq 0 \quad (17)$$

$$f_{\gamma_{n-1}}(\gamma_{n-1}) = \frac{m^{m(n-1)}}{\overline{\gamma}^{m(n-1)} \Gamma(mn-m)} \gamma_{n-1}^{m(n-1)-1} e^{-\frac{m\gamma_{n-1}}{\overline{\gamma}}} \quad (18)$$

where $m \in [1/2, \infty)$ is the shape parameter. $\overline{\gamma} = P_{tx} \cdot C \cdot r_i^{-\alpha}$ is the average received power determined by transmission power (P_{tx}), path-loss exponent (α), and a constant related to the antenna gains (C). Besides, for all vehicles in the network, the carrier frequency and the speed of light are both the same. According to our previous work in [52], the capture probability (i.e., Equation (16)) can be further expressed as

$$p_{cap}(n, z_0) = \frac{n}{\Gamma(m)\Gamma(mn-m)} \sum_{k=0}^{\infty} \frac{(-1)^k \Gamma(mn+k)}{k!(mn-m+k) z_0^{mn-m+k}} \quad (19)$$

According to the numerical method in [17], we can obtain τ_{tra} and p_c by figuring out the equation set (the non-linear system) composed of Equations (12) and (15) after submitting (13) into (12) and (19) into (15), respectively. Though the capture threshold (z_0) and the shape parameter (m) of this equation set are given in advance, the probability that there is at least one packet in the vehicle buffer (q) is still unknown, which is related to the service intensity ρ determined by the arrival rate of packets and the service rate. Here, we treat each vehicle as an $M/M/1/K$ queue with a first-in-first-out (FIFO) policy (as shown in Figure 3), where the packet arrival of each buffer from the upper layer is a Poission process with rate λ (in packets per second, pkts/s) and the interval of service time for each packet is exponentially distributed with mean value $1/\mu_{eff}$. In addition, for each buffer queue, the maximum length is K (including the packet in service).

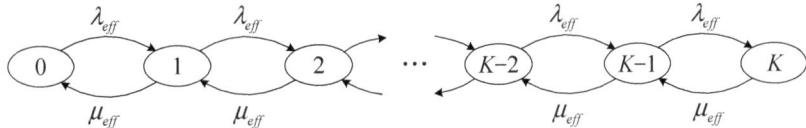

Figure 3. State transition diagram for an $M/M/1/K$ queue.

According to [53], the probability that the buffer queue of any vehicle is non-empty is given by

$$q = 1 - \frac{1 - \frac{\lambda_{pkt}}{\mu_{eff}}}{1 - \left(\frac{\lambda_{pkt}}{\mu_{eff}}\right)^{K+1}} \qquad (20)$$

where $\rho = \frac{\lambda_{pkt}}{\mu_{eff}} \neq 1$ and the effective packet service rate is given by

$$\mu_{eff} = \mu_{suc} + \mu_{dis} \qquad (21)$$

where μ_{suc} (i.e., the maximum service rate of packet successful transmission for a concerned vehicle) is given by

$$\mu_{suc} = \frac{\sum_{j=0}^{n-1} \frac{1}{j+1} p_{cap}(j+1, z_{th}) C_{n-1}^{j} (\tau_{tra}^{sat})^{j} (1 - \tau_{tra}^{sat})^{n-j-1}}{\sigma_{ave}^{sat}} \qquad (22)$$

where τ_{tra}^{sat} and σ_{ave}^{sat} are the average slot time and the transmission probability at saturation, respectively. Moreover, $p_{cap}(\cdot, \cdot)$ is the capture probability expressed as (19) with additionally $p_{cap}(1, \cdot) = 1$ (i.e., that only one vehicle transmits leading to a successful transmission). In addition, the rate at which packets are being discarded due to reaching the retry limit can be calculated by

$$\mu_{dis} = \frac{(p_c^{sat})^{M+f+1}}{\sigma_{ave}^{sat}} \qquad (23)$$

Since τ_{tra}^{sat}, p_c^{sat}, and σ_{ave}^{sat} are the values of transmission probability, collision probability and the average length of a virtual slot at saturation, we can substitute $\rho = a_i = a_b = 1$ into (12) and numerically solve a non-linear system for their values. The detailed steps for finding these two values can be found in our previous work in [43]. Therefore, after substituting (21), (22) and (23) into (20), the value of q can be obtained by a given value of λ. Finally, the equation set composed of (12) and (15) with unknown parameters τ_{tra} and p_c can be numerically solved with a unique solution.

3.3. Calculation of Normalized Throughput

Let η be the normalized throughput. Since it is the ratio of the duration of successful transmission of the packet payload (T_{L_p}) to the average length of a virtual slot (σ_{ave}), it can be calculated as

$$\eta = \frac{p_{suc} p_{tra} T_{L_p}}{\sigma_{ave}} \quad (24)$$

where $T_{L_p} = \frac{L_p}{R_t}$, L_p is the payload of the transmitted packet (which is usually assumed the same for all packets for simplicity) and R_t is the data transmission rate. p_{tra} is the probability that one or more of the vehicles transmit in a certain slot, p_{suc} denotes the probability that one vehicle successfully transmits in a certain slot on the conditioned that one or more of the vehicles transmit. Then, the average length of a virtual slot can be calculated as

$$\sigma_{ave} = (1 - p_{tra})\sigma + p_{suc} p_{tra} T_s + p_{tra}(1 - p_{suc}) T_c \quad (25)$$

where σ, T_s, T_c denote the average durations of an idle slot, successful transmission and collided transmission, respectively. Assume that there are n vehicles competing for transmission in the network; then p_{tra} can be computed by

$$p_{tra} = 1 - (1 - \tau_{tra})^n \quad (26)$$

Then, according to (26), p_{suc} can be calculated by

$$p_{suc} = \frac{p'_s}{p_{tra}} = \frac{\sum_{k=1}^{n}[n!/(k!(n-k)!)](\tau_{tra})^k(1-\tau_{tra})^{n-k} p_{cap}(i,z)}{1 - (1 - \tau_{tra})^n} \quad (27)$$

For the basic mode, the average durations of successful transmission and failed transmission are, respectively, computed as

$$\begin{cases} T_s^{bas} = T_H + T_{L_p} + T_{SIFS} + T_{ACK} + T_{DIFS} + 2T_{PD} \\ T_c^{bas} = T_H + T_{L_p} + T_{DIFS} + T_{PD} \end{cases} \quad (28)$$

where T_H is the transmission duration of the packet header including PHY header (PHY_{hdr}) and MAC header (MAC_{hdr}). T_{L_p}, T_{SIFS}, T_{DIFS}, T_{ACK} and T_{PD} are the durations of a successful transmission of the packet payload, SIFS, DIFS, a successful transmission of ACK and propagation delay, respectively.

For the RTS/CTS mode, the average durations of successful transmission and failed transmission can be computed as

$$\begin{cases} T_s^{rts} = T_{RTS} + T_{CTS} + T_H + T_{L_p} + T_{ACK} + T_{DIFS} + 3T_{SIFS} + 4T_{PD} \\ T_c^{rts} = T_{RTS} + T_{DIFS} + T_{PD} \end{cases} \quad (29)$$

where T_{RTS} and T_{CTS} are the durations of successful transmissions of RTS and CTS, respectively. Besides this, the other parameters are defined the same as those in Equation (28).

3.4. Calculation of Average Packet Delay

The average packet delay for successfully transmitting a packet is defined as the average time from the start when the packet enters the MAC buffer queue to the end when it is successfully received. Since the $M/M/1/K$ queue system is considered, if a packet from the upper layer is not discarded, it will enter the MAC buffer queue and wait to be transmitted (or be discarded by reaching the retry limit). As a result, it includes two parts, i.e., queue delay (D_{que}) and MAC delay (D_{MAC}). The former is the duration from the moment that this packet enters the MAC queue to the moment it becomes the head of the queue, and the latter is the duration from the moment it becomes the head of the queue to the moment it is successfully received. Therefore, the average packet delay can be calculated as

$$D_{ave} = D_{que} + D_{MAC} \quad (30)$$

According to the state transition diagram for an $M/M/1/K$ queue shown in Figure 3, we have

$$p_{k+1} = \rho p_k, 0 \leq k < K \qquad (31)$$

where service intensity ρ can be calculated by

$$\rho = \frac{\lambda_{pkt}}{\mu_{eff}} = \frac{\lambda_{pkt}}{\mu_{suc} + \mu_{dis}} \qquad (32)$$

Then, based on Equation (31), we can obtain

$$p_k = \rho^k p_0, 0 \leq k \leq K \qquad (33)$$

Subsequently, according to the normalization condition, i.e., $\sum_{j=0}^{K} p_j = 1$, the probability that the queue of any vehicle is empty is given by

$$p_0 = \begin{cases} \frac{1-\rho}{1-\rho^{K+1}}, \rho \neq 1 \\ \frac{1}{K+1}, \rho = 1 \end{cases} \qquad (34)$$

Therefore, combining Equations (33) and (34), the overflow probability of the MAC buffer queue can be expressed as

$$p_{of} = \begin{cases} \frac{\rho^K(1-\rho)}{1-\rho^{K+1}}, \rho \neq 1 \\ \frac{1}{K+1}, \rho = 1 \end{cases} \qquad (35)$$

In fact, for an $M/M/1/K$ queue system, the average number of packets in the queue can be calculated as

$$L_{ave} = \sum_{k=0}^{K} k p_k = \frac{1-\rho}{1-\rho^{K+1}} \sum_{k=1}^{K} k p_k = \frac{\rho \left[1 - (K+1)\rho^K + K\rho^{K+1}\right]}{(1-\rho)(1-\rho^{K+1})}, \rho \neq 1 \qquad (36)$$

where ρ can be given by (32) and K the given maximum of queue length. It is worth pointing out that $L_{ave} = \sum_{k=0}^{K} k p_k = \frac{1}{K+1} \sum_{k=1}^{K} k = \frac{K}{2}$ when $\rho = 1$. According to the Little's formula [49], the average waiting time for a packet in the buffer queue (i.e., queue delay) can be calculated by

$$D_{que} = \frac{L_{ave}}{\lambda_{pkt}(1 - p_{of})} \qquad (37)$$

where p_{of} and L_{ave} can be obtained by Equations (35) and (36).

For the calculation of MAC delay, it can be expressed as

$$D_{MAC} = \sum_{j=0}^{M+f} E[T^{(j)}] \cdot P^{(j)} = \sum_{j=0}^{M+f} \left(T_s + j T_c + \sigma_{ave} \sum_{i=0}^{j} \frac{W_i - 1}{2} \right) \cdot \frac{(p_c)^j (1 - p_c)}{1 - (p_c)^{M+f+1}} \qquad (38)$$

where $E\left[T^{(j)}\right]$ denotes the average delay of successfully transmitting a packet at backoff stage j and $P^{(j)}$ denotes the probability of the packet being successfully transmitted at backoff stage j under the premise of not being discarded. p_c and σ_{ave} can be obtained by Equations (15) and (25), respectively. Moreover, W_i can be obtained by Equation (1) and T_s and T_c can be calculated by Equation (28) or (29).

At last, after substituting Equations (37) and (38) into Equation (30), the average packet delay can be expressed as

$$D_{ave} = \frac{L_{ave}}{\lambda_{pkt}(1-p_{of})} + \sum_{j=0}^{M+f}\left(T_s + jT_c + \sigma_{ave}\sum_{i=0}^{j}\frac{W_i-1}{2}\right)\cdot\frac{(p_c)^j(1-p_c)}{1-(p_c)^{M+f+1}} \quad (39)$$

where p_c, σ_{ave}, T_s, T_c, p_{of}, and L_{ave} can be calculated by Equations (15), (25), (28) (or (29)), (35) and (36), respectively.

4. Model Valuation and Performance Evaluation

To validate the effectiveness of the proposed analytical model, the simulation results are given. For simplicity, the simulation scenario is that all vehicles, which are in the one-hop range of each other, communicate with an RSU, e.g., the V2I communication scenario as shown in Figure 4, like that in [54].

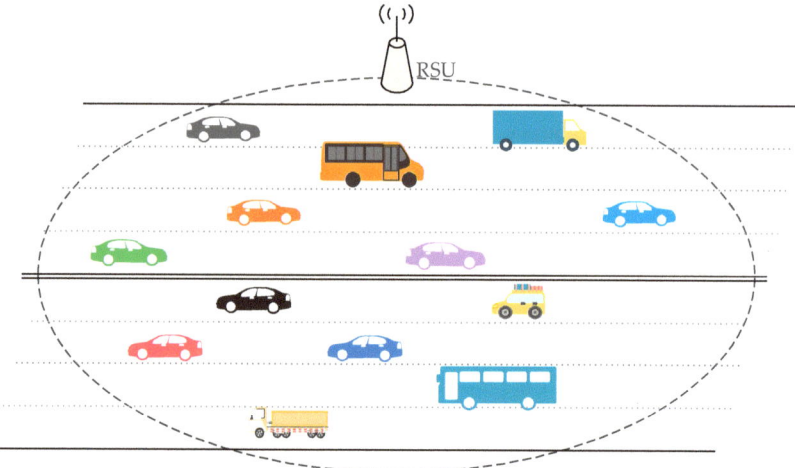

Figure 4. V2I communication scenario.

To verify the accuracy of the proposed model, it is compared with Zheng's model [15] and Malone's model [18] with one single AC queue for fairness. It is worth pointing out that Zheng's model is still adopted in their latest work in [55]. Since the transmission rates within the range of 3 and 27 Mbps are supported by IEEE 802.11p [29,50], a 3 Mbps transmission rate is chosen in the simulation. Like in [43], the capture threshold is set to $z_0 = 2$, because a smaller value of capture threshold means that the capture effect is much more likely to come up. Moreover, the packet arrival of each buffer, which is a Possion process, is set to $\lambda = 10$ pkts/s, because the performance analysis of the DCF is under the hypothesis of an unsaturated condition. The main parameters used are listed in Table 3.

Table 3. Simulation parameters.

Parameters	Setting
L_p	1024 bytes
MAC_{hdr}	224 bits
PHY_{hdr}	192 bits
ACK	304 bits
RTS	352 bits
CTS	304 bits

Table 3. *Cont.*

Parameters	Setting
T_{SIFS}	32 µs
T_{DIFS}	58 µs
σ	13 µs
T_{PD}	2 µs
R_t	3 Mbps
m	1.5
z_{th}	2
W_0	32
W_M	1024
M	5
f	2
λ_{pkt}	10 pkts/s
Simulation time	200 s

4.1. Transmission Probability and Collision Probability

According to Equations (12) and (15), the transmission probability of the vehicle (τ_{tra}) is related to the minimum contention window (W_0), maximum backoff stage (M), retransmission times in the maximum backoff stage (f), the number of vehicles (n), and the probability of a collided transmission under an unsaturated condition (p_c). As shown in Figures 5 and 6, with the increase of the number of vehicles, the transmission probability first increases, and then gradually decreases, while the probability of a collided transmission becomes larger and larger. Obviously, this is determined by the characteristics of the DCF protocol.

Under unsaturated conditions, when the number of vehicles is small, the possibility of a collided transmission is also small. Then, the probability of a successful transmission for packets in the buffer queue of vehicles is high, which also means that there are fewer packets waiting to be sent in the buffer queue, or even no packets waiting to be sent sometimes, resulting in a smaller transmission probability. With the increase in the number of vehicles in the network, the possibility of collided transmission increases. As a result, the vehicles need more time to successfully transmit packets, and the number of packets waiting to be transmitted in the buffer queue increases, resulting in the increase in the probability of vehicles transmitting. However, when the number of vehicles increases to a certain value, the transmission probability begins to decrease. The reason is that a high probability of collided transmissions leads to the increased possibility of delayed transmission of vehicles, which results in more time for vehicles to transmit packets, that is, the transmission probability begins to become smaller. As seen from Figure 5, compared with the theoretical results of transmission probabilities calculated by Zheng's model and Malone's model, the theoretical values calculated by the proposed model are much closer to the simulation results. The reason is that Malone's model does not consider the backoff freezing mechanism, which leads to a decrease in the waiting time before transmitting packets and then an increase in the collision probability. Besides, Zheng's model ignores the influence of capture effect, that is, the capture effect increases the transmission success rate and reduces the waiting time before transmitting packets, and then increases the transmission probability. Therefore, the transmission probabilities obtained by the proposed model, with consideration of the backoff freezing mechanism and the capture effect, are much closer to the actual transmission probabilities.

Similarly, as shown in Figure 6, since the proposed model takes the influence of the capture effect into account, one vehicle may successfully transmit among the collided vehicles. That is to say, the possibility of a collided transmission decreases. As a result, the theoretical probabilities of a collided transmission obtained by the proposed model are closer to the simulation results than that of the other two models. Therefore, when analyzing the performance of DCF (or EDCA), we should fully consider the characteristics of the protocol itself, and consider the influence of the capture effect on the protocol performance in the real IoV environment.

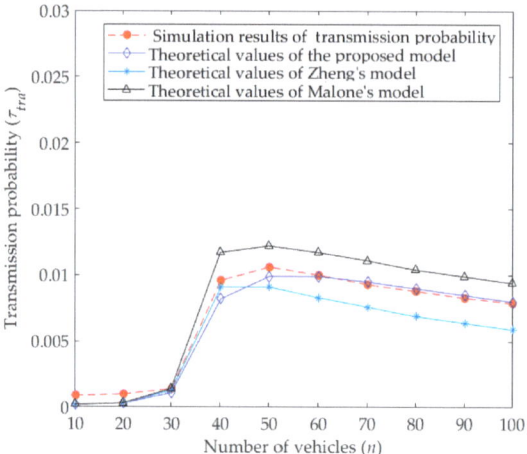

Figure 5. Transmission probability.

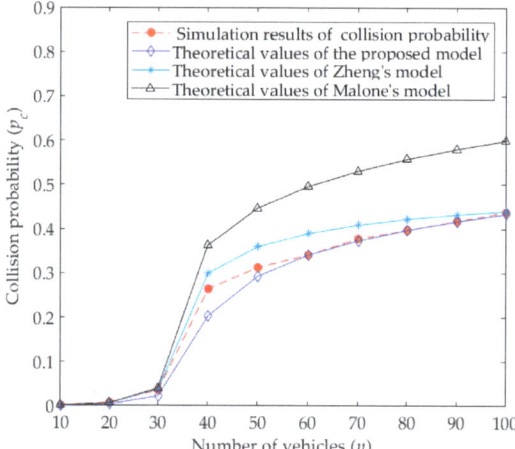

Figure 6. Collision probability.

4.2. Normalized Throughput

As shown in Figure 7, the values of normalized unsaturated throughput for the basic access mode under different numbers of vehicles are given. The theoretical values of normalized throughput for all models are calculated by Equation (24). As seen from Figure 7, the normalized throughput first increases with the increase in the number of vehicles, and then decreases with the increase in the number of vehicles after reaching the maximum value of normalized throughput. This is because the number of vehicles competing for channel resources is small at the beginning, and the normalized throughput gradually increases. As more and more vehicles compete for channel resources, the collision intensifies and the channel resource waste becomes more and more serious, which ultimately leads to the decrease in normalized throughput. In fact, the theoretical values of Zheng's model are much closer to the simulation results than that of Malone's model, because the former considers the backoff freezing mechanism. However, the theoretical values obtained by the proposed analytical model are much closer to the simulation results, and significantly higher than the theoretical values of other models when the number of vehicles is large. The reason is that the proposed analytical model not only fully considers the characteristics of the DCF protocol itself, but also considers the influence of the capture effect, thus improving the accuracy of the theoretical analysis.

As shown in Figure 8, the values of normalized unsaturated throughput for the RTS/CTS mode first gradually increases and then slowly decreases. In addition, the normalized throughput for the RTS/CTS mode is larger than that for the basic access mode. The reason is that the RTS/CTS mechanism limits the collision to smaller control frames (i.e., RTS and CTS), effectively avoiding the collision of larger packets, thus avoiding the sharp decline in the normalized throughput as the number of vehicles increases. Moreover, when the number of vehicles is large, the theoretical values of normalized throughput obtained by the proposed model are also slightly higher than that of the other models, and much closer to the simulation results. The reason is the same as that of basic access mode. In addition, according to Figures 7 and 8, it can be found that the normalized throughput for the RTS/CTS mode is less affected by the channel than for the basic mode, and with the increase in the number of vehicles, the normalized throughput of the former is significantly greater than that of the latter.

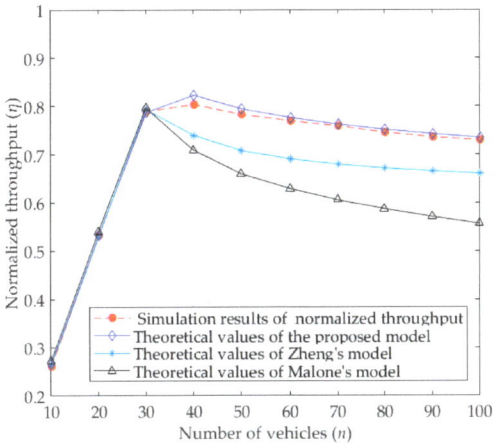

Figure 7. Normalized unsaturated throughput for basic access mode.

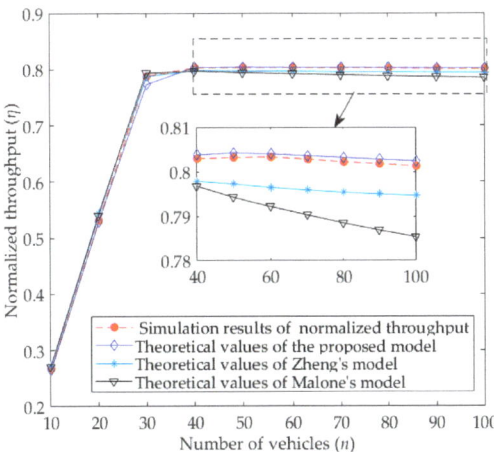

Figure 8. Normalized unsaturated throughput for RTS/CTS mode.

4.3. Average Packet Delay

Figures 9 and 10 give the comparisons between the theoretical values of average packet delay calculated by different analytical models and simulation results for the basic access mode and RTS/CTS mode, respectively. Among them, the theoretical values of the proposed model are calculated by Equation (37), while the theoretical values of Zheng's model and Malone's model are calculated by the calculation methods in the corresponding literatures [15,18], respectively. Due to the full consideration of the characteristics of the DCF protocol and the influence of capture effect, the theoretical values of average packet delay calculated by the proposed model are much closer to the simulation results.

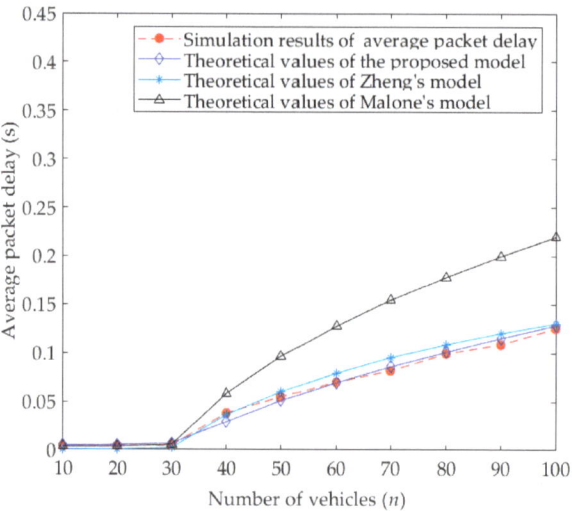

Figure 9. Average packet delay for basic access mode.

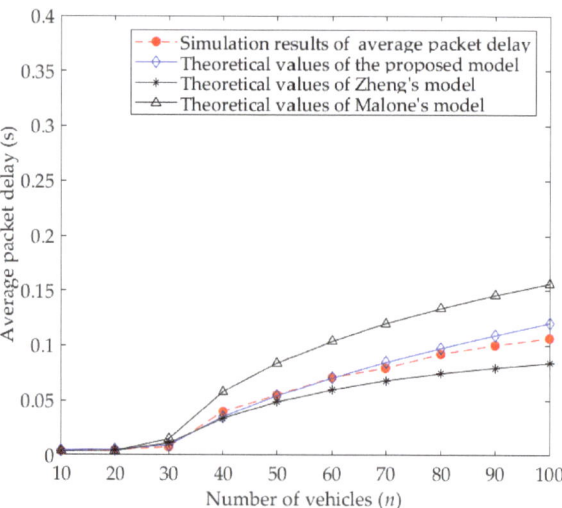

Figure 10. Average packet delay for RTS/CTS mode.

In Figure 9, when the number of vehicles increases in the network, more vehicles competing for the wireless channel leads to intensified collision. It makes the vehicles wait for longer to transmit packets successfully, which eventually results in an increase in the average packet delay. Since the backoff freezing mechanism is ignored in Malone's model, the collision probability calculated by this model increases, which means that the possibility of collided transmissions is amplified. Therefore, one vehicle needs more time to successfully transmit a packet, resulting in a larger average packet delay. Since both Zheng's model and the proposed model take the backoff freezing mechanism into account, the theoretical values of average packet delay calculated by these two models are smaller than that of Malone's model. However, compared with Zheng's model, the theoretical values obtained by the proposed model are much closer to the simulation results, because the proposed model not only considers the influence of capture effect, but also the queuing delay.

Similarly, as shown in Figure 10, in the RTS/CTS mode, the average packet delay gradually increases along with the increase in the number of vehicles. Moreover, the theoretical values of the proposed model are much closer to the simulation results than that of the other two models, which further shows the accuracy of the proposed model. In addition, combined with Figures 9 and 10, it can be found that the average packet delay for the RTS/CTS mode is lower than that for the basic access mode under the same simulation parameters. This is because the RTS/CTS mechanism limits collisions to smaller control frames (i.e., RTS and CTS), effectively avoiding collisions between larger packets.

5. Conclusions

In this paper, a novel analytical model of IEEE 802.11p/bd DCF with consideration of the capture effect under a Nakagami-m fading channel is proposed, which is more accurate than the existing analytical models and better suited to the IoV scenario. All the important characteristics of the DCF protocol and the capture effect under the Nakagami-m fading channel are considered in the proposed model. The accuracy of the proposed model is verified by comparison between the simulations and the analytical results, which show that the proposed model is more accurate than the existing ones, and the normalized unsaturated throughput with consideration of the capture effect is higher than that without a consideration of the capture effect. In addition, the average packet delay decreases, benefiting from the capture effect. As a result, when analyzing the DCF protocol in different communication scenarios or designing improved MAC protocols based on the DCF (or EDCA), the capture effect must be considered to make the MAC protocols more effective in real IoV scenarios. Furthermore, since the EDCA is based on the DCF with different ACs, the proposed analytical model can be easily extended to the performance analysis of the EDCA in IoV, which will be discussed in our future work.

Author Contributions: Conceptualization, Y.W. and J.S.; methodology, Y.W. and L.C.; software, Y.W.; validation, J.S. and L.C.; formal analysis, Y.W. and Z.F.; investigation, J.S.; writing—original draft preparation, Y.W.; writing—review and editing, Y.W. and Z.F.; supervision, J.S.; project administration, J.S.; funding acquisition, L.C. All authors have read and agreed to the published version of the manuscript.

Funding: This work was supported by the Science and Technology Key Project of Fujian Province (Grant No. 2021HZ021004).

Institutional Review Board Statement: Not applicable.

Informed Consent Statement: Not applicable.

Data Availability Statement: Data are contained within the article.

Conflicts of Interest: Author Zhiyuan Fang was employed by the company China Mobile Communications Group Shanxi Co., Ltd. The remaining authors declare that the research was conducted in the absence of any commercial or financial relationships that could be construed as a potential conflict of interest.

References

1. Ning, H.; An, Y.; Wei, Y.; Wu, N.; Mu, C.; Cheng, H.; Zhu, C. Modeling and analysis of traffic warning message dissemination system in VANETs. *Veh. Commun.* **2023**, *39*, 100566. [CrossRef]
2. Wu, Q.; Zheng, J. Performance modeling and analysis of the ADHOC MAC protocol for vehicular networks. *Wirel. Netw.* **2016**, *22*, 799–812. [CrossRef]
3. Wu, G.; Wang, H.; Zhang, H.; Zhao, Y.; Yu, S.; Shen, S. Computation offloading method using stochastic games for software-defined-network-based multiagent mobile edge computing. *IEEE Internet Things J.* **2023**, *10*, 17620–17634. [CrossRef]
4. Wu, G.; Xu, Z.; Zhang, H.; Shen, S.; Yu, S. Multi-agent DRL for joint completion delay and energy consumption with queuing theory in MEC-based IIoT. *J. Parallel Distrib. Comput.* **2023**, *176*, 80–94. [CrossRef]
5. Zhang, P.; Chen, N.; Shen, S.; Yu, S.; Kumar, N.; Hsu, C.H. AI-enabled space-air-ground integrated networks: Management and optimization. *IEEE Netw.* **2023**, early access. [CrossRef]
6. Wang, C.; Jiang, C.; Wang, J.; Shen, S.; Guo, S.; Zhang, P. Blockchain-aided network resource orchestration in intelligent internet of things. *IEEE Internet Things J.* **2022**, *10*, 6151–6163. [CrossRef]
7. Wu, G.; Chen, X.; Gao, Z.; Zhang, H.; Yu, S.; Shen, S. Privacy-preserving offloading scheme in multi-access mobile edge computing based on MADR. *J. Parallel Distrib. Comput.* **2024**, *183*, 104775. [CrossRef]
8. Guo, H.; Zhou, X.; Liu, J.; Zhang, Y. Vehicular intelligence in 6G: Networking, communications, and computing. *Veh. Commun.* **2022**, *33*, 100399. [CrossRef]
9. Yang, Y.; Wei, L.; Wu, J.; Long, C.; Li, B. A blockchain-based multidomain authentication scheme for conditional privacy preserving in vehicular ad-hoc network. *IEEE Internet Things J.* **2022**, *9*, 8078–8090. [CrossRef]
10. Wu, Q.; Wang, X.; Fan, Q.; Fan, P.; Zhang, C.; Li, Z. High stable and accurate vehicle selection scheme based on federated edge learning in vehicular networks. *China Commun.* **2023**, *20*, 1–17. [CrossRef]
11. Long, D.; Wu, Q.; Fan, Q.; Fan, P.; Li, Z.; Fan, J. A power allocation scheme for MIMO-NOMA and D2D vehicular edge computing based on decentralized DRL. *Sensors* **2023**, *23*, 3449. [CrossRef] [PubMed]
12. Ji, B.; Zhang, X.; Mumtaz, S.; Han, C.; Li, C.; Wen, H.; Wang, D. Survey on the internet of vehicles: Network architectures and applications. *IEEE Commun. Stand. Mag.* **2020**, *4*, 34–41. [CrossRef]
13. Wu, Q.; Wang, S.; Ge, H.; Fan, P.; Fan, Q.; Letaief, K.B. Delay-sensitive task offloading in vehicular fog computing-assisted platoons. *IEEE Trans. Netw. Serv. Man.* **2023**, early access. [CrossRef]
14. Wu, Q.; Zhao, Y.; Fan, Q.; Fan, P.; Wang, J.; Zhang, C. Mobility-aware cooperative caching in vehicular edge computing based on asynchronous federated and deep reinforcement learning. *IEEE J. Sel. Top. Signal Process.* **2023**, *17*, 66–81. [CrossRef]
15. Zheng, J.; Wu, Q. Performance modeling and analysis of the IEEE 802.11p EDCA mechanism for VANET. *IEEE Trans. Veh. Technol.* **2016**, *65*, 2673–2688. [CrossRef]
16. Wu, Z.; Bartoletti, S.; Martinez, V.; Todisco, V.; Bazzi, A. Analysis of co-channel coexistence mitigation methods applied to IEEE 802.11p and 5G NR-V2X sidelink. *Sensors* **2023**, *23*, 4337. [CrossRef] [PubMed]
17. Bianchi, G. Performance analysis of the IEEE 802.11 distributed coordination function. *IEEE J. Sel. Areas Commun.* **2000**, *18*, 535–547. [CrossRef]
18. Malone, D.; Duffy, K.; Leith, D. Modeling the 802.11 distributed coordination function in nonsaturated heterogeneous conditions. *IEEE/ACM Trans. Netw.* **2007**, *15*, 159–172. [CrossRef]
19. Madhavi, T.; Rao, G.S.B. Modelling collision alleviating DCF protocol with finite retry limits. *Electron. Lett.* **2015**, *51*, 185–187. [CrossRef]
20. Weng, C.E.; Chen, H.C. The performance evaluation of IEEE 802.11 DCF using Markov chain model for wireless LANs. *Comput. Stand. Inter.* **2016**, *44*, 144–149. [CrossRef]
21. Song, C.; Tan, G.; Yu, C. An efficient and QoS supported multichannel MAC protocol for vehicular ad hoc networks. *Sensors* **2017**, *17*, 2293. [CrossRef] [PubMed]
22. Vardakas, J.S.; Sidiropoulos, M.K.; Logothetis, M.D. Performance behaviour of IEEE 802.11 distributed coordination function. *IET Circ. Devices Syst.* **2008**, *2*, 50–59. [CrossRef]
23. Xie, Y.; Ho, I.W.H.; Magsino, E.R. The modeling and cross-layer optimization of 802.11p VANET unicast. *IEEE Access* **2017**, *6*, 171–186. [CrossRef]
24. Han, C.; Dianati, M.; Tafazolli, R.; Kernchen, R.; Shen, X.S. Analytical study of the IEEE 802.11p MAC sublayer in vehicular networks. *IEEE Trans. Intell. Transp. Syst.* **2012**, *13*, 873–886. [CrossRef]
25. Yao, Y.; Rao, L.; Liu, X. Performance and reliability analysis of IEEE 802.11p safety communication in a highway environment. *IEEE Trans. Veh. Technol.* **2013**, *62*, 4198–4212. [CrossRef]
26. Cao, S.; Lee, V.C.S. An accurate and complete performance modeling of the IEEE 802.11p MAC sublayer for VANET. *Comput. Commun.* **2020**, *149*, 107–120. [CrossRef]
27. Zhang, Y.; Li, S.; Shang, Z.; Zhang, Q. Performance analysis of IEEE 802.11 DCF under different channel conditions. In Proceedings of the IEEE 8th Joint International Information Technology and Artificial Intelligence Conference (ITAIC), Chongqing, China, 24–26 May 2019; pp. 1–4.

28. Peng, H.; Li, D.; Abboud, K.; Zhou, H.; Zhao, H.; Zhuang, W.; Shen, X. Performance analysis of IEEE 802.11p DCF for multiplatooning communications with autonomous vehicles. *IEEE Trans. Veh. Technol.* **2017**, *66*, 2485–2498. [CrossRef]
29. Almohammedi, A.A.; Shepelev, V. Saturation Throughput Analysis of Steganography in the IEEE 802.11p Protocol in the Presence of Non-Ideal Transmission Channel. *IEEE Access* **2021**, *9*, 14459–14469. [CrossRef]
30. Peng, J.; Li, S.; Dou, Z.; Yang, S. Optimization design and performance analysis of improved IEEE 802. 11p MAC mechanism based on high mobility of vehicle. *Math. Probl. Eng.* **2022**, *2022*, 8974673. [CrossRef]
31. Alshanyour, A.; Agarwal, A. Three-Dimensional Markov chain model for performance analysis of the IEEE 802.11 distributed coordination function. In Proceedings of the 2009 IEEE Global Telecommunications Conference (GLOBECOM), Honolulu, HL, USA, 30 November–4 December 2009; pp. 1–7.
32. Martorell, G.; Femenias, G.; Riera-Palou, F. Non-saturated IEEE 802.11 networks: A hierarchical 3D Markov model. *Comput. Netw.* **2015**, *80*, 27–50. [CrossRef]
33. Wang, N.; Hu, J. Performance analysis of the IEEE 802.11p EDCA for vehicular networks in imperfect channels. In Proceedings of the IEEE 20th International Conference on Ubiquitous Computing and Communications (IUCC/CIT/DSCI/SmartCNS), London, UK, 20–22 December 2021; pp. 535–540.
34. Harkat, Y.; Amrouche, A.; Lamini, E.S.; Kechadi, M.T. Modeling and performance analysis of the IEEE 802.11p EDCA mechanism for VANET under saturation traffic conditions and error-prone channel. *AEU-Int. J. Electron. Commun.* **2019**, *101*, 33–43. [CrossRef]
35. Li, S.; Li, H.; Gaber, J.; Yang, S.; Yang, Q. Performance analysis of IEEE 802.11p protocol in IoV under error-prone channel conditions. *Secur. Commun. Netw.* **2023**, *2023*, 5476836. [CrossRef]
36. Shah, A.F.M.S.; Ilhan, H.; Tureli, U. Modeling and performance analysis of the IEEE 802.11 MAC for VANETs under capture effect. In Proceedings of the IEEE 20th Wireless and Microwave Technology Conference (WAMICON), Sofia, Bulgaria, 8–9 April 2019; pp. 1–5.
37. Lei, L.; Zhang, T.; Zhou, L.; Song, X.; Cai, S. Saturation throughput analysis of IEEE 802.11 DCF with heterogeneous node transmit powers and capture effect. *Int. J. Ad Hoc Ubiquitous Comput.* **2017**, *26*, 1–11. [CrossRef]
38. Daneshgaran, F.; Laddomada, M.; Mesiti, F.; Mondin, M.; Zanolo, M. Saturation throughput analysis of IEEE 802.11 in the presence of non ideal transmission channel and capture effects. *IEEE Trans. Commun.* **2008**, *56*, 1178–1188. [CrossRef]
39. Daneshgaran, F.; Laddomada, M.; Mesiti, F.; Mondin, M. Unsaturated throughput analysis of IEEE 802.11 in presence of non ideal transmission channel and capture effects. *IEEE Trans. Wirel. Commun.* **2008**, *7*, 1276–1286. [CrossRef]
40. Han, H.; Pei, Z.; Zhu, W.; Li, N. Saturation throughput analysis of IEEE 802.11b DCF considering capture effects. In Proceedings of the 2012 8th International Conference on Wireless Communications, Networking and Mobile Computing (WiCOM), Shanghai, China, 21–23 September 2012; pp. 1–4.
41. Sutton, G.J.; Liu, R.; Collings, I.B. Modelling IEEE 802.11 DCF heterogeneous networks with Rayleigh fading and capture. *IEEE Trans. Commun.* **2013**, *61*, 3336–3348. [CrossRef]
42. Leonardo, E.J.; Yacoub, M.D. Exact formulations for the throughput of IEEE 802.11 DCF in Hoyt, Rice, and Nakagami-m fading channels. *IEEE Trans. Wirel. Commun.* **2013**, *12*, 2261–2271. [CrossRef]
43. Wang, Y.; Shi, J.; Chen, L. Performance Analysis of IEEE 802.11p MAC with considering capture effect under Nakagami-m fading channel in VANETs. *Entropy* **2023**, *25*, 218. [CrossRef]
44. Zhao, H.; Garcia-Palacios, E.; Wang, S.; Wei, J.; Ma, D. Evaluating the impact of network density, hidden nodes and capture effect for throughput guarantee in multi-hop wireless networks. *Ad Hoc Netw.* **2013**, *11*, 54–69. [CrossRef]
45. Sun, X. Maximum throughput of CSMA networks with capture. *IEEE Wirel. Commun. Lett.* **2016**, *6*, 86–89. [CrossRef]
46. Cheng, L.; Henty, B.E.; Stancil, D.D.; Bai, F.; Mudalige, P. Mobile vehicle-to-vehicle narrow-band channel measurement and characterization of the 5.9 GHz dedicated short range communication (DSRC) frequency band. *IEEE J. Sel. Areas Commun.* **2007**, *25*, 1501–1516. [CrossRef]
47. Bharati, S.; Zhuang, W.; Thanayankizil, L.V.; Bai, F. Link-layer cooperation based on distributed TDMA MAC for vehicular networks. *IEEE Trans. Veh. Technol.* **2016**, *66*, 6415–6427. [CrossRef]
48. Ravi, B.; Thangaraj, J.; Petale, S. Data traffic forwarding for inter-vehicular communication in VANETs using stochastic method. *Wirel. Pers. Commun.* **2019**, *106*, 1591–1607. [CrossRef]
49. Ghosh, A.; Paranthaman, V.V.; Mapp, G.; Gemikonakli, O.; Loo, J. Enabling seamless V2I communications: Toward developing cooperative automotive applications in VANET systems. *IEEE Commun. Mag.* **2015**, *53*, 80–86. [CrossRef]
50. Bazzi, A.; Masini, B.M.; Zanella, A.; Thibault, I. On the performance of IEEE 802.11p and LTE-V2V for the cooperative awareness of connected vehicles. *IEEE Trans. Veh. Technol.* **2017**, *66*, 10419–10432. [CrossRef]
51. Sanchez-Garcia, J.; Smith, D.R. Capture probability in Rician fading channels with power control in the transmitters. *IEEE Trans. Commun.* **2002**, *50*, 1889–1891. [CrossRef]
52. Wang, Y.; Shi, J.; Chen, L.; Lu, B.; Yang, Q. A novel capture-aware TDMA-based MAC protocol for safety messages broadcast in vehicular ad hoc networks. *IEEE Access* **2019**, *7*, 116542–116554. [CrossRef]
53. Kleinrock, L. *Queueing Systems, Vol. 1: Theory*; John Wiley & Sons: Hoboken, NJ, USA, 1975; ISBN 9780471555971.

54. Wu, Q.; Shi, S.; Wan, Z.; Fan, Q.; Fan, P.; Zhang, C. Towards V2I age-aware fairness access: A DQN based intelligent vehicular node training and test method. *Chinese J. Electron.* **2023**, *32*, 1230–1244.
55. Wang, B.; Zheng, J.; Ren, Q.; Li, C. Analysis of IEEE 802.11p-based intra-platoon message broadcast delay in a platoon of vehicles. *IEEE Trans. Veh. Technol.* **2023**, *72*, 13417–13429. [CrossRef]

Disclaimer/Publisher's Note: The statements, opinions and data contained in all publications are solely those of the individual author(s) and contributor(s) and not of MDPI and/or the editor(s). MDPI and/or the editor(s) disclaim responsibility for any injury to people or property resulting from any ideas, methods, instructions or products referred to in the content.

Article

Heuristic Path Search and Multi-Attribute Decision-Making-Based Routing Method for Vehicular Safety Messages

Lei Nie [1,2,*], Junjie Zhang [1,2], Haizhou Bao [1,2] and Yiming Huo [3]

1. School of Computer Science and Technology, Wuhan University of Science and Technology, Wuhan 430065, China; zhangjunjie0726@126.com (J.Z.); baohaizhou@wust.edu.cn (H.B.)
2. Hubei Province Key Laboratory of Intelligent Information Processing and Real-Time Industrial System, Wuhan University of Science and Technology, Wuhan 430065, China
3. Department of Electrical and Computer Engineering, University of Victoria, Victoria, BC V8P 5C2, Canada; yhuo@ieee.org
* Correspondence: lnie@wust.edu.cn

Abstract: Efficient routing in urban vehicular networks is essential for timely and reliable safety message transmission, and the selection of paths and relays greatly affects the quality of routing. However, existing routing methods usually face difficulty in finding the globally optimal transmission path due to their greedy search strategies or the lack of effective ways to accurately evaluate relay performance in intricate traffic scenarios. Therefore, we present a vehicular safety message routing method based on heuristic path search and multi-attribute decision-making (HMDR). Initially, HMDR utilizes a heuristic path search, focusing on road section connectivity, to pinpoint the most favorable routing path. Subsequently, it employs a multi-attribute decision-making (MADM) technique to evaluate candidate relay performance. The subjective and objective weights of the candidate relays are determined using ordinal relationship analysis and the Criteria Importance Through Intercriteria Correlation (CRITIC) weighting methods, respectively. Finally, the comprehensive utility values of the candidate relays are calculated in combination with the link time and the optimal relay is selected. In summary, the proposed HMDR method is capable of selecting the globally optimal transmission path, and it comprehensively considers multiple metrics and their relationships when evaluating relays, which is conducive to finding the optimal relay. The experimental results show that even if the path length is long, the proposed HMDR method gives preference to the path with better connectivity, resulting in a shorter total transmission delay for safety messages; in addition, HMDR demonstrates faster propagation speed than the other evaluated methods while ensuring better one-hop distance and one-hop delay. Therefore, it helps to improve the performance of vehicular safety message transmission in intricate traffic scenarios, thus providing timely data support for secure driving.

Keywords: heuristic path search; multi-attribute decision-making; safety message transmission; routing method; vehicular network

1. Introduction

Over the past few decades, the surge in urban vehicle numbers has intensified challenges in city transportation systems, notably in traffic safety and efficiency. The Vehicular Ad Hoc Network (VANET) has emerged as a cornerstone for Intelligent Transportation Systems (ITSs), offering a promising solution to these challenges [1]. VANET has garnered extensive research attention and has found applications in signal control [2,3], route optimization [4,5], cooperative tasks [6–8], and safety communication [9,10], among others. Particularly, VANET has significant research merit and practical value in safety communication. It paves the way for a streamlined and secure driving landscape, minimizing accidents and curtailing economic costs.

Owing to the dynamic nature of the VANET topology, vehicles often grapple with inconsistent communication links amongst themselves. Such instability can quickly escalate to network congestion, packet loss, and significant end-to-end delays. In crisis situations like traffic accidents, it is imperative that safety messages are dispatched promptly and reliably to the intended destination or vehicle [11,12]. The intricate urban traffic landscape further complicates matters; disjointed road sections can disrupt safety message transmission, and suboptimal message relays hamper swift message propagation. Thus, efficient VANET routing strategies should prioritize both path and relay selection to ensure the promptness and dependability of safety message delivery.

As shown in Figure 1, three routing methods with different path selection strategies are presented in a vehicular safety message transmission scenario. Method 1 tends to find the shortest path, Method 2 tends to find a path with the highest connectivity, and Method 3 forms a path based on constantly selecting relays, which leads them to select different routing paths. In addition, even if the same paths are selected, the performance of the routing methods may vary due to different relay selection strategies.

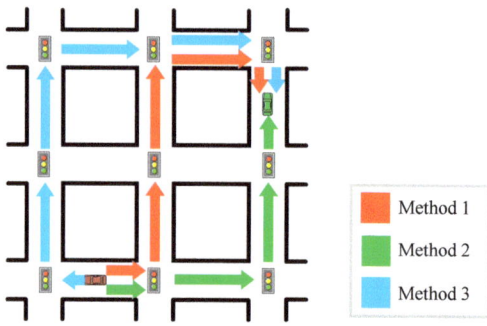

Figure 1. Examples of different path selection strategies.

For path selection, many current routing approaches tend to concentrate solely on identifying the next best intersection from a given point. This narrow focus can result in localized optimization issues. Regarding relay selection, these methods often prioritize one-hop distance or link quality, neglecting the interplay and significance of multiple evaluation metrics. This oversight complicates the accurate evaluation of relay performance in intricate traffic scenarios. Existing routing methods, therefore, face difficulty in determining a globally optimal transmission path or are unable to accurately assess relay performance, leading to poor performance in routing in intricate traffic scenarios, which affects timely data support for secure driving. It is, therefore, necessary to design an efficient routing method that takes into account the selection of globally optimal transmission paths and relays for timely and reliable transmission.

In this study, we design a heuristic path search and multi-attribute decision-making-based routing (HMDR) method for vehicular safety messages. Our main contributions are as follows:

1. For the identification of a global optimal transmission path, HMDR initiates by constructing a path tree model spanning from the destination intersections back to the source intersection. It then pinpoints the most efficient route through a heuristic path search that emphasizes road section connectivity;
2. To accurately evaluate the performance of candidate relays, the HDMR method employs a multi-attribute decision-making-based approach. This calculates the subjective and objective weights of candidate relays based on the ordinal relationship analysis method and the Criteria Importance Through Intercriteria Correlation (CRITIC) weighting method, respectively;
3. The innovative HMDR strategy synergistically blends path search with relay selection, equipping it to manage safety message transmission in multifaceted traffic scenarios.

Our experimental results underscore the effectiveness and proficiency of HMDR in this domain.

The remainder of this paper is organized as follows: Section 2 reviews some related work. A detailed description of the proposed HDMR method is given in Section 3. The performance of the HDMR method is evaluated in Section 4. Finally, we conclude the study in Section 5.

2. Related Work

Existing VANET routing methods are mainly categorized into network-topology-based, cluster-based, and geographic-location-based types. Network-topology-based routing methods store the routing information of the network nodes in a routing table and periodically update the routing table to maintain its availability. Perkins et al. proposed a hop-by-hop Destination Sequenced Distance Vector (DSDV) routing method [13], wherein each node maintains a routing table to other nodes; the route with a large sequence number and a small number of hops is the optimal route, and the routing table is periodically updated network-wide or partially to maintain the effectiveness of communication. Since then, Perkins et al. proposed an Ad Hoc On-demand Distance Vector (AODV) routing method based on DSDV [14], wherein the source node initiates the route lookup process only when a packet is sent to the destination node. If a link is found to be broken, the node sends an "error" message for route repair. Kanani et al. built a collision-avoidance system using the DSDV routing protocol and successfully forwarded emergency messages over VANET [15]. The DSDV routing protocol is used to provide a reliable and efficient routing solution and maintains a routing table for each node. Malnar et al. proposed a Neighbourhood Density AODV (ND-AODV) routing method to reduce routing overhead in large-scale dynamic wireless ad hoc networks [16]. The proposed routing method uses an expected transmission count (ETX)-based metric called Power Light Reflection ETX (PLRE) instead of the hop count metric, which greatly improves the reliability of the AODV protocol. However, in a vehicular network environment where the network topology changes frequently, the network topology-based routing method tends to consume and waste network resources because it must constantly interact with packets to update the routing table.

Cluster-based routing approaches use hierarchical network organization by cluster head nodes in order to reduce network communication and, thus, network resource overhead. Knowing how to select cluster head nodes and optimize the data transmission path to balance the network load and reduce the network communication overhead are urgent optimization problems. Zhang et al. proposed a new algorithm—AODV-MEC—for AODV clustering based on an edge computing strategy [17], which considers vehicle node energy and travel speed to select stable multi-hop links. It solves the problem of resource bandwidth consumption and additional network delay caused by offloading computing tasks to the cloud core network and improves the routing efficiency in the vehicular network. Kandali et al. proposed a new cluster-based vehicular routing protocol that combines a modified k-means algorithm with a continuous Hopfield network and Maximum Stable Set Problem (KMRP) [18]; the cluster head is selected via the weight function according to the amount of free buffer space, the speed, and the node degree. The experimental results show that it performs better in highway vehicular environments. Raja proposed a perspective on road safety by adopting a routing protocol for hybrid VANET-WSN communication (PRAVN) [19]; the proposed routing protocol uses the Improved Water Wave Optimization (IWWO) algorithm for clustering, which minimizes energy consumption while maintaining balanced clusters to monitor bandwidth. However, knowing how to reasonably select cluster head and forwarding nodes and maintain the stability of the cluster structure is a common problem faced by cluster-based routing methods.

Location-based routing methods use the location information between vehicles to decide the relay of route transmission without consideration of route discovery, route maintenance, and network topology, which is more suitable for high-mobility vehicular networks. Karp et al. proposed a greedy peripheral stateless routing protocol named Greedy Perimeter Stateless Routing (GPSR) [20], which combines greedy routing with surface routing and uses surface routing to avoid local minima where greedy forwarding fails. However, due to the presence of obstacles, packets may travel through longer paths and result in higher latency. Zhang et al. proposed a Weight-based Path-aware Greedy Perimeter Stateless Routing (W-PAGPSR) protocol [21]. At the routing establishment stage, the proposed protocol integrates the node distance, reliable node density, cumulative communication duration, and node movement direction to evaluate the communication reliability of the nodes, and the next-hop node is selected using a greedy weight forwarding strategy. Chen et al. proposed an Artificial Spider Geographic Routing (ASGR) algorithm for urban environments [22]. From the perspective of bionics, a spider-web-based network topology is constructed; a connection quality model and transmission delay model are established to select the optimal path from all feasible paths; and, finally, a selective forwarding scheme is proposed based on the node motion and signal propagation characteristics. Rana et al. [23] proposed a novel routing model named Fuzzy-logic-based Multi-hop Directional Location Routing (FLMDLR) in VANET. FLMDLR selects outstanding next hops that help to establish a stable routing path from the source to the destination. In intricate traffic scenarios, the location-based routing method is currently the most reasonable and effective method.

The key to the safety message routing problem in vehicular network environments lies in knowing how to select appropriate routing paths and relays and is reflected in the transmission delay, speed, and coverage of safety messages. However, combined with the above description and summary, existing routing methods often fail to solve the key problem described above, which affects timely data support for secure driving. In this study, we intend to study and propose the method of combining path and relay selection, which integrates the effects of transmission paths and relay nodes on the quality of communication links and message propagation speed to adapt to intricate traffic scenarios.

3. Heuristic Path Search and Multi-Attribute Decision-Making-Based Routing Method

3.1. System Model and Assumptions

The system model and related assumptions are given in this section, and the meaning of the main variables frequently used in this paper are listed in Table 1.

Table 1. The Meaning of the Main Variables.

Variable	Description
o_i	$O = \{o_1, o_2, \cdots, o_i\}$ denotes road intersections, o_s is the source, and o_{d1} and o_{d2} are the targets.
$l_{i,j}, \lvert l_{i,j} \rvert$	The road section that joins two intersections $o_i, o_j \in O$, and its length is $\lvert l_{i,j} \rvert$
q_j	$Q = \{q_1, q_2, \cdots, q_m\}$ denotes the candidate relays of the current forwarding vehicle.
f_j	$F = \{f_1, f_2, \cdots, f_n\}$ denotes the evaluation metrics of candidate relays.
(x_0, y_0)	The location of the current forwarding vehicle.
(x_j, y_j)	The location of q_j.
R	The transmission radius of the vehicles.
v_0	The speed of the current forwarding vehicle.
v_j	The speed of q_j.
v_{max}	The maximum speed limit of vehicles.
$\theta_{0,j}$	The direction angle between the current forwarding vehicle and q_j.
rss_j	The RRS value of q_j recorded by the current forwarding vehicle.
rss_{min}, rss_{max}	The minimum and maximum thresholds of the RRS value.
ρ_j	The area density of q_j.
ρ_{max}	The maximum area density of vehicles.
$con_{i,j}$	The connectivity of $l_{i,j}$.

In our study, an urban vehicular network environment within a certain geographical range is analyzed as a whole. A system model of our study is shown in Figure 2, where vehicles are traveling in a multi-lane urban traffic scenario. The safety messages are generated via the yellow vehicle and transmitted to the green target vehicle using the blue vehicles as relays, while the other red vehicles are ordinary nodes that only receive safety messages. The system model satisfies the following assumptions:

1. Initially, the vehicles with Poisson distribution travel in the urban traffic scenario with multiple lanes in both directions;
2. Each vehicle is equipped with an On-board Unit (OBU) that utilizes a Dedicated Short-range Communication (DSRC) interface for Vehicle-to-vehicle (V2V) communication and obtains location information based on GPS and electronic map;
3. Each vehicle receives road section information from the equipped electronic map, as well as periodical Beacon messages (including vehicle ID, speed, position, timestamp, etc.) from its neighbors, and maintains a road section information table and a neighbor information table;
4. Both buildings and trees attenuate the RSS of safety messages, and each vehicle records the RSS values of monitored neighboring vehicles in the neighbor information table.

Figure 2. System model.

3.2. Overview of HDMR

The main process of HMDR is shown in Figure 3. In a V2V communication-based vehicular network environment, each vehicle obtains information about relevant vehicles and road sections through periodic Beacon messages as well as in-vehicle GPS and electronic maps, while, at the same time, the current forwarding node of the safety message interacts with the candidate relays and selects the optimal relay by executing the HMDR method. In this process, the main problems that require a solution are searching for the optimal path and selecting the optimal relay, which are two important phases of HMDR.

In the optimal path search phase, the source and destination intersections are firstly determined based on the geographical location of the source node and the target node; then, the path model, based on the connectivity of road sections, is established according to road-section-related information, and the least costly path is finally selected as the optimal routing path.

In the optimal relay selection phase, the current forwarding vehicle first calculates the utility values of the four evaluation metrics of candidate delays by using corresponding utility functions and constructs the decision matrix. Then, it calculates the subjective and objective weights of the metrics of candidate relays by using ordinal relationship analysis and CRITIC weighting methods, respectively, and obtains the comprehensive utility values reflecting the performance of candidate relays. Finally, it calculates the node with the biggest comprehensive utility value combined with the link time as the next forwarding node of the safety message.

When a candidate relay is selected as the optimal relay and receives the safety message, it becomes a new forwarding node and continues to select a new relay based on HDMR. The above steps are repeated until the safety message reaches the destination after multi-hop forwarding.

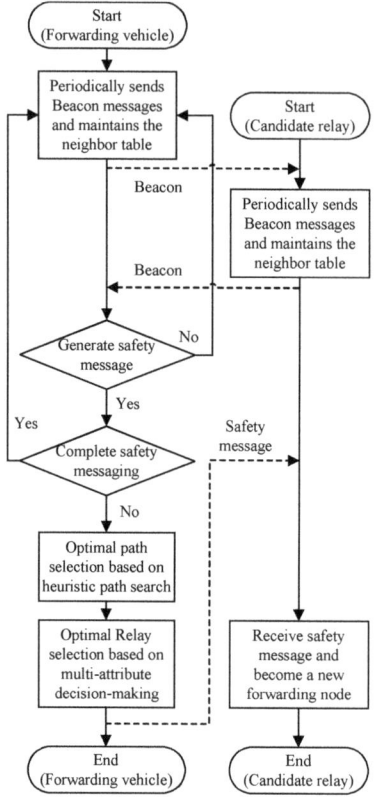

Figure 3. The main process of HDMR.

3.3. Optimal Path Search

Intersections in urban environments are usually regarded as key nodes for finding routing paths. In the first optimal path search phase, the HMDR method determines the source and destination intersections based on the geographical locations of the source and target vehicles and constructs a path tree model to obtain all possible transmission paths; then, the connectivity value of each road section is calculated. Finally, the optimal path with the least cost is determined using a heuristic search method.

3.3.1. Path Tree Model

In this subsection, we use a tree structure to search for routing paths, called the path tree model. The path tree stores the relationship between intersections in the form of an adjacency table and adopts the depth-first search method for traversing the routing paths. The traversal time complexity of a path tree is $\mathcal{O}(V + E)$, where V is the number of vertices (i.e., intersections) and E is the number of edges (i.e., road segments). As the number of intersections increases, the path tree still traverses all of them quickly. In addition, the path tree avoids path loops so that there are no duplicate segments in the routing path.

The selection of source and destination intersections is related to the geographic location of the source and target vehicles. If the source or target vehicle is located at an intersection, the intersection is considered the source or destination intersection; if the source vehicle is located at a road section, the intersection closer to the destination intersection is considered the source intersection; and if the target vehicle is located at a road section, the intersections at both ends of the road section are considered the destination intersections. As shown in Figure 4, after selecting the source intersection o_s and the destination intersections o_{d1} and o_{d2}, extension lines are drawn along the road direction from these intersections, and the maximum quadrilateral area formed by the intersection of the four extension lines is the search area of the optimal path.

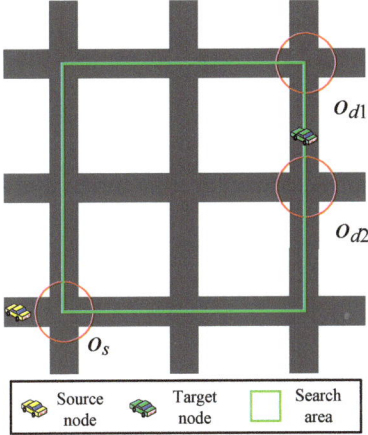

Figure 4. Example of an optimal path search area.

After determining the optimal path search area, the path model can be created by extending from the source node to the destination node. The source node is recorded as layer 0 nodes, its adjacent intersections as layer 1 nodes, the subsequent adjacent intersections as layer 2 nodes, etc., until the destination intersection is covered, and a completed path model is shown in Figure 5.

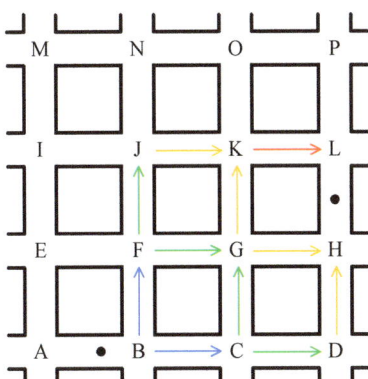

Figure 5. A case of a completed path model.

After completing the path model, path trees are constructed using the following rules:

1. Each path tree starts at a destination intersection and ends at the source intersection;

2. When an intersection is added to a path tree, it is removed from the set of intersections to which it belongs;
3. The children of an intersection in layer $i+1$ can only be its neighboring intersections in layer i;
4. The construction of all path trees is completed when the intersection set is $NULL$.

According to the above rules, path trees with the destination intersection as the root are generated. As shown in Figure 5, the destination intersections are L and H. A depth-first traversal algorithm is used to find all available paths starting from the root and ending at the leaf nodes, the reverse paths of which are feasible paths from the destination to the source node, including $LKJFB$, $LKGFB$, $LKGCB$, $HGFB$, $HGCB$, and $HDCB$, as shown in Figure 6. In this way, it is possible to find the optimal one among these paths to satisfy specific requirements.

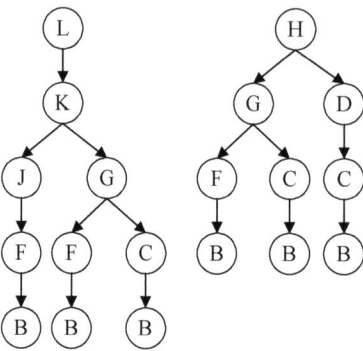

Figure 6. A case of constructing path trees.

3.3.2. Connectivity Analysis of Road Sections

In addition to the length of the routing path, the connectivity of the routing path, also known as the connectivity of the communication link, is an important factor that affects the message transmission delay, and since the routing path is composed of multiple road sections, it is necessary to analyze the connectivity of road sections in order to select a more reliable routing path.

In a V2V communication-based vehicular network environment, each vehicle obtains road and vehicle-related information based on an installed electronic map and receives periodic Beacon messages from its neighbors. Referring to the literature [24], the special density $\rho_{i,j}$ of candidate relays over the road section $l_{i,j}$ is defined and calculated as shown below:

$$\rho_{ij} = \frac{R \cdot N_{ij}}{|l_{i,j}|}, \tag{1}$$

where R is the communication radius of vehicles and $|l_{i,j}|$ and N_{ij} are the length of l_{ij} and the number of vehicles on l_{ij}, respectively.

According to the Poisson distribution, the probability of k vehicles occurring in the counting interval on l_{ij} is denoted as $P_{ij}(k)$ as follows:

$$P_{ij}(k) = \frac{\rho_{ij}^k \cdot e^{-\rho_{ij}}}{k!}. \tag{2}$$

To calculate the connectivity of l_{ij}, the road section is divided into $2|l_{ij}|/R$ sub-segments of length $R/2$. The probability that there is at least one vehicle on a sub-segment is shown in Equation (3):

$$P_{ij}(k>0) = 1 - P_{ij}(0) = 1 - e^{-\rho_{i,j}}. \tag{3}$$

The fact that a road section is connected indicates that all its sub-sections are connected, i.e., there is at least one vehicle in each sub-section; thus, the connectivity of l_{ij} is shown in Equation (4):

$$con_{i,j} = (1 - e^{-\rho_{i,j}})^{|l_{ij}|/R}. \qquad (4)$$

3.3.3. Heuristic Path Search

Combining the path tree model and connectivity analysis of road sections, the optimal routing path from the source node to the target node is selected using the following criteria. The specific selection criterion for the optimal path is that the path has connectivity above a certain level and is of the shortest length. The selection of the optimal path can be formalized as follows:

$$\min \sum |l_{ij}| \cdot g_{ij}, \\ \text{s.t.} \quad con_{ij} \cdot g_{ij} \geq \lambda, \qquad (5)$$

where the value of the binary variable g_{ij} is 1 when the routing path selects the intersections o_i and o_j; otherwise, the value of g_{ij} is 0, and s and d are the start and end of the selected segment, respectively. λ is the connectivity threshold of the road section.

3.4. Optimal Relay Selection

During the multi-hop transmission of safety messages on the optimal routing path, the optimal relay node is selected at each hop. The HMDR method selects four important metrics for relay selection and analyzes the subjective and objective weights of each metric to obtain a table for the comprehensive utility values of all candidate relays. After excluding the "edge nodes" that are about to leave the communication range, the optimal relay node is selected as the node that forwards the safety message, i.e., the node with the first rank in the comprehensive utility value transmits the message.

To facilitate the description and solution of the problem, it is assumed that m vehicles are recorded in the neighbor information table of the current forwarding vehicle and constitute the set of candidate relays $T = \{T_1, T_2, \cdots, T_m\}$; each candidate relay has n metrics for evaluating its performance and constitutes the set of evaluation metrics $F = \{F_1, F_2, \cdots, F_n\}$. The current forwarding vehicle updates the metric values of candidate relays in real time and constructs the initial decision evaluation matrix H for relay selection after data preprocessing based on related utility functions.

3.4.1. Data Preprocessing

The proposed HDMR method selects four important metrics, i.e., the relative distance and relative speed between the current forwarding vehicle and the candidate relay, the RSS of the candidate relay, and area density, to evaluate the performance of the candidate relay. First, the data need to be preprocessed using the corresponding utility functions for the four detected metrics.

(1) Relative distance

The relative distance between the current forwarding vehicle and the candidate relay is the next hop distance for safety message forwarding. If the relative distance is shorter, the safety message requires more hops to cover the target area or reach the destination. Assuming that the location of the current forwarding vehicle is (x_0, y_0), the location of its jth candidate relay is (x_j, y_j); then, the relative distance between them can be expressed as follows:

$$\Delta d_j = \sqrt{(x_0 - x_j)^2 - (y_0 - y_j)^2}, \qquad (6)$$

The utility value of the relative distance Δd_j is as follows:

$$U(\Delta d_j) = \frac{\Delta d_j}{R}, \qquad (7)$$

where R is the transmission radius.

(2) Relative speed

The relative speed between the current forwarding vehicle and the candidate relay is another important factor that affects the link stability. The smaller the relative speed, the less likely the candidate relay will easily leave the communication range of the current forwarding vehicle, which also means the higher the link stability. Assuming that the speed of the current forwarding vehicle is v_0, the speed of the jth candidate relay is v_j, and the direction angle between the two vehicles is $\theta_{0,j}$; the relative speeds of the current forwarding vehicle i and the candidate relay j can be expressed as follows:

$$\Delta v_j = \sqrt{v_0^2 + v_j^2 - 2v_0 v_j \cos(\theta_{0,j})}. \tag{8}$$

The utility value of the relative speed Δv_j is:

$$U(\Delta v_j) = \frac{\Delta v_j}{2 v_{max}}, \tag{9}$$

where v_{max} is the maximum speed limit of vehicles.

(3) RSS

The RSS of the candidate relay is a key factor in ensuring whether the safety message can be successfully received or not, and a larger RSS value indicates that the safety message is more likely to be successfully received. The value of RSS depends on the distance between vehicles; the further the distance, the smaller the RSS value. Assuming that the RSS value of the jth candidate relay recorded in the neighbor information table of the current forwarding vehicle is rss_j, the utility value of rss_j is as follows:

$$U(rss_j) = \begin{cases} 0, rss_j \leq rss_{min}, \\ \frac{rss_j - rss_{min}}{rss_{max} - rss_{min}}, rss_{min} < rss_j < rss_{max}, \\ 1, rss_j \geq rss_{max}, \end{cases} \tag{10}$$

where rss_{min} and rss_{max} are the minimum and maximum thresholds of RSS for not being able to receive and guaranteed to receive safety messages, respectively.

(4) Area density

The area density of a candidate relay is the number of neighboring vehicles per unit distance length, which reflects the degree of traffic congestion on the lanes; too small an area density may lead to link interruptions, while too large an area density is more likely to generate message conflicts. Only if the density is appropriate can it be guaranteed that the selected relay will achieve a higher quality of routing. The area density of candidate relay j and its utility value are as follows:

$$\rho_j = \frac{m}{\pi R^2}, \tag{11}$$

$$U(\rho_j) = \sin \frac{\pi \cdot \rho_j}{\rho_{max}}, \tag{12}$$

respectively, where ρ_{max} is the maximum area density.

(5) Decision matrix

The initial evaluation metrics of candidate relays are normalized using the above utility functions, and the initial standardized decision matrix H is obtained as follows:

$$H = \begin{bmatrix} h_{11} & h_{12} & \cdots & h_{1n} \\ h_{21} & h_{22} & \cdots & h_{2n} \\ \vdots & \vdots & \ddots & \vdots \\ h_{m1} & h_{m2} & \cdots & h_{mn} \end{bmatrix} = (h_{ij})_{m \times n}, \tag{13}$$

where h_{ij} represents the utility value corresponding to the j-th evaluation metric of the i-th candidate relay.

3.4.2. Relay Evaluation

In this section, we select four evaluation metrics and use a multi-attribute decision-making-based method to evaluate the performance of candidate relays. In order to obtain comprehensive utility values that can more accurately reflect the performance of candidate relays, the subjective and objective weights of the candidate relays are comprehensively considered and calculated based on ordinal relationship analysis and CRITIC weighting methods, respectively.

(1) Subjective weights

Ordinal relationship analysis is an improved Analytic Hierarchy Process (AHP) method without a consistency test, which is adopted to calculate the subjective weights of evaluation metrics of candidate relays in this section.

Assuming that there is an ordinal relationship $f_1 \succ f_2 \succ \cdots \succ f_n$ between the n evaluation metrics, which indicates that the adjacent evaluation metric f_i is more important than the evaluation metric f_{i-1}, the ratio of importance between them is denoted as r_k, and r_k is related to the subjective weights of evaluation metrics, as shown below:

$$r_k = \frac{w_{sub,k-1}}{w_{sub,k}}, \tag{14}$$

where $w_{sub,k-1}$ and $w_{sub,k}$ are the subjective weights of f_{k-1} and f_k, respectively. r_k is illustrated in Table 2. It should be noted that the evaluation metrics involved in this study have the following ordinal relationship: $relativedistance \succ RSS \succ areadensity \succ relativespeed$.

Table 2. Description of the values of r_k.

The Value of r_k	Description
1.0	f_{k-1} is as equally important as f_k.
2.0	f_{k-1} is slightly more important than f_k.
3.0	f_{k-1} is significantly more important than f_k.
4.0	f_{k-1} is more important than f_k.
5.0	f_{k-1} is extremely more important than f_k.

The subjective weight of metric f_n is calculated as follows:

$$w_{sub,n} = (1 + \sum_{j=2}^{n} \prod_{i=j}^{n} r_i)^{-1}. \tag{15}$$

According to the proportional relationship between adjacent evaluation metrics, the subjective weight of metric f_j can be derived based on Formula (19), as shown in Equation (20):

$$w_{sub,j} = w_{sub,n} \cdot \prod_{i=j+1}^{n} r_i. \tag{16}$$

(2) Objective weights

The CRITIC weighting method is an objective empowerment method based on the volatility and the correlation of data, which is utilized to calculate the objective weights of evaluation metrics of candidate relays in this section.

The volatility of data is expressed in the form of standard deviation. The larger the standard deviation of a metric, the more information it reflects and the more weight it should be assigned. Let S_j denote the standard deviation of the j-th metric; then, calculate S_j as follows:

$$S_j = \sqrt{\frac{\sum_{i=1}^{n}(h_{ij} - \overline{h}_j)}{n-1}}, \tag{17}$$

$$\overline{h}_j = \frac{1}{n}\sum_{i=1}^{n} h_{ij}, \tag{18}$$

where $\overline{h}_j$ is the mean value of the j-th metric of candidate relays.

The correlation of data is expressed by the correlation coefficient. The larger the correlation coefficient of a metric, the more it reflects the same information and the less weight it should be assigned. Let c_{ij} represent the correlation coefficient between the i-th and j-th metrics; then, the correlation coefficient for metric f_j is as follows:

$$C_j = \sum_{i=1}^{n}(1 - c_{ij}). \tag{19}$$

Considering the above two aspects together, the objective weight of metric f_j is calculated as shown below:

$$w_{obj,j} = \frac{S_j \cdot C_j}{\sum_{i=1}^{n}(S_j \cdot C_j)}. \tag{20}$$

(3) Comprehensive weights and utility values

We evaluate the performance of candidate relays from both subjective and objective aspects. After obtaining the subjective and objective weights of the metrics, the adjustment coefficient β ($0 < \beta < 1$) is introduced and the simple weighting method is used to calculate the comprehensive weights of metrics. The comprehensive weight of metric F_j is as follows:

$$w_{sum,j} = (1-\beta) \cdot w_{sub,j} + \beta \cdot w_{obj,j}. \tag{21}$$

Similarly, the comprehensive weights of other metrics can be calculated. Finally, the comprehensive utility values of candidate relays are calculated by combing the comprehensive weights and the decision matrix through the simple weighting method, and the comprehensive utility value C_i of any candidate relay $q_i (1 \leq i \leq m)$ can be obtained, which is calculated as follows:

$$C_i = \sum_{j=1}^{n} w_{sum,j} \cdot h_{ij}. \tag{22}$$

Candidate relays with higher comprehensive utility values exhibit better performance, as evidenced by better link quality while maintaining longer one-hop distances.

3.4.3. Relay Selection

Theoretically, selecting the candidate relay with the higher comprehensive utility value can result in greater gains; however, due to the high-speed mobility of vehicles, candidate relays at the edge of the current forwarding vehicle's communication range may leave that range in a short period of time, resulting in the interruption of the communication link between them and routing failures. Therefore, these "edge" vehicles need to be removed before selecting a relay. Here, the method of the literature [25,26] is utilized to estimate the link duration between vehicles, and a relay selection strategy combining the link duration estimation is proposed.

Suppose that we now need to estimate the link time between vehicles q_a and q_b, the locations of which are (x_a, y_a) and (x_b, y_b), the speeds of which are v_a and v_b, and of which the angles of direction are θ_a and θ_b, respectively. Then, the remaining connected link time for the two vehicles is as follows:

$$Linktime_{a,b} = \frac{-(AB+CD) + \sqrt{(A^2+C^2) \cdot R^2 - (AD-BC)^2}}{A^2+C^2}, \quad (23)$$

where A, B, C, and D are calculated as follows:

$$\begin{aligned} A &= v_a \cos\theta_a - v_b \cos\theta_b; \\ B &= x_a - x_b; \\ C &= v_a \sin\theta_a - v_b \sin\theta_b; \\ D &= y_a - y_b. \end{aligned} \quad (24)$$

The current forwarding vehicle q_0 estimates the link time with all candidate relays based on the above strategy, and if its link time with candidate relay q_i is less than the threshold value Δt, i.e., $Linktime_{q_0,q_i} \leq \Delta t$, then the candidate relay q_i is removed from the optional relays, thus minimizing the link with the "edge" vehicles, which is conducive to guaranteeing the stability of the communication link. Finally, among the remaining candidate relays that satisfy the link time, the vehicle with the largest integrated utility value is selected as the next hop relay.

4. Experiments and Analysis

4.1. Experimental Scenario and Parameters

In this section, we implemented and evaluated our proposed HDMR method and comparative methods in a simulated vehicular environment built via MATLAB, an experimental urban traffic scenario with multiple lanes in both directions, as shown in Figure 7. Initially, the vehicles obey Poisson distribution, vehicle q_s is the source node of the safety message, vehicle q_d is the target node, safety messages are transmitted in multiple hops along the lanes via V2V communication, and the experimental data are processed using Python. The main simulation parameters related to the experiment are shown in Table 3.

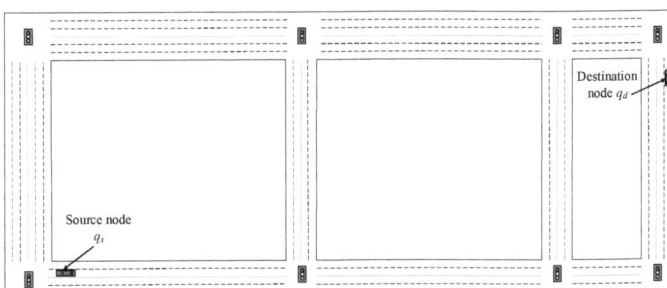

Figure 7. Experimental scenario.

Table 3. Main simulation parameters.

Parameter	Value
Transmission radius R/m	270
Vehicle speed v/ms^{-1}	8~16
Maximum vehicle density ρ_{max}/vehs per km	250
Minimum RSS threshold rss_{min}/dBm	−85
Beacon sending interval/s	1
Minimum distance between vehicles/m	1
The value of β	0.4

4.2. Experimental Results and Analysis

4.2.1. Sensitivity Analysis of the β Value

In this study, the subjective and objective weights of four metrics are comprehensively considered in evaluating candidate relays and weighted via the adjustment factor β. Therefore, the value of β is adjusted at intervals of 0.1 under different traffic density conditions, and the changes in the average propagation speed of the safety messages of the proposed HDMR method are observed. As shown in Figure 8, the average propagation speed is positively correlated with the value of β at the beginning, and the speed is the fastest when β exhibits values from 0.3 to 0.5; it then shows a rapidly decreasing trend, so the value of β in the comparison experiments is taken as 0.4.

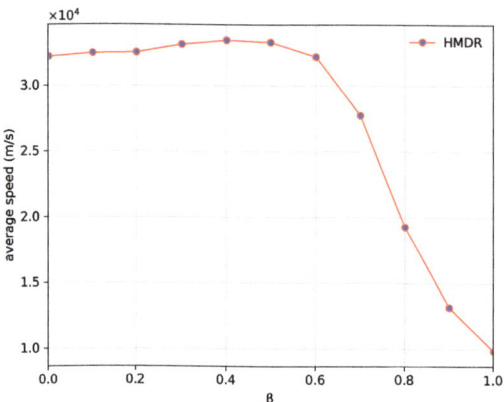

Figure 8. Effect of different β values on the propagation speed of safety messages.

It can be seen that the subjective weight calculation method based on the ordinal relationship method combined with expert experience can better reflect the performance of candidate relays and effectively improve the performance of HDMR as its proportion increases. The objective weight calculation method based on CRITIC cannot effectively reflect the importance of the evaluation metrics of candidate relays and has difficulty in accurately selecting relays with better overall performance when their proportion is higher; however, it can effectively differentiate between the metrics that have a certain degree of correlation, and when the objective weight obtained based on it is combined with the subjective weight, it can, to a certain extent, improve the accuracy of the evaluation of the candidate relay, which is conducive to the selection of the best relay.

4.2.2. Road Connectivity Analysis

From Section 3.3.2, it can be seen that the defined special density determines the connectivity of the road section, which is related to the number of vehicles, the transmission radius, and the length of the road section. This section verifies its rationality through two experimental scenarios. In the first case, the number of vehicles grows from 1 to 100 and the transmission radius is 100, 200, and 300 m. From Figure 9, it can be seen that the connectivity of the road section is proportional to both the number of vehicles and the transmission radius for a certain length of the road section.

In the second case, the length of the road section ranges from 2500 to 6500 m and the transmission radius is 100, 200, and 300 m. From Figure 10, it can be seen that for a given number of vehicles, the road connectivity is inversely proportional to the length of the road section and directly proportional to the transmission radius. Therefore, the two experimental results described above allow us to verify the justification of the special density defined.

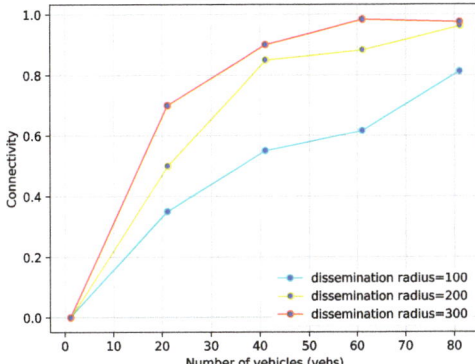

Figure 9. Impact of number of vehicles and transmission radius on connectivity.

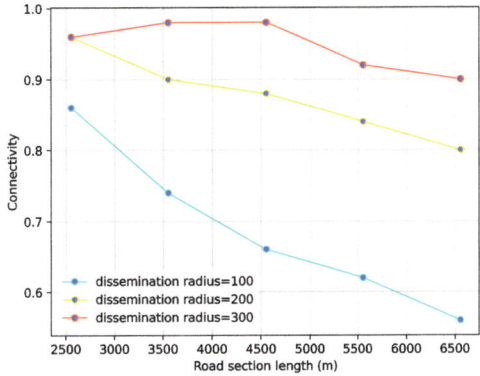

Figure 10. Impact of length of road section and transmission radius on connectivity.

4.2.3. Comparative Results and Analysis

The selection of paths and the selection of relays are the two main factors that affect the performance of routing methods; therefore, several typical methods that can improve the selection strategies of both are selected for comparison and, thus, used to verify the effectiveness of the proposed HMDR. Specifically, four methods, namely, BRAVE [27]; MISR [28]; IGCR [29]; OPBR [30]; and simplified HMDR, which ignores the link time estimation in the optimal relay selection phase, are selected as comparison models in the experimental section. BRAVE uses Dijkstra's algorithm to find the shortest routing path and a distance-based random back-off time to select the relay. MISR selects routing paths taking into account the connectivity of road segments and gives preference to wider paths; it selects relays based on distance and link quality. IGCR also considers the connectivity of road segments when selecting routing paths, particularly in relation to vehicle mobility, direction, and traffic density. In addition, it selects vehicles that are closer to the destination and faster as relays. OPBR determines the routing path by continuously selecting relays; specifically, it selects vehicles as relays that are closer to the destination, not obscured by buildings, and maintain the communication link.

Figure 11 shows the routing paths of all the methods in the experimental scenario, and it can be seen that they select different routing paths due to the differences in path selection strategies. In short, BRAVE and OPBR tend to select the shortest paths, MISR tends to select wider paths with guaranteed connectivity, and our proposed method and IGCR tend to select paths with higher connectivity of road segments. Then, several performance metrics of the six methods are compared by determining the average value through multiple experiments under different traffic densities.

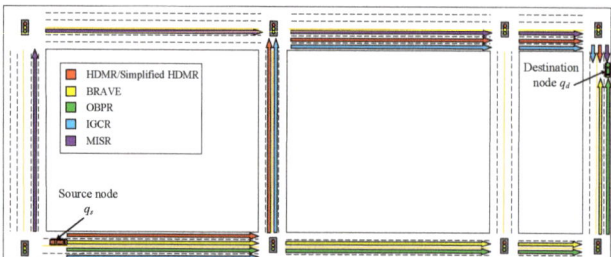

Figure 11. Routing paths of methods.

The transmission of safety messages is a multi-hop forwarding process; first of all, the total hops, the total transmission distance, and the total transmission delay of the safety message from the source node to the destination node are compared under different traffic density conditions. The comparison results are shown in Figures 12–14.

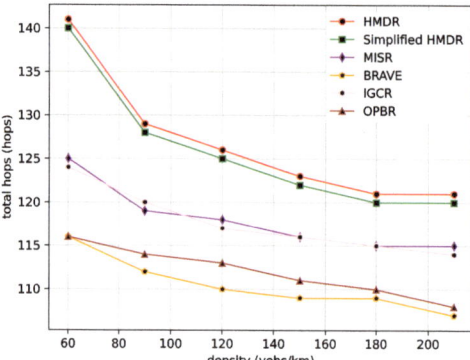

Figure 12. Total hops under different traffic densities.

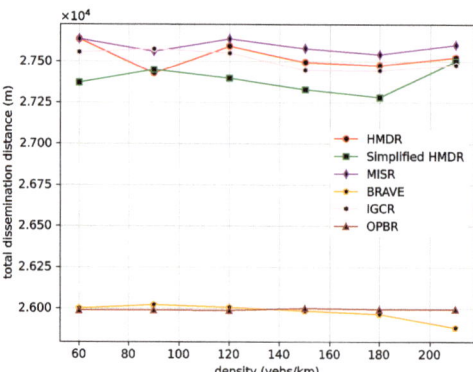

Figure 13. Total transmission distance under different traffic densities.

BRAVE always selects the shortest routing path as it uses Dijkstra's algorithm, and OPBR is also able to find the shortest path in our experimental scenario by continuously selecting relays. Both of them try to select a candidate relay with a longer distance within the one-hop range; thus, they have smaller total hops and transmission distances. By considering the connectivity of road segments, the other four methods select relatively long routing paths. In addition, compared with MISR and IGCR, our proposed HMDR and simplified HMDR methods further enhance the transmission reliability at the expense of reducing the one-hop distance, which is beneficial to reducing transmission delay, despite having the maximum number of hops. Specifically, the total number of hops for the proposed HMDR method is similar to that of the simplified HMDR method, while it is approximately 10.00% to 21.55% more than BRAVE and OPBR and approximately 5.22% to 13.71% more than MISR and IGCR. Moreover, the total transmission distance of HMDR is also similar to that of the simplified HMDR, MISR, and IGCR, while it is approximately 5.40% to 6.34% more than BRAVE and OPBR. Moreover, as can be seen in Figure 14, HDMR performs slightly worse than IGCR and OPBR in the beginning, which is due to our selection strategy sacrificing a larger one-hop distance to guarantee the connectivity of road segments, and this strategy does not bring as much benefit as selecting a longer one-hop distance relay when the traffic density is small. When the traffic density increases, the advantage of HDMR is demonstrated. Particularly, at higher traffic densities, HDMR reduces the total transmission delay by about 7.49% compared to the better-performing OPBR and by about 17.05% to 18.76% compared to the three methods, namely, BRAVE, MISR, and IGCR. Due to the lack of link time estimation, simplified HMDR still has a small probability of selecting "edge" vehicles and, therefore, experiences lower performance than HMDR.

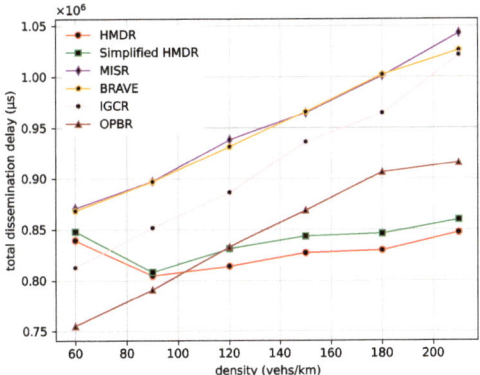

Figure 14. Total transmission delay under different traffic densities.

The average one-hop distance, the average one-hop delay, and the average speed of the safety message from the source node to the destination node are then compared under different traffic density conditions. The comparison results are shown in Figures 15–17.

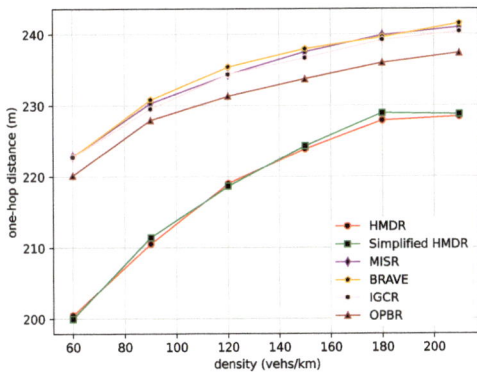

Figure 15. Average one-hop distance under different traffic densities.

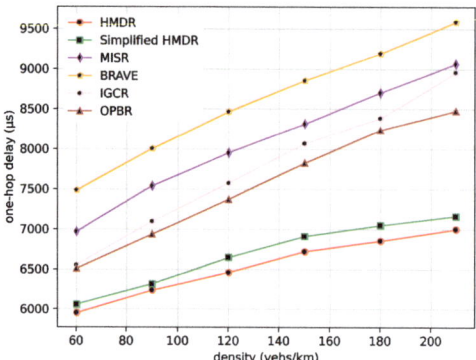

Figure 16. Average one-hop delay under different traffic densities.

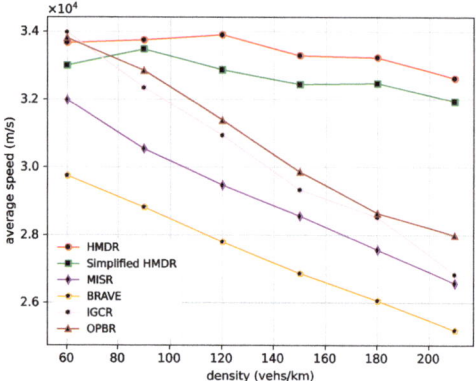

Figure 17. Average transmission speed under different traffic densities.

As traffic density increases, the average one-hop distance increases for all methods. Different from other methods, our proposed HMDR and simplified HMDR simultaneously consider multiple evaluation metrics when selecting relays and use a combination of subjective and objective methods to evaluate the comprehensive weights of candidate relays; thus, the link stability is higher, although the one-hop distance of the selected relay is shorter. The other methods also simultaneously consider multiple factors to select a path but lack a predictive method for link stability and, therefore, have a higher probability of link interruption. Specifically, the average one-hop distance and one-hop delay of HMDR are similar to simplified HMDR; the average one-hop distance is approximately 3.43% to 10.03% shorter than the other methods, and the average one-hop delay of HMDR is approximately 8.55% to 26.98% lower than the other methods. Furthermore, as traffic density increases, the average transmission speed of each method decreases, which is mainly due to the increase in transmission delay. As can be seen in Figure 17, HDMR performs comparably to IGCR and OPBR in sparse traffic. However, since the average one-hop delay of our proposed method has the slowest upward trend, it can always transmit at a high speed. Furthermore, when the traffic density increases, the advantage of our method is gradually highlighted, and the average transmission speed of our method is 16.56% to 29.54% faster than the other methods when in dense traffic scenarios.

5. Conclusions

Currently, urban traffic safety is an important research topic, and safety message routing for vehicular networks is of great research interest and application value. The

V2V-based routing method is more suitable for application in urban scenarios and does not have to rely on the construction of infrastructure, such as RSUs, reflecting its advantages of low latency and low cost. In this study, we proposed a vehicular safety message routing method HMDR that combines path search and relay selection for timely and reliable transmission in intricate traffic scenarios. It uses a heuristic path search method based on road connectivity to select the optimal global routing path to avoid the local optimal problem of path searching. It selects the optimal relay node based on multi-attribute decision-making to accurately evaluate the relay performance in intricate traffic scenarios. The experimental results show that the proposed HMDR tends to choose the path with better connectivity and shorter length, with the characteristics of a lower delay and a higher speed, which is superior to other methods. Therefore, it can be said that HMDR helps to improve the performance of vehicular safety message transmission in intricate traffic scenarios and provides timely data support for secure driving.

However, V2V communication-based routing methods face the problem of poor road connectivity when in sparse traffic and message collisions when in dense traffic, which is less suitable for long-distance transmission in urban scenarios; while V2I-based communication effectively improves communication quality and transmission distance, V2I-assisted safety message routing methods will be studied in future work.

Author Contributions: Conceptualization, L.N. and J.Z.; methodology, L.N. and J.Z.; software, J.Z.; validation, L.N. and J.Z.; formal analysis, L.N. and J.Z.; data curation, J.Z.; writing—original draft preparation, J.Z.; writing—review and editing, L.N., H.B. and Y.H.; supervision, L.N., H.B. and Y.H.; project administration, L.N.; and funding acquisition, L.N. and H.B. All authors have read and agreed to the published version of the manuscript

Funding: This work was partially supported by the National Natural Science Foundation of China (61802286), the Scientific Research Program Youth Project of Hubei Provincial Department of Education (Q20221108), and the Hubei Province Key Laboratory of Intelligent Information Processing and Real-time Industrial System Fund (ZNXX2022009).

Institutional Review Board Statement: Not applicable.

Informed Consent Statement: Not applicable.

Data Availability Statement: Data are contained within the article.

Conflicts of Interest: The authors declare no conflict of interest.

Abbreviations

The following abbreviations are used in this manuscript:

HDMR	Heuristic path search and multi-attribute decision-making-based routing method
RSS	Received signal strength
CRITIC	Criteria Importance through Intercriteria Correlation
VANET	Vehicular Ad Hoc Network
V2V	Vehicle-to-vehicle
V2I	Vehicle-to-infrastructure

References

1. Rafter, C.B.; Anvari, B.; Box, S.; Cherrett, T. Augmenting traffic signal control systems for urban road networks with connected vehicles. *IEEE Trans. Intell. Transp. Syst.* **2020**, *21*, 1728–1740. [CrossRef]
2. Nguyen, C.H.P.; Hoang, N.H.; Vu, H.L. A joint trajectory planning and signal control framework for a network of connected and autonomous vehicles. *IEEE Trans. Intell. Transp. Syst.* **2023**, *24*, 5052–5068. [CrossRef]
3. Finkelberg, I.; Petrov, T.; Gal-Tzur, A.; Zarkhin, N.; Počta, P.; Kováčiková, T.; Buzna, L.; Dado, M.; Toledo, T. The effects of vehicle-to-infrastructure communication reliability on performance of signalized intersection traffic control *IEEE Trans. Intell. Transp. Syst.* **2022**, *23*, 15450–15461. [CrossRef]
4. Wu, H.; Zhou, H.; Zhao, J.; Xu, Y.; Qian, B.; Shen, X. Deep learning enabled fine-grained path planning for connected vehicular networks. *IEEE Trans. Veh. Technol.* **2022**, *71*, 10303–10315. [CrossRef]

5. Zhao, J.; Ma, X.; Yang, B.; Chen, Y.; Zhou, Z.; Xiao, P. Global path planning of unmanned vehicle based on fusion of A* algorithm and Voronoi field. *J. Intell. Connect. Veh.* **2022**, *5*, 250–259. [CrossRef]
6. Wu, Q.; Wang, S.; Ge, H.; Fan, P.; Fan, Q.; Letaief, K.B. Delay-sensitive task offloading in vehicular fog computing-assisted platoons. *IEEE Trans. Netw. Serv. Manag.* **2023**, 1–16. . [CrossRef]
7. Wu, Q.; Zhao, Y.; Fan, Q.; Fan, P.; Wang, J.; Zhang, C. Mobility-aware cooperative caching in vehicular edge computing based on asynchronous federated and deep reinforcement learning. *IEEE J. Sel. Top. Signal Process.* **2023**, *17*, 66–81. [CrossRef]
8. Wu, Q.; Shi, S.; Wan, Z.; Fan, Q.; Fan, P.; Zhang, C. Towards V2I age-aware fairness access: A DQN based intelligent vehicular node training and test method. *Chin. J. Electron.* **2023**, *32*, 1230–1244.
9. Takahashi, K.; Shioda, S. Distributed congestion control method for sending safety messages to vehicles at a set target distance. *Ann. Telecommun.* **2023**, 1–15. . [CrossRef]
10. St. Amour, B.; Jaekel, A. Data rate selection strategies for periodic transmission of safety messages in VANET. *Electronics* **2023**, *12*, 3790. [CrossRef]
11. Han, R.; Shi, J.; Guan, Q.; Banoori, F.; Shen, W. Speed and position aware dynamic routing for emergency message dissemination in VANETs. *IEEE Access* **2022**, *10*, 1376–1385. [CrossRef]
12. Chen, X.; Yang, A.; Tong, Y.; Weng, J.; Weng, J.; Li, T. A multisignature-based secure and OBU-friendly emergency reporting scheme in VANET. *IEEE Internet Things J.* **2022**, *9*, 23130–23141. [CrossRef]
13. Perkins, C.E.; Bhagwat, P. Highly dynamic destination-sequenced distance-vector routing (DSDV) for mobile computers. *Comput. Commun. Rev.* **1994**, *24*, 234–244. [CrossRef]
14. Perkins, C.E.; Royer, E.M. Ad-hoc on-demand distance vector routing. In Proceedings of the Second IEEE Workshop on Mobile Computing Systems and Applicationss, New Orleans, LA, USA, 25–26 February 1999.
15. Kanani, P.; Patil, N.; Shelke, V.; Salot, K.; Nanavati, A.; Damodaran, N.; Desai, S. Improving QoS of DSDV protocol to deliver a successful collision avoidance message in case of an emergency in VANET. *Soft Comput.* **2023**, 1–11. . [CrossRef]
16. Malnar, M.; Jevtic, N. An improvement of AODV protocol for the overhead reduction in scalable dynamic wireless ad hoc networks. *Wirel. Netw.* **2022**, *28*, 1039–1051. [CrossRef]
17. Zhang, D; Gong, C; Zhang, T; Zhang, J; Piao, M. A new algorithm of clustering AODV based on edge computing strategy in IOV. *Wirel. Netw.* **2021**, *27*, 2891–2908. [CrossRef]
18. Kandali, K.; Bennis, L.; Bennis, H. A new hybrid routing protocol using a modified k-means clustering algorithm and continuous hopfield network for VANET. *IEEE Access* **2021**, *9*, 47169–47183. [CrossRef]
19. Raja, M. PRAVN: Perspective on road safety adopted routing protocol for hybrid VANET-WSN communication using balanced clustering and optimal neighborhood selection. *Soft Comput.* **2021**, *25*, 4053–4072. [CrossRef]
20. Karp, B.; Kung, H.T. GPSR: Greedy perimeter stateless routing for wireless networks. In Proceedings of the 6th International Conference on Mobile Computing and Networking (Mobicom 2000), Boston, MA, USA, 6–11 August 2000.
21. Zhang, W.; Jiang, L.; Song, X.; Shao, Z. Weight-Based PA-GPSR Protocol Improvement Method in VANET. *Sensors* **2023**, *23*, 5991. [CrossRef]
22. Chen, C.; Liu, L.; Qiu, T.; Yang, K.; Gong, F.; Song, H. ASGR: An artificial spider-web-based geographic routing in heterogeneous vehicular networks. *IEEE Trans. Intell. Transp. Syst.* **2019**, *20*, 1604–1620. [CrossRef]
23. Rana, K.K.; Sharma, V.; Tiwari, G. Fuzzy logic-based multi-hop directional location routing in vehicular ad hoc network. *Wirel. Pers. Commun.* **2021**, *121*, 831–855. [CrossRef]
24. Hawbani, A.; Wang, X.; Al-Dubai, A.; Zhao, L.; Busaileh, O.; Liu, P.; Al-Qaness, M.A.A. A novel heuristic data routing for urban vehicular ad hoc networks. *IEEE Internet Things J.* **2021**, *8*, 8976–8989. [CrossRef]
25. Chaib, N.; Oubbati, O.S.; Bensaad, M.L.; Lakas, A.; Lorenz, P.; Jamalipour, A. BRT: Bus-based routing technique in urban vehicular networks. *IEEE Trans. Intell. Transp. Syst.* **2020**, *21*, 4550–4562. [CrossRef]
26. Su, W.; Lee, S.-J.; Gerla, M. Mobility prediction and routing in ad hoc wireless networks. *Int. J. Netw. Manag.* **2001**, *11*, 3–30. [CrossRef]
27. Ruiz, P.M.; Cabrera, V.; Martinez, J.A.; Ros, F.J. BRAVE: Beacon-less routing algorithm for vehicular environments. In Proceedings of the 7th IEEE International Conference on Mobile Ad-hoc and Sensor Systems (IEEE MASS 2010), San Francisco, CA, USA, 8–12 November 2010.
28. Zhou, S.; Li, D.; Tang, Q.; Fu, Y.; Guo, C.; Chen, X. Multiple intersection selection routing protocol based on road section connectivity probability for urban VANETs. *Comput. Commun.* **2021**, *177*, 255–264. [CrossRef]
29. Qureshi, K.N.; Ahmed, M.; Jeon, G. An enhanced multi-hop intersection-based geographical routing protocol for the internet of connected vehicles network. *IEEE Trans. Intell. Transp. Syst.* **2021**, *22*, 3850–3858. [CrossRef]
30. Diaa, M.K.; Mohamed, I.S.; Hassan, M.A. OPBRP-obstacle prediction based routing protocol in VANETs. *Ain Shams Eng. J.* **2023**, *14*, 101989. [CrossRef]

Disclaimer/Publisher's Note: The statements, opinions and data contained in all publications are solely those of the individual author(s) and contributor(s) and not of MDPI and/or the editor(s). MDPI and/or the editor(s) disclaim responsibility for any injury to people or property resulting from any ideas, methods, instructions or products referred to in the content.

Article

Computing Offloading Based on TD3 Algorithm in Cache-Assisted Vehicular NOMA–MEC Networks

Tianqing Zhou [1], Ming Xu [1], Dong Qin [2,*], Xuefang Nie [1], Xuan Li [1] and Chunguo Li [3]

[1] School of Information Engineering, East China Jiaotong University, Nanchang 330013, China; zhoutian930@163.com (T.Z.); xm1020487915@163.com (M.X.); xuefangnie@163.com (X.N.); lixuan@ecjtu.edu.cn (X.L.)
[2] School of Information Engineering, Nanchang University, Nanchang 330031, China
[3] School of Information Science and Engineering, Southeast University, Nanjing 210096, China; chunguoli@seu.edu.cn
* Correspondence: qindong@ncu.edu.cn; Tel.: +86-157-9789-6518

Abstract: In this paper, in order to reduce the energy consumption and time of data transmission, the non-orthogonal multiple access (NOMA) and mobile edge caching technologies are jointly considered in mobile edge computing (MEC) networks. As for the cache-assisted vehicular NOMA–MEC networks, a problem of minimizing the energy consumed by vehicles (mobile devices, MDs) is formulated under time and resource constraints, which jointly optimize the computing resource allocation, subchannel selection, device association, offloading and caching decisions. To solve the formulated problem, we develop an effective joint computation offloading and task-caching algorithm based on the twin-delayed deep deterministic policy gradient (TD3) algorithm. Such a TD3-based offloading (TD3O) algorithm includes a designed action transformation (AT) algorithm used for transforming continuous action space into a discrete one. In addition, to solve the formulated problem in a non-iterative manner, an effective heuristic algorithm (HA) is also designed. As for the designed algorithms, we provide some detailed analyses of computation complexity and convergence, and give some meaningful insights through simulation. Simulation results show that the TD3O algorithm could achieve lower local energy consumption than several benchmark algorithms, and HA could achieve lower consumption than the completely offloading algorithm and local execution algorithm.

Keywords: TD3; MEC; NOMA; vehicular networks; edge cache; computation offloading; resource allocation

Citation: Zhou, T.; Xu, M.; Qin, D.; Nie, X.; Li, X.; Li, C. Computing Offloading Based on TD3 Algorithm in Cache-Assisted Vehicular NOMA–MEC Networks. *Sensors* **2023**, *23*, 9064. https://doi.org/10.3390/s23229064

Academic Editors: Pingyi Fan and Qiong Wu

Received: 10 October 2023
Revised: 2 November 2023
Accepted: 7 November 2023
Published: 9 November 2023

Copyright: © 2023 by the authors. Licensee MDPI, Basel, Switzerland. This article is an open access article distributed under the terms and conditions of the Creative Commons Attribution (CC BY) license (https://creativecommons.org/licenses/by/4.0/).

1. Introduction

With the rapid development of information and communication technologies, the data traffic generated by vehicles (mobile devices, MDs) has also significantly increased [1]. For wireless communication networks, more spectrum resources are required for data traffic transmission [2–5]. In addition, higher computing power is required by MDs for supporting large amounts of task calculation. However, due to the limited battery capacity of MDs, it may be challenging to process these computation tasks for them. By deploying edge computing servers at base stations (BSs), mobile edge computing (MEC) can support MDs in processing tasks at the adjacent edge servers [6,7]. Compared with cloud computing (CC), which requires tasks to be uploaded to a remote cloud, MEC can provide additional computing resources for MDs within its coverage area and thus reduce their computing overhead [8–14].

Although the edge servers can reduce the computing overhead of MDs by providing more computing resources, the extra time and energy consumption caused by offloading tasks through wireless channels cannot be ignored, especially for high-size computation tasks. In order to further reduce the time and energy consumption caused by offloading tasks, edge caching technology is also introduced into MEC networks. By caching tasks of

MDs at edge servers in advance, the overhead caused by offloading tasks could be greatly reduced [15–19].

To upload tasks from MDs to edge servers, orthogonal multiple access (OMA) is often used, but it may be greatly challenging to provide a high transmission rate and support massive connections. As another type of resource utilization management, non-orthogonal multiple access (NOMA) technologies can let multiple users share the same frequency bands, achieve higher spectral efficiency and support extensive connections [20–23]. It is evident that NOMA is a good type of resource utilization management for reducing the cost of task transmission in MEC networks.

Although the application of caching and NOMA technologies in MEC networks can reduce time and energy consumption, such a framework will make the design of computation offloading and edge caching schemes more complex. To the best of our knowledge, until now, how to jointly perform the device association, computation offloading, edge caching, subchannel selection and resource allocation is still an important and open topic in cache-assisted NOMA–MEC networks.

1.1. Related Work

So far, a lot of work has been conducted on joint computation offloading and resource optimization in NOMA–MEC networks. In [20], joint radio and computation resource allocation was optimized to maximize the offloading energy efficiency in NOMA–MEC-enabled IoT networks, and a solution based on a multi-layer iterative algorithm was proposed. In [21], local computation resource, offloading ratio, uplink transmission time and power and subcarrier assignment were jointly optimized to minimize the sum of weighted energy consumed by users in NOMA–MEC networks, and some effective iterative algorithms were designed for single-user and multi-user cases. In [24], joint task offloading, power allocation and computing resource allocation were optimized to achieve delay minimization using a deep reinforcement learning (DRL) algorithm in NOMA–MEC networks. In [25], the joint optimization of offloading decisions, local and edge computing resource allocation and power and subchannel allocation were realized to minimize energy consumption in heterogeneous NOMA–MEC networks, and an effective iterative algorithm was designed. In [26], power and computation resource allocations were jointly optimized to minimize overall computation and transmission delay for massive MIMO and NOMA-assisted MEC systems, and a solution based on an interior-point algorithm was given. In [27], the channel resource allocation and computation offloading policy were jointly optimized to minimize the sum of weighted energy and latency in NOMA–MEC networks, and some efficient solutions were found using a DRL algorithm based on actor–critic and deep Q-network (DQN) methods.

To further reduce the offloading time and energy consumption, edge caching technology is introduced into conventional MEC networks. Such a framework has attracted more and more attention. In [28], the offloading and caching decisions, uplink power and edge computing resources were jointly optimized to minimize the sum of weighted local processing time and energy consumption in two-tier cache-assisted MEC networks, and a distributed collaborative iterative algorithm was proposed. In [29], a problem of adaptive request scheduling and cooperative service caching was studied in cache-assisted MEC networks. After formulating the optimization problems as partially observable Markov decision process (MDP) problems, an online DRL algorithm was proposed to improve the service hitting ratio and latency reduction rate. In [30], optimal offloading and caching strategies were established to minimize overall delay and energy consumption of all regions using a deep deterministic policy gradient (DDPG) framework in cache-assisted multi-region MEC networks. In [31], joint MD association and resource allocation were performed to minimize the sum of MDs' weighted delay in heterogeneous cellular networks with MEC and edge caching functions, and an effective iterative algorithm was developed using coalitional game and convex optimization theorems. In [32], to minimize the content transmission delay in vehicular edge computing networks, a cooperative vehicular edge computing and caching scheme based on asynchronous federated and deep reinforcement learning was proposed to predict the popular content and the optimal cooperative caching location of the content. In [33], to reduce the cost of the cloud service center through the asynchronous advantage actor–critic algorithm, the offloading decision, service caching and resource allocation strategies were jointly optimized in the three-tier mobile cloud–edge computing structure combining computation offloading and service caching mechanisms. In [34], a logistic function-based service reliability probability (SRP) estimation model was built, and the average SRP maximization problem of a virtual machine-based edge computing server was studied for such a model. At last, a low-complexity heuristic alternative optimization algorithm was proposed.

To enhance spectral efficiency and support massive connections, NOMA technology has attracted increasing attention in cache-assisted MEC networks. In [35], the multi-agent deep deterministic policy gradient method was used to dynamically optimize the user association, power control and cache placement of BSs and satellites to improve the network energy efficiency in a NOMA-enabled satellite integrated with a terrestrial network scenario. In [36], joint optimization of offloading and caching decisions and computation resource allocation was performed to maximize long-term reward in cache-assisted NOMA–MEC networks under the predicted task popularity, and single-agent and multi-agent Q-learning algorithms were proposed to find feasible solutions. In [37], joint optimization of offloading and caching decisions was performed to minimize the system delay in cache-assisted NOMA–MEC networks, and a multi-agent DQN algorithm was used for finding efficient solutions under the predicted popularity. In [38], local task processing time was minimized by jointly optimizing offloading and caching decisions and the allocation of edge computing resources and uplink power in cache-assisted NOMA–MEC networks with single BS, and the blocking successive upper-bound minimization method was utilized to achieve efficient solutions.

Although the framework of cache-assisted (vehicular) NOMA–MEC networks can greatly reduce the task processing time and energy consumption and support massive connections, there exist very few relevant efforts. Unlike the above-mentioned work, we jointly optimize the edge computing resource allocation, subchannel selection, device association, offloading and caching decisions for the cache-assisted vehicular NOMA–MEC networks with multiple BSs, minimizing the energy consumed by MDs under time and resource constraints. In addition, unlike existing efforts, we develop an effective dynamic joint computation offloading and task-caching algorithm based on the twin-delayed deep deterministic policy gradient algorithm (TD3) to find efficient solutions, named the TD3-based offloading (TD3O) algorithm.

1.2. Contribution and Organization

In this paper, we jointly optimize the edge computing resource allocation, subchannel selection, device association, offloading and caching decisions in cache-assisted vehicular NOMA–MEC networks, minimizing the energy consumed by MDs under time and resource constraints. Specifically, the main contributions and work of this paper can be listed as follows.

- Edge computing resource allocation, subchannel selection, device association, computation offloading and edge caching are jointly performed in cache-assisted vehicular NOMA–MEC networks. To the best of our knowledge, work that concerns subchannel selection is a new investigation for cache-assisted vehicular NOMA–MEC networks with multi-server scenarios. Meanwhile, as far as this problem is concerned, the goal is to minimize the energy consumed by MDs under the constraints of time, computing resources, caching capacity, the number of MDs associated with each BS and the number of MDs associated with each subchannel. As far as we know, such an optimization problem is a new concentration in cache-assisted vehicular NOMA–MEC networks.
- We design effective algorithms to find feasible solutions to the formulated problem. Considering that the formulated problem is in a mixed-integer, nonlinear, multi-

constraint form, a simple map between actions and actual policies in a conventional twin-delayed deep deterministic policy gradient (TD3) algorithm cannot be well applied. In addition, too large an action space will cause the TD3 algorithm to fail to search for correct actions and thus fail to converge. In view of these concerns, we develop an effective TD3O algorithm integrating with the AT algorithm to solve the formulated problem. Moreover, in order to solve this problem in a non-iterative manner, an effective heuristic algorithm (HA) is also designed.
- Performance analyses of the designed algorithms. Some analyses are made for the computation complexity and convergence of the designed algorithms in detail. In addition, some meaningful simulation analyses are also made by introducing other benchmark algorithms for comparison, and some good results and insights are achieved.

The rest of the paper is organized as follows. Section 2 introduces the system model. Section 3 formulates a problem of minimizing local energy consumption in cache-assisted vehicular NOMA–MEC networks. Section 4 designs the HA and TD3O algorithm. Section 5 gives the computation complexity and convergence analyses for the designed algorithms. Section 6 investigates the performance of the designed algorithms through simulation. Section 7 gives conclusions and discussions.

2. System Model

2.1. Network Model

Figure 1 shows the cache-assisted vehicular NOMA–MEC networks. In such network, there exist M MDs, and the index set of them is denoted as $\mathcal{M} = \{1, 2, \cdots, M\}$; B BSs are deployed, and the index set of them is given by $\mathcal{I} = \{1, 2, \cdots, I\}$. In addition, each BS is equipped with one edge computing server and one edge caching server, and these BSs connect to each other through wired links. We assume that each MD has one computation task at any timeslot, which can be processed by itself, its associated BS or another auxiliary BS selected by this associated BS. When tasks have been cached at the BSs used for processing them, they do not need to be uploaded to these BSs; when the associated BSs have not cached tasks, MDs need to upload tasks to these BSs; when the auxiliary BSs have not cached tasks, the associated BSs need to upload tasks to their selected auxiliary BSs.

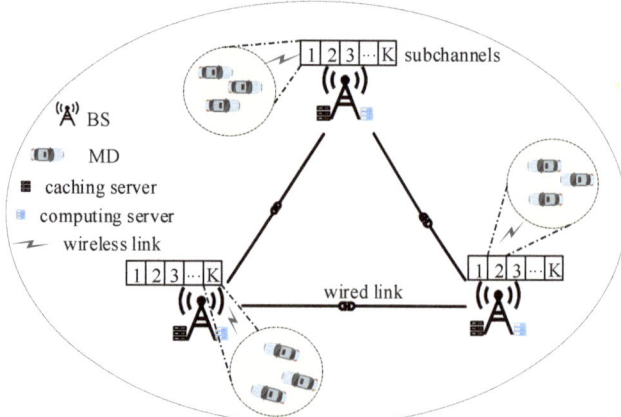

Figure 1. Cache-assisted vehicular NOMA–MEC networks.

Assume that the association index between MD m and BS i is $x_{m,i} \in \{0, 1\}$, where $\mathbf{X} = \{x_{m,i} | \forall m \in \mathcal{M}, \forall i \in \mathcal{I}\}$. $x_{m,i} = 1$ if MD m is associated with BS i, otherwise $x_{m,i} = 0$. In addition, we assume that the caching index of the task of MD m at BS i is denoted as $y_{m,i} \in \{0, 1\}$, where $\mathbf{Y} = \{y_{m,i} | \forall m \in \mathcal{M}, \forall i \in \mathcal{I}\}$. $y_{m,i} = 1$ if the task of MD m is cached at BS i, otherwise $y_{m,i} = 0$. We also assume that the offloading (execution) index of the task of MD m at BS i is denoted as $u_{m,i}$, where $\mathbf{U} = \{u_{m,i} | \forall m \in \mathcal{M}, \forall i \in \mathcal{I}\}$. $u_{m,i} = 1$ if the task of MD m is executed at BS i, otherwise $u_{m,i} = 0$. At last, we assume that the association index between MD m and subchannel k of BS i is denoted as $z_{m,i,k}$, where $\mathbf{Z} = \{z_{m,i,k} | \forall m \in \mathcal{M}, \forall i \in \mathcal{I}, \forall k \in \mathcal{K}\}$. If $x_{m,i}(1-y_{m,i})(1-y_{m,\bar{i}})u_{m,\bar{i}} = 1$ or $x_{m,i}(1-y_{m,i})u_{m,i} = 1$ under $\bar{i} \neq i$, MD m can select (be associated with) some subchannel k of BS i, which means $z_{m,i,k} = 1$. Otherwise, the subchannel k of BS i cannot be selected by MD m, which means $z_{m,i,k} = 0$.

2.2. Communication Model

In this paper, the system bandwidth W is divided into K subchannels with equal bandwidth, which are indexed by $\mathcal{K} = \{1, 2, \cdots, K\}$. These subchannels can be shared by different MDs through NOMA. Significantly, each MD can occupy at most one subchannel, the number of MDs selecting each subchannel cannot exceed the upper limit ρ, and the number of MDs associated with any BS that need to upload tasks should be less than or equal to the number of subchannels K [37].

As revealed in [39], the successive interference cancellation (SIC) technology in NOMA technology can effectively reduce the interference between MDs in the same subchannel. The channel gains of MDs sharing the same subchannel of a BS should be sorted in descending order at first, and then the uplink NOMA signals received by this BS can be decoded in this order. We assume that $\mathcal{M}_k^{SC}$ is the set of MDs selecting subchannel k, and $o_{m,i,k}$ represents the sequence number of channel gain between MD m and BS i on subchannel k. When MD i and MD m access the subchannel k of BS i simultaneously and the channel gain $h_{j,i,k}$ between MD j and BS i on subchannel k is lower than the channel gain $h_{m,i,k}$ between MD m and BS i on subchannel k, $o_{j,i,k} < o_{m,i,k}$ is satisfied. Then, the signal of MD m is decoded but the signal of MD j will be treated as noise. Therefore, when MD m selects subchannel k of BS i, its uplink data rate $r_{m,i,k}$ can be given by

$$r_{m,i,k} = W \log_2 \left(1 + p_m h_{m,i,k} / \left(\Gamma_{m,i,k} + \sigma^2\right)\right) / K, \quad (1)$$

where $\Gamma_{m,i,k} = \sum_{j \in \mathcal{M}_k^{sc}/\{m\}: o_{j,i,k} < o_{m,i,k}} p_j h_{j,i,k}$ is the interference caused by other MDs (excluding MD m) sharing subchannel k of BS i through NOMA; p_m is the transmission power of MD m; σ^2 is the noise power. When MD m is decoded, it is no longer regarded as interference in the subchannel, and the device with the maximum channel gain among the remaining MDs in the current subchannel is decoded in the same way until all MDs of the current subchannel are decoded.

2.3. Caching and Offloading Models

In this paper, we assume that any MD m has a time-sensitive task denoted as $\mathcal{L}_m = \{d_m, c_m, \tau_m^{\max}\}$ at each timeslot, where d_m is the data size of the task of MD m, c_m is the number of CPU cycles required to complete a one-bit task, and $\tau_m^{\max}$ is the maximum task processing time of MD m.

Figure 2 illustrates the caching and offloading models. At each timeslot, BSs precache the tasks for processing at the next timeslot. When MD m is associated with BS i, it first checks whether the associated BS has cached the corresponding task. If $\sum_{i \in \mathcal{I}} u_{m,i} = 0$, the task of MD m is calculated by itself, e.g., MD 1 in Figure 2. If $x_{m,i} y_{m,i} u_{m,i} = 1$, the task of MD m can be directly calculated at its associated BS i, and the results will be fed back from BS i to MD m, e.g., MD 2 in Figure 2. If $x_{m,i} y_{m,i}(1-y_{m,\bar{i}})u_{m,\bar{i}} = 1$, the task of MD m is offloaded from its associated BS i to another auxiliary BS $\bar{i} \neq i$ for computing through a wired link, e.g., MD 3 in Figure 2. If $x_{m,i} y_{m,i} y_{m,\bar{i}} u_{m,\bar{i}} = 1$, the task of MD m can be directly calculated at auxiliary BS $\bar{i} \neq i$, e.g., MD 4 in Figure 2. If $x_{m,i}(1-y_{m,i})u_{m,i} = 1$, the task of MD m will be offloaded to its associated BS i for computing, e.g., MD 5 in Figure 2. If $x_{m,i}(1-y_{m,i})(1-y_{m,\bar{i}})u_{m,\bar{i}} = 1$, the task of MD m first needs to be offloaded to its associated BS i, and then it is transmitted from this BS to another auxiliary BS $\bar{i} \neq i$

for computing through a wired link, e.g., MD 6 in Figure 2. If $x_{m,i}(1-y_{m,i})y_{m,\bar{i}}u_{m,\bar{i}} = 1$, the task of MD m can be directly calculated at an auxiliary BS $\bar{i} \neq i$, e.g., MD 4 in Figure 2.

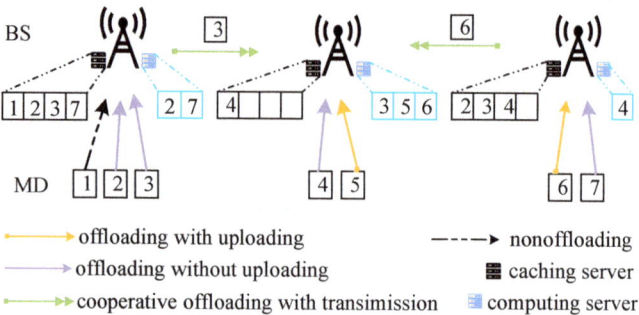

Figure 2. Caching and offloading models.

2.3.1. Local Computing

If $\sum_{i \in \mathcal{I}} u_{m,i} = 0$ is satisfied, the task of MD m should be executed locally, and the processing time and energy consumption are, respectively, given by

$$\tau_m^{\text{loc}} = c_m d_m / f_m^{\text{loc}}, \qquad (2)$$

$$\varepsilon_m^{\text{loc}} = \zeta c_m d_m \left(f_m^{\text{loc}}\right)^2, \qquad (3)$$

where f_m^{loc} is the computing capacity of MD m, and ζ is an energy-consumption coefficient depending on the hardware architecture.

2.3.2. Task Transmission

If $x_{m,i}(1-y_{m,i})(1-y_{m,\bar{i}})u_{m,\bar{i}} = 1$ or $x_{m,i}(1-y_{m,i})u_{m,i} = 1$ are satisfied under $\bar{i} \neq i$, the task of MD m should be, respectively, uploaded to BS $\bar{i}$ or i for execution through NOMA. Then, the uploading time and energy consumption of MD m, respectively, are given by

$$\tau_m^{\text{trs}} = \sum_{i \in \mathcal{I}} \sum_{k \in \mathcal{K}} z_{m,i,k} d_m / r_{m,i,k}, \qquad (4)$$

$$\varepsilon_m^{\text{trs}} = p_m \tau_m^{\text{trs}}. \qquad (5)$$

In addition, if $x_{m,i}(1-y_{m,i})(1-y_{m,\bar{i}})u_{m,\bar{i}} = 1$ or $x_{m,i}y_{m,i}(1-y_{m,\bar{i}})u_{m,\bar{i}} = 1$ is satisfied under $\bar{i} \neq i$, the task of MD m should be transmitted from its associated BS i to an auxiliary BS $\bar{i}$ through a wired link, and the corresponding time is given by

$$\tau_m^{\text{bh}} = d_m / r^{\text{bh}}, \qquad (6)$$

where r^{bh} is the backhualing rate between any two BSs.

In this paper, we mainly concentrate on the energy consumption of MDs but not the energy consumed by BSs. In addition, the downlink transferring time of results is often ignored since they are fairly small [40].

2.3.3. Edge Computing

When MD m executes its task at BS i, the task processing time at this BS can be given by

$$\tau_{m,i}^{\text{exe}} = c_m d_m / f_{m,i}, \qquad (7)$$

where $f_{m,i}$ is the computing capacity allocated to MD m by BS i.

2.3.4. Task Processing Time and Energy Consumption

Then, the total time used for processing the task of MD m can be given by

$$\begin{aligned}
\tau_m^{tot} = \sum_{i \in \mathcal{I}} &\Big(\big(1 - \sum_{i \in \mathcal{I}} u_{m,i}\big) \tau_m^{loc} \\
&+ x_{m,i}(1 - y_{m,i}) \sum_{\bar{i} \in \mathcal{I} \setminus \{i\}} u_{m,\bar{i}}(1 - y_{m,\bar{i}}) \tau_m^{trs} \\
&+ x_{m,i}(1 - y_{m,i}) \sum_{\bar{i} \in \mathcal{I} \setminus \{i\}} u_{m,\bar{i}}(1 - y_{m,\bar{i}}) \tau_m^{bh} \\
&+ x_{m,i}(1 - y_{m,i}) \sum_{\bar{i} \in \mathcal{I} \setminus \{i\}} u_{m,\bar{i}} \tau_{m,\bar{i}}^{exe} \\
&+ x_{m,i} y_{m,i} \sum_{\bar{i} \in \mathcal{I} \setminus \{i\}} u_{m,\bar{i}}(1 - y_{m,\bar{i}}) \tau_m^{bh} \\
&+ x_{m,i} y_{m,i} \sum_{\bar{i} \in \mathcal{I} \setminus \{i\}} u_{m,\bar{i}} \tau_{m,\bar{i}}^{exe} \\
&+ x_{m,i}(1 - y_{m,i}) u_{m,i} (\tau_m^{trs} + \tau_{m,i}^{exe}) \\
&+ x_{m,i} y_{m,i} u_{m,i} \tau_{m,i}^{exe} \Big).
\end{aligned} \quad (8)$$

On the right side of the equality sign in (8), the first item represents the local executing time; the second item is the time used for uploading the task from MD m to the associated BS i, which does not cache this task and further transmits it to auxiliary BS for computing; the third item is the time used for transmitting the task from the associated BS i to another auxiliary BS, where these two BSs do not cache this task; the fourth item is the time used for executing the task of MD m at an auxiliary BS, where the associated BS does not cache this task; the fifth item is the time used for transmitting the task from the associated BS i to another auxiliary BS, where the associated BS caches this task but the auxiliary BS does not; the sixth item is the time used for executing the task of MD m at an auxiliary BS, where the associated BS caches this task; the seventh item includes the time used for transmitting the task from MD m to the associated BS i, which does not cache this task, and the time used for executing the task of MD m at this BS; the eighth item is the time used for executing the task of MD m at the associated BS i, which caches this task.

Then, the total local energy consumption used for processing the task of MD m can be given by

$$\begin{aligned}
\varepsilon_m^{tot} = \sum_{i \in \mathcal{I}} &\Big(\big(1 - \sum_{i \in \mathcal{I}} u_{m,i}\big) \varepsilon_m^{loc} \\
&+ x_{m,i}(1 - y_{m,i}) \sum_{\bar{i} \in \mathcal{I} \setminus \{i\}} u_{m,\bar{i}}(1 - y_{m,\bar{i}}) \varepsilon_m^{trs} \\
&+ x_{m,i}(1 - y_{m,i}) u_{m,i} \varepsilon_m^{trs} \Big),
\end{aligned} \quad (9)$$

on the right side of equality sign in (9), the first item represents the local executing energy consumption; the second item is the energy consumption caused by offloading the task from MD m to its associated BS i, which further transmits this task to auxiliary BS $\bar{i} \neq i$ for computing; the third item is the energy consumption caused by transmitting the task from MD m to the associated BS i, which does not cache this task.

3. Problem Formulation

Until now, we have formulated a problem of minimizing local energy consumption at each given period. Specifically, under the constraints of time, computing resources, caching capacity, the number of MDs associated with each BS and the number of MDs associated with each subchannel, we jointly optimized the edge computing resource allocation, subchannel selection, device association, offloading and caching decisions to minimize the energy consumed by MDs in cache-assisted vehicular NOMA–MEC networks. Mathematically, this is formulated as

$$P1: \min_{\mathbf{X},\mathbf{Y},\mathbf{U},\mathbf{Z},\mathbf{F}} \sum_{m \in \mathcal{M}} \varepsilon_m^{tot}$$
$$\text{s.t. } C_1: \tau_m^{tot} \leq \tau_m, \forall m \in \mathcal{M},$$
$$C_2: \sum_{i \in \mathcal{I}} x_{m,i} = 1, \forall m \in \mathcal{M},$$
$$C_3: \sum_{i \in \mathcal{I}} \sum_{k \in \mathcal{K}} z_{m,i,k} \leq 1, \forall m \in \mathcal{M},$$
$$C_4: \sum_{m \in \mathcal{M}} \sum_{k \in \mathcal{K}} z_{m,i,k} \leq K, \forall i \in \mathcal{I},$$
$$C_5: \sum_{i \in \mathcal{I}} u_{m,i} \leq 1, \forall m \in \mathcal{M},$$
$$C_6: x_{m,i} \in \{0,1\}, \forall m \in \mathcal{M}, \forall i \in \mathcal{I}, \quad (10)$$
$$C_7: y_{m,i} \in \{0,1\}, \forall m \in \mathcal{M}, \forall i \in \mathcal{I},$$
$$C_8: z_{m,i,k} \in \{0,1\}, \forall m \in \mathcal{M}, \forall i \in \mathcal{I}, \forall k \in \mathcal{K},$$
$$C_9: u_{m,i} \in \{0,1\}, \forall m \in \mathcal{M}, \forall i \in \mathcal{I},$$
$$C_{10}: \sum_{m \in \mathcal{M}} y_{m,i} d_m \leq \vartheta_i, \forall i \in \mathcal{I},$$
$$C_{11}: \sum_{m \in \mathcal{M}} \sum_{i \in \mathcal{I}} z_{m,i,k} \leq \rho, \forall k \in \mathcal{K},$$
$$C_{12}: \sum_{m \in \mathcal{M}} u_{m,i} f_{m,i} \leq f_i^{BS}, \forall i \in \mathcal{I},$$

where $\mathbf{F} = \{f_{m,i} | \forall m \in \mathcal{M}, \forall i \in \mathcal{I}\}$; the constraint C_1 gives the maximum task processing time of MD m; C_2 and C_6 indicate that any MD m can select only one BS; C_3 and C_8 indicate that any MD m can occupy at most one subchannel; C_4 and C_8 mean that the number of MDs selecting any BS that needs to upload tasks should be less than or equal to the number of subchannels; C_5 and C_9 mean that any MD m can select at most one BS to execute its task; C_7 and C_{10} indicate that the data size of tasks cached at BS i does not exceed the caching capacity ϑ_i of this BS; C_8 and C_{11} show that the number of MDs selecting a subchannel cannot exceed its upper limit; C_7 and C_{12} reveal that the total computing capacity allocated to MDs by BS i cannot exceed the computing capacity of this BS.

4. Algorithm Design

As previously mentioned, the optimization problem P1 refers to minimizing local energy consumption within a given period. In view of this, we adopt the DRL algorithm to solve it. DRL is based on MDP, which implements the environment-based output of agent policy in MDP through neural networks, maximizing certain rewards. Considering that the overestimation of some conventional DRL algorithms (e.g., DQN and DDPG), the TD3 algorithm has been widely advocated because it can overcome well the problems of the above algorithms and achieve more stable output [41,42]. The main features of the TD3 algorithm are adding a new neural network and reducing the training frequency of the network based on DDPG.

The problem P1 has both continuous and discrete variables, and the solution space formed by the combination of all variables is very large, which is not a suitable scenario for the DQN algorithm to solve discrete space problems. Therefore, we use the TD3 algorithm, which can solve the continuous solution space problems. Considering that a simple mapping between only the decision of the algorithm and the actual strategy will fail to achieve convergence because of there being too many feasible strategies and the inability to search for the correct one, we develop an effective TD3O algorithm integrated with the AT algorithm to solve the problem P1.

4.1. MDP Used for Describing Problem P1

Considering that the optimization problem P1 needs to be tackled within a given period, in order to apply TD3O to the problem P1, such a period is divided into T timeslots and denoted as $\mathcal{T} = \{1, 2, \cdots, T\}$. Furthermore, the problem of joint computing offloading, task caching and resource allocation is described as a MDP, the state space, action space and reward function are defined as follows.

❶ State space: At each timeslot, the state space contains the information used for decisions made by the network. Here, the state s_t at timeslot t can be denoted as $s_t = \{\tilde{\mathbf{D}}(t+1), \tilde{\mathbf{Y}}(t)\}$. The detailed definitions can be found as follows.

- $\tilde{\mathbf{D}}(t+1) = \{\tilde{d}_m(t+1)|\forall m \in \mathcal{M}\}$ are the standardized data sizes of tasks of MDs at timeslot $t+1$, where

$$\tilde{d}_m(t) = \frac{d_m(t) - d^{\min}(t)}{d^{\max}(t) - d^{\min}(t)}, \quad (11)$$

$d^{\min}(t)$ is the minimum data size of the tasks of all MDs at timeslot t, and $d^{\max}$ is the maximum data size of the tasks of all MDs at timeslot t.

- $\tilde{\mathbf{Y}}(t) = \{\tilde{y}_m(t)|\forall m \in \mathcal{M}\}$ are the task caching decision factors at BSs at timeslot t, where $\tilde{y}_m \in [0,1]$.

❷ Action space: At each timeslot, the action space refers to the decisions made by the network according to the state s_t. The action a_t at timeslot t can be denoted as $a_t = \{\tilde{\mathbf{X}}(t), \tilde{\mathbf{Y}}(t+1), \tilde{\mathbf{Z}}(t), \tilde{\mathbf{U}}(t), \tilde{\mathbf{F}}(t)\}$. Specifically:

- $\tilde{\mathbf{X}}(t) = \{\tilde{x}_m(t)|\forall m \in \mathcal{M}\}$ are the association decision factors of MDs at timeslot t, where $\tilde{x}_m \in [0,1]$.
- $\tilde{\mathbf{Y}}(t+1) = \{\tilde{y}_m(t+1)|\forall m \in \mathcal{M}\}$ are the caching decision factors at timeslot t for the next timeslot.
- $\tilde{\mathbf{Z}}(t) = \{\tilde{z}_i(t)|\forall i \in \mathcal{I}\}$ are the subchannel allocation decision factors of BSs at timeslot t, where $\tilde{z}_i \in [0,1]$.
- $\tilde{\mathbf{U}}(t) = \{\tilde{u}_m(t)|\forall m \in \mathcal{M}\}$ are the offloading decision factors of MDs at timeslot t, where $\tilde{u}_m \in [0,1]$.
- $\tilde{\mathbf{F}}(t) = \{\tilde{f}_m(t)|\forall m \in \mathcal{M}\}$ are the computing resource allocation factors of MDs at timeslot t, where $\tilde{f}_m \in [0,1]$.

It is noteworthy that the dimensions of the above-mentioned state and action spaces have been greatly reduced compared to the actual ones. The actual state and action spaces can be achieved by executing an AT algorithm in the following parts.

❸ Reward: Considering that the goal of problem P1 is to minimize local energy consumption and the constraints C_1 and C_{10} cannot be strictly satisfied in the DRL-based iteration procedure, the reward w_t at timeslot t is given by

$$w_t = -\omega_1 \sum_{m \in \mathcal{M}} \varepsilon_m^{\text{tot}}(t) - \omega_2 \phi(t) - \omega_3 \varphi(t), \quad (12)$$

where $\phi(t) = \sum_{m \in \mathcal{M}} \max(\tau_m^{\text{tot}}(t) - \tau_m, 0)$ is the penalty function added for guaranteeing the constraint C_1; $\varphi(t) = \sum_{i \in \mathcal{I}} \max(\sum_{m \in \mathcal{M}} y_{m,i}(t) d_m(t) - \vartheta_i(t), 0)$ is the penalty function introduced for guaranteeing the constraint C_{10}; ω_1 is the energy-consumption discount factor; ω_2 and ω_3 are penalty coefficients.

When the network obtains action a_t according to the state s_t, the state space will obtain the next state s_{t+1} according to the action a_t. Specifically, the task-caching decisions of BSs can be directly achieved from $\mathbf{Y}(t+1)$ in a_t. Therefore, the total return of minimizing long-term local energy consumption within T timeslots can be given by

$$R = \sum_{t \in \mathcal{T}} \gamma w_t, \quad (13)$$

where γ is the reward discount factor satisfying $\gamma \in (0,1)$.

4.2. TD3O Algorithm

The TD3 algorithm is an actor-critic-based framework that comprises the policy (μ) network, critic (Q) network and their corresponding target networks and updates the network parameters using gradient algorithms. It is characterized by using two critic networks and two critic target networks in the design of critic networks. The TD3 algorithm is often divided into two parts consisting of experience collection and training. In the phase of collecting experience, a new action a_t can be generated by adding random Gaussian noise into the output of the policy network at the state s_t, i.e.,

$$a_t = \mu(s_t, \theta^\mu) + \tilde{\sigma}^2. \tag{14}$$

where θ^μ is the parameter of the policy network and $\tilde{\sigma}^2$ is the additive Gaussian noise.

After that, the environment is rewarded with w_t and the next state s_{t+1} can be achieved according to the state and action (s_t, a_t). To enable the algorithm to obtain better decisions through past experience-assisted training, we put the quadruple (s_t, a_t, w_t, s_{t+1}) into the experience replay buffer as a historical experience. In the training process, a certain number of quadruples are randomly selected from the experience replay buffer for training. Since the TD3 algorithm consists of policy and critic networks, the training part of the network is relatively independent, so it is divided into the following two parts.

4.2.1. Training Policy Network

The training process of the policy network is shown in Figure 3. In the training phase, N quadruples are extracted from the experience replay buffer and denoted as $\mathcal{N} = \{1, 2, \cdots, N\}$. For any quadruple $n \in \mathcal{N}$, the policy network outputs a new action $a'_n = \mu(s_n, \theta^\mu)$ according to the state s_n. It should be noted that the policy a'_n is different from a_n existing in the experience replay buffer. After s_n and a'_n are inputted into any critic network (e.g., critic Q_1 network), such network outputs $q_n = Q_1(s_n, \mu(s_n, \theta^\mu), \theta^{Q_1})$, where θ^{Q_1} is the parameter of the critic Q_1 network. After achieving all q_n, their mathematical expectation is given by

$$J(\theta^\mu) = \mathbb{E}\left[Q_1\left(\mathcal{S}, \mu(\mathcal{S}, \theta^\mu), \theta^{Q_1}\right)\right], \tag{15}$$

where $\mathcal{S} = \{s_n | n \in \mathcal{N}\}$. Then, the policy gradient of function J with respect to θ^μ can be given by

$$\nabla_{\theta^\mu} J = \mathbb{E}\left[\nabla_{\mathcal{A}} Q_1\left(\mathcal{S}, \mathcal{A}, \theta^{Q_1}\right) \nabla_{\theta^\mu} \mu(\mathcal{S}, \theta^\mu)\right], \tag{16}$$

where $\mathcal{A} = \{a_n | n \in \mathcal{N}\}$.

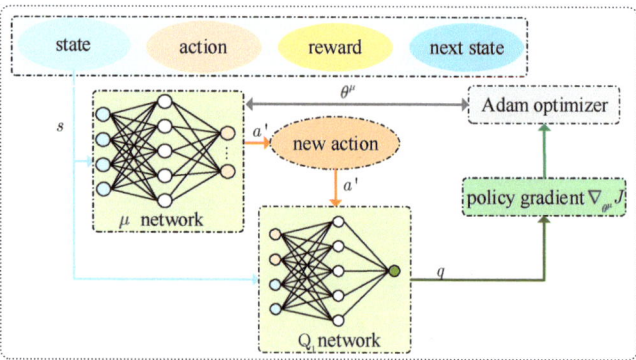

Figure 3. Training policy network.

Significantly, the calculated gradient requires gradient clipping, which can avoid skipping the optimal solution because the gradient is too large. The calculated policy gradients will be used to update the parameters of the policy networks. We assume that the learning rate of the policy network is β^μ, and use the adaptive moment (Adam) estimation commonly used in DRL to obtain the optimal θ^μ [43].

4.2.2. Training Critic Network

Figure 4 shows the training process of the critic network. During the critic network training, the policy at the next time is first estimated through the state at the next time by the policy target(μ^-) network, i.e., $a'_n = \mu^-(s'_n, \theta^{\mu^-}) + \hat{\sigma}^2$, where $\hat{\sigma}^2$ is policy noise, that is, trimmed additive Gaussian noise. Then, the action a'_n and the state s'_n are used as the input of the critic target (Q_1^-) network and critic target (Q_2^-) network, where $\theta^{Q_1^-}$ and $\theta^{Q_2^-}$ are their parameters. After that, these two networks output $\tilde{q}_{n,1}$ and $\tilde{q}_{n,2}$, respectively. Next, the approximation of Q value is $\tilde{q}_n = r_n + \gamma \tilde{q}_n$ achieved using Behrman equation, where $\tilde{q}_n = \min(\tilde{q}_{n,1}, \tilde{q}_{n,2})$. At the same time, the action a_n and the state s_n are used as the input of the critic Q_1 network and critic Q_2 network, where θ^{Q_1} and θ^{Q_2} are their parameters. After that, these two networks output $q_{n,1}$ and $q_{n,2}$. At last, for all $\tilde{q}_n$, according to the theorem of mean squared error (MSE), the expectation function of the squared loss between $Q_1(S, A, \theta^{Q_1})$ and $\bar{Q}$ is

$$L_1\left(\theta^{Q_1}\right) = 0.5 \mathbb{E}\left[\left(Q_1\left(S, A, \theta^{Q_1}\right) - \bar{Q}\right)^2\right], \quad (17)$$

and the expectation function of the squared loss between $Q_2(S, A, \theta^{Q_2})$ and $\bar{Q}$ is given by

$$L_2\left(\theta^{Q_2}\right) = 0.5 \mathbb{E}\left[\left(Q_2\left(S, A, \theta^{Q_2}\right) - \bar{Q}\right)^2\right], \quad (18)$$

where $\bar{Q} = \{\tilde{q}_n | n \in \mathcal{N}\}$. Then, the gradient of the loss function $L_1(\theta^{Q_1})$ with respect to the parameter θ^{Q_1} is

$$\nabla_{\theta^{Q_1}} L_1 = \mathbb{E}\left[\left(Q_1\left(S, A, \theta^{Q_1}\right) - \bar{Q}\right)\nabla_{\theta^{Q_1}} Q_1\left(S, A, \theta^{Q_1}\right)\right], \quad (19)$$

and the gradient of the loss function $L_2(\theta^{Q_2})$ with respect to the parameter θ^{Q_2} is given by

$$\nabla_{\theta^{Q_2}} L_2 = \mathbb{E}\left[\left(Q_2\left(S, A, \theta^{Q_2}\right) - \bar{Q}\right)\nabla_{\theta^{Q_2}} Q_2\left(S, A, \theta^{Q_2}\right)\right]. \quad (20)$$

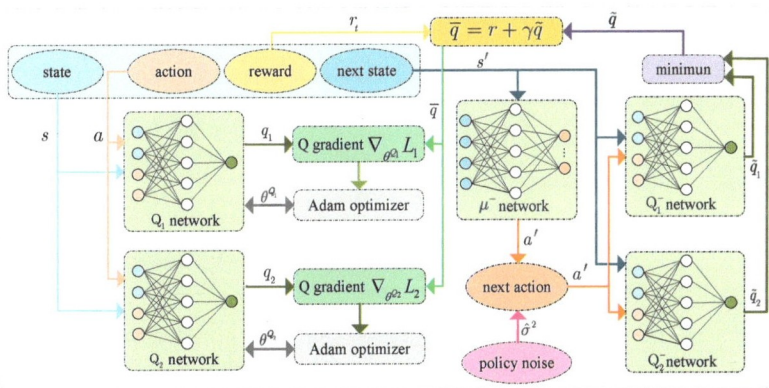

Figure 4. Training critic network.

Similar to calculating the policy gradient, the gradient clipping needs to be performed after calculating the gradients using (19) and (20). In addition, β^Q is the learning rate of the critic network, and the parameters of the two critic networks are updated using the Adam algorithm. Certainly, the parameters of critic target networks also need to be updated using the soft update method, i.e.,

$$\theta^{\mu^-} = \lambda \theta^\mu + (1-\lambda)\theta^{\mu^-}, \quad (21)$$

$$\theta^{Q_1^-} = \lambda \theta^{Q_1} + (1-\lambda)\theta^{Q_1^-}, \quad (22)$$

$$\theta^{Q_2^-} = \lambda \theta^{Q_2} + (1-\lambda)\theta^{Q_2^-}, \quad (23)$$

where λ is the learning rate of target networks.

It is noteworthy that a lower network updating frequency is adopted in this paper. We assume that the update interval of the critic network is t^{cti} and the update interval between the policy and critic networks is t^{pti}. The critic networks are trained many times to ensure the stability of Q value. After that, the policy network can be updated. The detailed procedure of the TD3O algorithm is summarized in Algorithm 1, where t^{mep} is the maximal number of epochs.

Algorithm 1: TD3-based offloading (TD3O)

1: Initialization: $\theta^{Q_1}, \theta^{Q_2}, \theta^{\mu}, \theta^{Q_1^-}, \theta^{Q_2^-}, \theta^{\mu^-}, t^{step} = 0, t^{epoch} = 0$.
2: **While** $t^{epoch} < t^{mep}$
3:    Let $t = 0$, state s_t and reward $R = 0$.
4:    **While** $t < T$
5:      Generate action a_t using (14).
6:      Achieve actual action by executing Algorithm 2.
7:      Calculate reward w_t using (12) and obtain the state s_{t+1}.
8:      **If** $t^{step} \geq \kappa$
9:        Replace the previous quadruple with (s_t, a_t, w_t, s_{t+1}).
10:      **Else**
11:        Put the quadruple (s_t, a_t, w_t, s_{t+1}) into the queue.
12:      **EndIf**
13:      Update state $s_t = s_{t+1}$.
14:      **If** $t^{step}\%t^{cti} = 0$ and $t^{step} > N$
15:        Extract N quadruples for training.
16:        For any sample n, Q_1^- and Q_2^- networks output $\tilde{q}_{n,1}$ and
17:        $\tilde{q}_{n,2}$, respectively, and obtain the minimum value $\tilde{q}_n$.
18:        Calculate $L(\theta_1^Q)$ and $L(\theta_2^Q)$ using (17) and (18), respectively.
19:        Calculate Q gradient using (19) and (20), and clip it.
20:        Find θ^{Q_1} and θ^{Q_2} using Adam optimizer.
21:        **If** $t^{step}\%t^{pti} = 0$
22:          Calculate q through Q_1.
23:          Calculate policy gradient using (16), and clip it.
24:          Find θ^{μ} using Adam optimizer.
25:        **EndIf**
26:        Calculate $\theta^{Q_1^-}, \theta^{Q_2^-}$ and θ^{μ^-} using (21)–(23), respectively.
27:      **EndIf**
28:      $R = R + \gamma w_t$.
29:      $t^{step} = t^{step} + 1; t = t + 1$.
30:    **EndWhile**
31:    $t^{epoch} = t^{epoch} + 1$.
32: **EndWhile**

4.3. AT Algorithm

In order to apply the TD3O algorithm to solve the problem P1, it is necessary to convert the achieved continuous action $a_t = \{\tilde{\mathbf{X}}(t), \tilde{\mathbf{Y}}(t+1), \tilde{\mathbf{Z}}(t), \tilde{\mathbf{U}}(t), \tilde{\mathbf{F}}(t)\}$ into a discrete one [44]. To this end, we consider the following transformations for a_t.

4.3.1. The Discretization of Device Association Array

In $\tilde{\mathbf{X}} = \{\tilde{x}_m | \forall m \in \mathcal{M}\}$, $\tilde{x}_m$ is the non-integer association index of MD m, which is the continuous action achieved by the TD3 algorithm. Then, it is converted into an integer form, i.e.,

$$\begin{cases} x_{m,\text{ceil}(I\tilde{x}_m)} = 1, & \text{if } I\tilde{x}_m \neq 0, \\ x_{m,1} = 1, & \text{otherwise,} \end{cases} \quad (24)$$

where ceil(b) is an upward rounding function with respect to b. Such a transformation can ensure that each MD can be associated with one BS.

Algorithm 2: Action transformation (AT)

1: **For** each MD $m \in \mathcal{M}$
2:   Achieve MD association matrix **X** using discretization rule.
3:   Achieve task caching matrix **Y** using discretization rule.
4:   Achieve task offloading matrix **U** using discretization rule.
5: **EndFor**
6: **For** each BS $i \in \mathcal{I}$
7:   Returns the set $\mathcal{K}_i$ of available subchannels and the set $\mathcal{M}_i$ of
8:   offloading MDs.
9:   **If** $M_i > K_i$
10:    $M_i - K_i$ associated MDs are randomly selected, disassociated
11:    and execute tasks locally.
12: **EndIf**
13: Achieve subchannel allocation matrix **Z** using discretization rule.
14: **EndFor**
15: **For** each MD $m \in \mathcal{M}$
16: **If** $\sum_{i \in \mathcal{I}} u_{m,i} = 1$
17:   **If** $\tilde{f}_m = 0$
18:    Assign small enough computing capacity to MD m to avoid
19:    zero division.
20:   **Else**
21:    Allocate computing resources to MD m using (28).
22:   **EndIf**
23: **EndIf**
24: **EndFor**

4.3.2. The Discretization of Task-Caching Array

In $\tilde{\mathbf{Y}}$, $\tilde{y}_m$ represents the non-integer caching index of MD m, which is the continuous action achieved by the TD3 algorithm. Since each MD can store its task at all BSs, there exist 2^I storage options for it. Consequently, in order to convert $\tilde{y}_m$ into a discrete form, we first need to perform

$$\begin{cases} \hat{y}_m = \text{floor}\left(2^I \tilde{y}_m\right), & \text{if } 2^I \tilde{y}_m \neq 0, \\ \hat{y}_m = 0, & \text{otherwise,} \end{cases} \quad (25)$$

where floor(b) is a downward rounding function with respect to b. Then, in order to achieve the binary caching index, the decimal $\hat{y}_m$ needs to be converted into a binary number of I 0–1 digits, which is given by bin$(\tilde{y}_m)$. In it, bin(b) is a function used for calculating the binary number of decimal b. Then, $y_{m,i} = \text{bin}(\tilde{y}_m)_i$, where bin$(\tilde{y}_m)_i$ represents the i-th digit of the binary number bin$(\tilde{y}_m)$.

4.3.3. The Discretization of Task Offloading Array

In $\tilde{\mathbf{U}}$, $\tilde{u}_m$ is the non-integer offloading index of MD m, which is the continuous action achieved by the TD3 algorithm. Considering that each MD can offload its task to at most one BS, $\tilde{u}_m$ is converted into an integer form, i.e.,

$$\begin{cases} u_{m,\text{ceil}(I\tilde{u}_m)} = 1, & \text{if } I\tilde{u}_m \neq 0, \\ u_{m,i} = 0, \forall i \in \mathcal{I}, & \text{otherwise.} \end{cases} \quad (26)$$

4.3.4. The Discretization of Subchannel Allocation Array

In $\tilde{\mathbf{Z}}$, $\tilde{z}_i$ is the non-integer index of the subchannels allocated by BS i to its associated MDs who need to offload tasks, which is the continuous action achieved by the TD3 algorithm. To achieve the integer form of $\tilde{z}_i$, we first need to perform

$$\begin{cases} \hat{z}_i = \text{ceil}(C(M_i, K_i)\tilde{z}_i), & \text{if } C(M_i, K_i)\tilde{z}_i \neq 0, \\ \hat{z}_i = 1, & \text{otherwise,} \end{cases} \quad (27)$$

where M_i is the number of MDs that are associated with BS i and need to offload tasks; K_i is the number of available subchannels at BS i; $C(M_i, K_i) = \text{fac}(K_i)/\text{fac}(M_i)\text{fac}(K_i - M_i)$ is a function with respect to M_i and K_i and is used for calculating the number of feasible subchannel allocation policies between M_i MDs and K_i subchannels at BS i; $\text{fac}(b)$ is a factorial function with respect to b; $M_i \leq K_i$ shall be satisfied.

Then, we assume that $\mathcal{Z}_i = \{1, 2, \cdots, C(M_i, K_i)\}$ is the set of $C(M_i, K_i)$ feasible subchannel allocation policies between M_i MDs and K_i subchannels at BS i. After that, the subchannel allocation policy $\hat{z}_i$ in the set $\mathcal{Z}_i$ is selected according to the Equation (27). It is noteworthy that $C(M_i, K_i)$ feasible subchannel allocation policies are generated in advance. That is to say, in the policy $\hat{z}_i$, we can easily know the utilized indices of K_i subchannels for M_i MDs. According to these rules, we can easily find the subchannel allocation index $\mathbf{Z}$.

4.3.5. The Transformation of Computing Resource Allocation Array

In $\bar{\mathbf{F}} = \{\bar{f}_m | \forall m \in M\}$, $\bar{f}_m$ represents the computing resource score of MD m at the target BS that is executing its task. If $\sum_{i \in I} u_{m,i} = 1$ is satisfied between MD m and BS i, according to the proportional allocation of computing resources, the computing resources allocated to MD m by BS i can be given by

$$f_{m,i} = u_{m,i} f_i^{BS} \bar{f}_m / \sum_{j \in \mathcal{M}} u_{j,i} \bar{f}_j. \tag{28}$$

Based on the above-mentioned operations, the output action $a_t = \{\bar{\mathbf{X}}, \bar{\mathbf{Y}}, \bar{\mathbf{Z}}, \bar{\mathbf{U}}, \bar{\mathbf{F}}\}$ of the TD3O algorithm can be effectively converted into an actual decision, which is summarized as Algorithm 2.

4.4. HA

To solve the problem P1 in a non-iterative manner, we design an effective heuristic algorithm, which is summarized in Algorithm 3. In such an algorithm, to reduce the uplink transmission time and energy consumption, some MDs are associated with the nearest BSs, and the BSs randomly cache the tasks of their associated MDs until the cache space cannot cache more tasks. Then, the uncached MDs randomly select a BS as the offloading target, and to guarantee time constraints, the BS will evenly distribute the computing resources according to the computation amount of the task. Finally, a part of the MDs are disassociated from BSs without sufficient subchannels and execute tasks by themselves.

Algorithm 3: Heuristic algorithm (HA)

1: Initialization: energy consumption $\bar{\varepsilon}^{tot} = 0$.
2: Each MD selects (is associated with) the nearest BS.
3: **For** each BS $i \in \mathcal{I}$
4:    **If** $M_i > K_i$
5:      $M_i - K_i$ associated MDs are randomly selected, disassociated
6:        and execute tasks locally.
7:    **EndIf**
8:    Randomly select the tasks of MDs associated with BS i for caching
9:      until the caching space is full.
10: **EndFor**
11: **For** $t \in \mathcal{T}$
12:    Randomly select a target BS for each MD without cached task.
13:    Randomly allocate subchannels to MDs associated with each BS.
14:    **If** subchannels are insufficient
15:      Extra MDs are randomly selected to execute tasks locally.
16:    **EndIf**
17:    Proportionally allocate computing resources to MDs associated with
18:      each BS according to the CPU cycles required by tasks.
19:    Calculate the total local energy consumption $\bar{\varepsilon}$.
20:    $\bar{\varepsilon}^{tot} = \bar{\varepsilon}^{tot} + \bar{\varepsilon}$.
21: **EndFor**

5. Algorithm Analysis

5.1. Computation Complexity Analysis

In this section, the computation complexity of proposed algorithms are analysed as follows.

Proposition 1. *The computation complexity of Algorithm 2 is $\mathcal{O}(MIK)$ in the worst case.*

Proof. In Algorithm 2, the computation complexity of steps 1–5 is $\mathcal{O}(M)$, the computation complexity of steps 6–14 is $\mathcal{O}(MIK)$ in the worst case, and the computation complexity of steps 15–24 is $\mathcal{O}(MI)$. In general, the computation complexity of Algorithm 2 is $\mathcal{O}(MIK)$ in the worst case. □

Proposition 2. *The computation complexity of Algorithm 1 is $\mathcal{O}\left(\max\left(\sum_{l=0}^{L^Q}\psi_l^Q\psi_{l+1}^Q, \sum_{l=0}^{L^\mu}\psi_l^\mu\psi_{l+1}^\mu\right)\right)$ at each timeslot, where L^μ is the number of layers of the policy network, L^Q is the number of layers of the critic network, ψ_l^μ is the number of neurons at the l-th layer of the policy network and ψ_l^Q is the number of neurons at l-th layer of the critic network.*

Proof. In Algorithm 1, the computation complexity is mainly related to the action transformation, the calculation of the reward and task processing time and the structure of the neural network. As previously mentioned, the computation complexity of the action transformation should be $\mathcal{O}(MIK)$ in the worst case. As seen from Formulas (8) and (12), the computation complexity of the calculation of the reward and task processing time is $\mathcal{O}(MIK)$.

In Algorithm 1, there exist four critic networks and two policy networks. We assume that the structure of the policy network and its target network is the same, and the structure of the two critic networks and its target network is the same. Then, we can easily deduce that the computation complexity of establishing policy networks is $\mathcal{O}\left(\sum_{l=0}^{L^\mu}\psi_l^\mu\psi_{l+1}^\mu\right)$ and the computation complexity of establishing critic networks is $\mathcal{O}\left(\sum_{l=0}^{L^Q}\psi_l^Q\psi_{l+1}^Q\right)$. Therefore, the computation complexity of establishing neural networks is $\mathcal{O}\left(\max\left(\sum_{l=0}^{L^Q}\psi_l^Q\psi_{l+1}^Q, \sum_{l=0}^{L^\mu}\psi_l^\mu\psi_{l+1}^\mu\right)\right)$.

Since the computation complexity of establishing neural networks is much higher than that of the other operations in Algorithm 2. In general, the computation complexity of Algorithm 2 is $\mathcal{O}\left(\max\left(\sum_{l=0}^{L^Q}\psi_l^Q\psi_{l+1}^Q, \sum_{l=0}^{L^\mu}\psi_l^\mu\psi_{l+1}^\mu\right)\right)$ at each timeslot. □

Proposition 3. *The computation complexity of Algorithm 3 is $\mathcal{O}(MI)$ at each timeslot.*

Proof. In Algorithm 3, the computation complexity of step 2 is $\mathcal{O}(MI)$, the computation complexity of steps 3–10 is $\mathcal{O}(I)$, the computation complexity of steps 12–16 is $\mathcal{O}(M)$, the computation complexity of steps 17–19 is $\mathcal{O}(MI)$. In general, the computation complexity of Algorithm 3 is $\mathcal{O}(MI)$ at each timeslot. □

5.2. Convergence Analysis

Since Algorithm 2 is a part of Algorithms 1 and 3 and is non-iterative, we just need to concentrate on the convergence of Algorithm 1. In detail, this is established as follows.

Theorem 1. *Algorithm 1 can be guaranteed to converge after finite iterations.*

Proof. In Algorithm 1, the neural networks are updated by the gradient descent method used in the Adam optimizer. This utilizes the gradient information of the functions $J(\theta^\mu)$, $L_1(\theta^{Q_1})$ and $L_2(\theta^{Q_2})$ to guide the updating directions of the parameters θ^μ, θ^{Q_1} and θ^{Q_2}, so that the objective functions can reach the optimal or suboptimal values. When these values tend to be stable, the parameters θ^μ, θ^{Q_1} and θ^{Q_2} also tend to be stable. At this time, Algorithm 1 is deemed convergent. □

6. Performance Evaluation

In order to verify the performance of the designed algorithms, we introduce the following algorithms for comparison.

DDPG-based offloading (DDPGO): DDPG is a classical DRL algorithm [45]. Compared with the TD3 algorithm, the DDPG algorithm reduces the critic network and the critic target network. In addition, both the critic network and policy network are updated at each timeslot in the DDPG algorithm. In this paper, the DDPG algorithm used to solve the problem P1 is named the DDPG-based offloading (DDPGO) algorithm. The difference between the two algorithms is that the state input and action output are the same as the mode used by TD3O, and it also uses the AT algorithm to convert continuous actions.

Completely offloading (CO): In the CO algorithm, the task of each MD is offloaded to the nearest BS for computing. Such BS proportionally allocates the computing capacity to its associated MDs according to the CPU cycles required by the tasks of these MDs.

Completely local executing (CLE): In the CLE algorithm, the tasks of all MDs can be executed by themselves.

In this paper, we consider that each BS is deployed in a non-overlapping area with a radius of 400 m and a power spectral density of -174 dBm/Hz. In addition, $I = 3$, $f_m^{loc} = 1$ GHz, $f_i = 8$ GHz, $W = 40$ MHz, $K = 4$, $d_m = 2 \sim 5$ MB, $c_m = 50$ cycles/bit, $\xi = 10^{-27}$, $\tau_m = 10$ s, $\rho = 2$, $r^{bh} = 1$ Gbps, $p_m = 23$ dBm, $\kappa = 80,000$, $N = 128$, $\gamma = 0.94$ and $\lambda = 0.04$. In the DRL algorithm, we consider that both the policy network and the critic network are composed of three-layer fully connected neural networks, where the numbers of neurons in three-layer neural networks in the policy network are 300, 200 and 128, respectively, and the corresponding target network has the same structure with this policy network; the number of neurons in three-layer neural networks in the critic network are 300, 128 and 32, respectively, and the corresponding target network has the same structure as this critic network. Significantly, the first-layer fully connected neural network of the policy network and the critic network utilizes the rectified linear unit 6 (RELU6), which suppresses the maximum value as the activation function, while other layers use RELU as the activation function.

Figure 5 shows the convergence of the TD3O and DDPGO algorithms. As shown in Figure 5, DDPGO may have a higher convergence rate than TD3O, but the former may have worse convergence stability than the latter. The reason for this may be that the critic network and the policy network are updated synchronously in DDPGO. In DDPGO, the network parameters are updated in each training phase, which speeds up the convergence. Synchronously, the policy network parameters are updated in the training, which results in the instability of the long-term reward value and training bias. As we know, TD3O is composed of two sets of critic networks. Consequently, it could be trained in a relatively stable Q value so that the algorithm can converge stably. In the simulation, it is also easy to find that TD3O could achieve a more stable and better solution to the problem P1 than DDPGO in general.

Figure 6 shows the impact of the training interval t^{pti} on the convergence of the TD3O algorithm. As we know, under the same number of iterations, a larger t^{pti} can effectively reduce the overall training time of the network. However, it will reduce the total learning times of the policy network and its target network. As illustrated in Figure 6, the convergence rate of TD3O may decrease with t^{pti} in general.

Figure 7 shows the impacts of learning rates β^Q and β^μ on the convergence of the TD3O algorithm. As we know, when the learning rate β^Q of the critic network increases, the parameters of such network will be updated at a larger scale, which speeds up the convergence of TD3O. However, it may lead to the failure of stable evaluation of environmental information, which weakens the convergence stability of TD3O. As illustrated in Figure 7, when $\beta^Q = 0.001$, the convergence rate of TD3O is relatively high, but the achieved long-term reward dramatically fluctuates at this moment. On the other hand, the learning rate β^μ of the policy network can affect the optimization capability of TD3O.

Specifically, a lower β^μ means a smaller amplitude of updating the policy network, which is better for finding better solutions. As seen from Figure 7, TD3O can achieve a better long-term reward when $\beta^Q = 0.0001$ and $\beta^\mu = 0.0001$.

Figure 8 shows the impact of the number of MDs on the long-term local energy consumption ε^{MD}, where ε^{MD} is the sum of the total local energy consumption in T timeslots. In general, the ε^{MD} increases with the number of MDs since a greater energy consumption is used when tackling more tasks of more MDs. Since CLE executes tasks in maximal computation capacity, it could achieve the highest ε^{MD} among all algorithms. In CO, MDs are associated with the nearest BSs, which may result in a relatively imbalanced load distribution. Then, some overloaded BSs cannot provide good services for their associated MDs because of limited resources, which may result in high ε^{MD}. Consequently, CO could achieve higher ε^{MD} than other algorithms excluding CLE. As illustrated in Figure 8, TD3O could achieve lower ε^{MD} than DDPGO since the former can effectively mitigate the overestimation existing in the latter. Although HA lets MDs be associated with the nearest BSs, some MDs associated with overloaded BSs will disassociate and execute tasks locally. Such an operation may result in relatively low ε^{MD}.

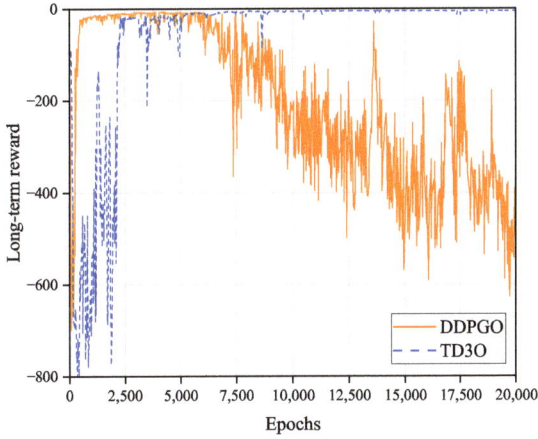

Figure 5. The convergence of DDPGO and TD3O algorithms.

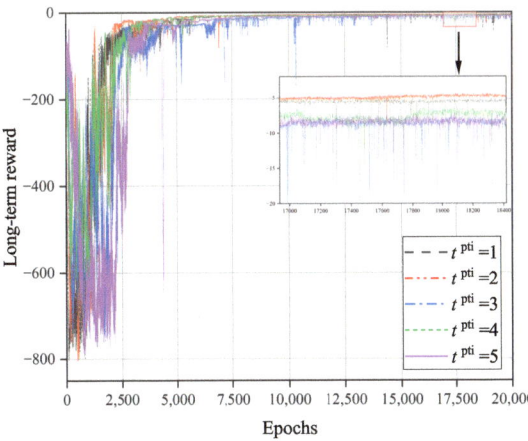

Figure 6. The impact of training interval t^{pti} on the convergence of TD3O algorithm.

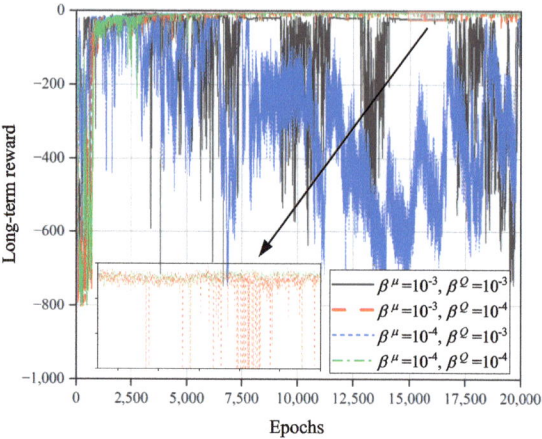

Figure 7. The impacts of learning rates β^Q and β^μ on the convergence of TD3O algorithm.

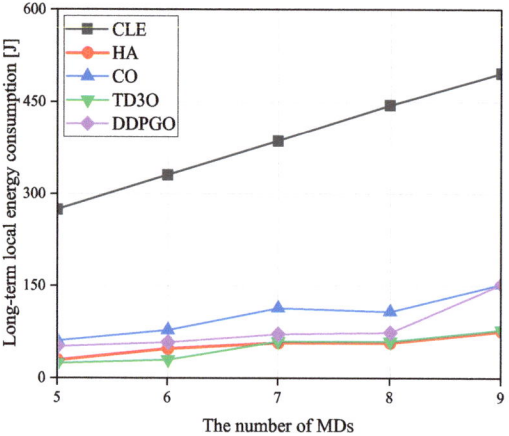

Figure 8. The impacts of the number of MDs on the long-term local energy consumption ε^{MD}.

Figure 9 shows the impacts of the number of MDs on the long-term reward (R). As illustrated in Figure 9, R may decrease with the number of MDs since more MDs result in a higher energy consumption. Since both TD3O and DDPGO try to maximize the reward but in other algorithms this is not the case, the former could achieve higher R than other algorithms in general. Since TD3O could achieve lower ε^{MD} than DDPGO, the former could achieve a higher R than the latter. In view of the unstable convergence of DDPGO, its reward may dramatically fluctuate. Since CLE could achieve the highest ε^{MD} among all algorithms, it could achieve the lowest R in general. In addition, CO could achieve a lower R than HA since the former consumes more energy than the latter.

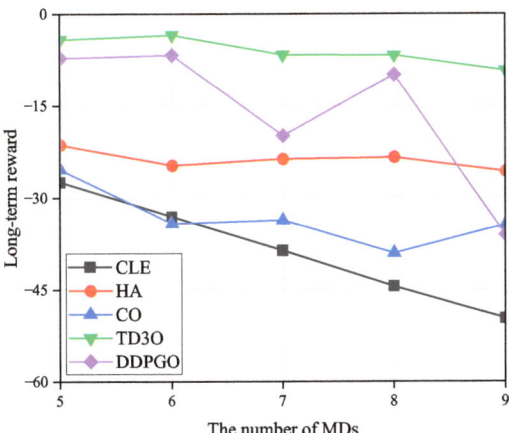

Figure 9. The impacts of the number of MDs on the long-term reward R.

Figure 10 shows the impacts of the size of caching space of each BS on the relative long-term energy consumption η^ε denoted as the ratio of ε^{MD} achieved by an offloading algorithm to the one attained by CLE. Evidently, a smaller η^ε means a higher energy gain caused by offloading tasks. As illustrated in Figure 10, in addition to DDPGO and CO, η^ε in other algorithms decreases with the size of the caching space of each BS in general. The reason for this may be that a larger caching space can hold more tasks to reduce the transmission energy consumption. However, as revealed in Figures 5 and 9, the unstable convergence of DDPGO may result in a dramatically fluctuating performance. Therefore, η^ε in DDPGO may evidently be fluctuating. In addition, η^ε in CO may not change with the size of caching space of each BS since it does not utilize caching space. By minimizing the ε^{MD}, TD3O and DDPGO could achieve a lower η^ε than other algorithms in general. In addition, TD3O could achieve a lower η^ε than DDPGO since the former mitigates the overestimation existing in DDPGO. As seen from Figure 10, CO could achieve the highest η^ε among all algorithms since it has no sufficient resources to provide for MDs associated with overloaded BSs.

Figure 11 shows the impacts of the size of the caching space of each BS on the relative long-term reward η^R denoted as the ratio of the long-term reward achieved by an offloading algorithm to the one attained by CLE. Evidently, a smaller η^R means a higher reward gain caused by offloading tasks. As illustrated in Figure 11, η^R in TD3O and HA decreases with the size of the caching space of each BS in general. The reason for this may be that a larger caching space can hold more tasks to reduce the transmission energy consumption, and then bring a higher reward. However, due to the unstable convergence of DDPGO, η^R in DDPGO may be fluctuating. Moreover, η^R in CO may not change with the size of caching space of each BS since it does not utilize caching space. By minimizing the ε^{MD} and thus increasing the reward, TD3O and DDPGO could achieve a lower η^R than other algorithms in general. In addition, TD3O could achieve a lower η^R than DDPGO since the former mitigates the overestimation existing in DDPGO. As can be seen from Figure 11, CO could achieve the highest η^R because of the high energy consumed by MDs associated with overloaded BSs.

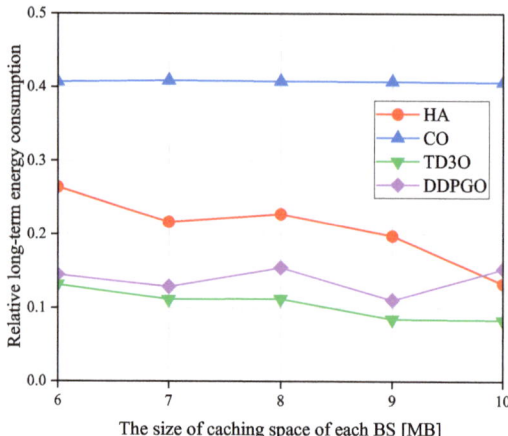

Figure 10. The impacts of the size of caching space of each BS on the relative long-term energy consumption η^ε.

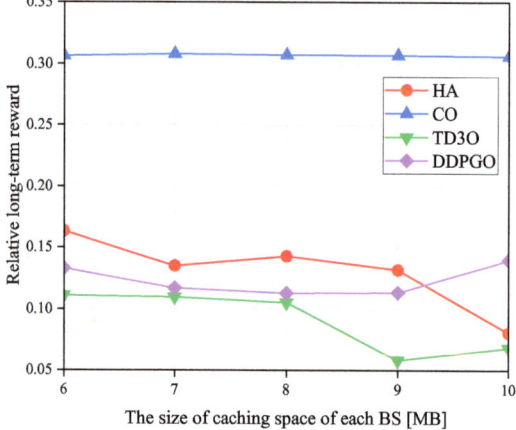

Figure 11. The impacts of the size of caching space of each BS on the relative long-term reward η^R.

As seen from the above-mentioned simulation figures, although HA is a non-iterative algorithm, it could sometimes achieve better performance than DDPGO. In addition, it may always achieve fairly better performance than CO and CLE.

7. Conclusions

In this paper, the problem of minimizing the local energy consumption is concentrated in the cache-assisted vehicular NOMA–MEC networks under time and resource constraints, which refers to the joint optimization of the computing resource allocation, subchannel selection, device association, offloading and caching decisions. To solve the formulated problem, we developed an effective TD3O algorithm that was integrated with the AT algorithm and designed HA simultaneously. As for the designed algorithms, we have given some analyses of the convergence and computation complexity. Simulation results show that TD3O could achieve lower local energy consumption than several benchmark algorithms, and HA could achieve lower local energy consumption than the CO and CLE algorithms. Future work can include power allocation and secure communications, such as optimizing the transmission power of MDs for task offloading, and how to encrypt part of the task data in the network at a low cost to achieve secure communication, etc.

Author Contributions: Conceptualization, T.Z. and M.X.; methodology, T.Z. and M.X.; software, M.X.; investigation, X.N., X.L. and C.L.; writing—original draft preparation, T.Z. and M.X.; writing—review and editing, D.Q. All authors have read and agreed to the published version of the manuscript.

Funding: This work was supported by the National Natural Science Foundation of China under grant Nos. 62261020, 62171119, 62062034, 62361026, 61961020, 62001201 and 61963017; the National Key Research and Development Program of China under grant No. 2020YFB1807201; the Jiangxi Provincial Natural Science Foundation under grant Nos. 20232ACB212005, 20224BAB202001, 20232BAB202019, 20212BAB202004 and 20212BAB212001; the key research and development plan of Jiangsu Province under grant No. BE2021013-3.

Institutional Review Board Statement: Not applicable.

Informed Consent Statement: Not applicable.

Data Availability Statement: Data are contained within the article.

Conflicts of Interest: The authors declare no conflict of interest.

References

1. Yao, Y.; Shu, F.; Li, Z.; Cheng X.; Wu, L. Secure transmission scheme based on joint radar and communication in mobile vehicular networks. *IEEE Trans. Intell. Transport. Syst.* **2023**, *24*, 10027–10037. [CrossRef]
2. Yao, Y.; Zhao, J.; Li, Z.; Cheng X.; Wu, L. Jamming and eavesdropping defense scheme based on deep reinforcement learning in autonomous vehicle networks. *IEEE Trans. Inf. Forens. Secur.* **2023**, *18*, 1211–1224. [CrossRef]
3. Wu, Q.; Shi, S.; Wan, Z.; Fan, Q.; Fan, P.; Zhang, C. Towards V2I age-aware fairness access: A dqn based intelligent vehicular node training and test method. *Chin. J. Electron.* **2023**, *32*, 1230–1244.
4. Khan, S.; Luo, F.; Zhang, Z.; Ullah, F.; Amin, F.; Qadri, S.; Heyat, M.; Ruby, R.; Wang, L.; Ullah, S.; et al. A survey on X.509 public-key infrastructure, certificate revocation, and their modern implementation on blockchain and ledger technologies. *IEEE Commun. Surv. Tutor.* **2023**; early access. [CrossRef]
5. Khan, S.; Luo, F.; Zhang, Z.; Rahim, M.; Ahmad, M.; Wu, K. Survey on issues and recent advances in vehicular public-key infrastructure (VPKI). *IEEE Commun. Surv. Tutor.* **2023**, *24*, 1574–1601. [CrossRef]
6. Wang, D.; Tian, X.; Cui, H.; Liu, Z. Reinforcement learning-based joint task offloading and migration schemes optimization in mobility-aware MEC network. *China Commun.* **2020**, *17*, 31–44. [CrossRef]
7. Spinelli, F.; Mancuso, V. Toward enabled industrial verticals in 5G: A survey on MEC-based approaches to provisioning and flexibility. *IEEE Commun. Surv. Tutor.* **2021**, *23*, 596–630. [CrossRef]
8. Zhang, Y.; Di, B.; Zheng, Z.; Lin, J.; Song, L. Distributed multi-cloud multi-access edge computing by multi-agent reinforcement learning. *IEEE Trans. Wirel. Commun.* **2021**, *20*, 2565–2578. [CrossRef]
9. Zhou, T.; Yue, Y.; Qin, D.; Nie, X.; Li, X.; Li, C. Joint device association, resource allocation, and computation offloading in ultradense multidevice and multitask IoT networks. *IEEE Internet Things J.* **2022**, *9*, 18695–18709. [CrossRef]
10. Zhang, W.; Zhang, G.; Mao, S. Joint parallel offloading and load balancing for cooperative-MEC systems with delay constraints. *IEEE Trans. Veh. Technol.* **2022**, *71*, 4249–4263. [CrossRef]
11. Malik, R.; Vu, M. Energy-efficient joint wireless charging and computation offloading in MEC systems. *IEEE J. Sel. Top. Signal Proces.* **2021**, *15*, 1110–1126. [CrossRef]
12. Hu, H.; Song, W.; Wang, Q.; Hu, R.Q.; Zhu, H. Energy efficiency and delay tradeoff in an MEC-enabled mobile IoT network. *IEEE Internet Things J.* **2022**, *9*, 15942–15956. [CrossRef]
13. Wu, Q.; Wang, S.; Ge, H.; Fan, P.; Fan, Q.; Letaief, K. Delay-sensitive task offloading in vehicular fog computing-assisted platoons. *IEEE Trans. Netw. Service Manag.* **2023**; early access. [CrossRef]
14. Bai, Y.; Zhao, H.; Zhang, X.; Chang, Z.; Jäntti, R.; Yang, K. Towards autonomous multiuav wireless network: A survey of reinforcement learning-based approaches. *IEEE Commun. Surv. Tutor.* **2023**; early access. [CrossRef]
15. Liu, Y.; Liu, J.; Argyriou, A.; Wang, L.; Xu, Z. Rendering-aware VR video caching over multi-cell MEC networks. *IEEE Trans. Veh. Technol.* **2021**, *70*, 2728–2742. [CrossRef]
16. Lekharu, A.; Jain, M.; Sur, A.; Sarkar, A. Deep learning model for content aware caching at MEC servers. *IEEE Trans. Netw. Service Manag.* **2022**, *19*, 1413–1425. [CrossRef]
17. Huang, X.; He, L.; Wang, L.; Li, F. Towards 5G: Joint optimization of video segment caching, transcoding and resource allocation for adaptive video streaming in a multi-access edge computing network. *IEEE Trans. Veh. Technol.* **2021**, *70*, 10909–10924. [CrossRef]
18. Zheng, G.; Xu, C.; Wen, M.; Zhao, X. Service caching based aerial cooperative computing and resource allocation in multi-UAV enabled MEC systems. *IEEE Trans. Veh. Technol.* **2022**, *71*, 10934–10947. [CrossRef]

19. Wu, Q.; Wang, X.; Fan, Q.; Fan, P.; Zhang, C.; Li, Z. High stable and accurate vehicle selection scheme based on federated edge learning in vehicular networks. *China Commun.* **2023**, *20*, 1–17. [CrossRef]
20. Liu, B.; Liu, C.; Peng, M. Resource allocation for energy-efficient MEC in NOMA-enabled massive IoT networks. *IEEE J. Sel. Areas Commun.* **2021**, *39*, 1015–1027. [CrossRef]
21. Song, Z.; Liu, Y.; Sun, X. Joint task offloading and resource allocation for NOMA-enabled multi-access mobile edge computing. *IEEE Trans. Commun.* **2021**, *69*, 1548–1564. [CrossRef]
22. Ding, Z.; Xu, D.; Schober, R.; Poor, H.V. Hybrid NOMA offloading in multi-user MEC networks. *IEEE Trans. Wirel. Commun.* **2022**, *21*, 5377–5391. [CrossRef]
23. Farha, Y.A.; Ismail, M.H. Design and optimization of a UAV-enabled non-orthogonal multiple access edge computing IoT system. *IEEE Access* **2022**, *10*, 117385–117398. [CrossRef]
24. Wang, K.; Li, H.; Ding, Z.; Xiao, P. Reinforcement learning based latency minimization in secure NOMA-MEC systems with hybrid SIC. *IEEE Trans. Wirel. Commun.* **2023**, *22*, 408–422. [CrossRef]
25. Xu, C.; Zheng, G.; Zhao, X. Energy-minimization task offloading and resource allocation for mobile edge computing in NOMA heterogeneous networks. *IEEE Trans. Veh. Technol.* **2020**, *69*, 16001–16016. [CrossRef]
26. Ylmaz, S.S.; Zbek, B. Massive MIMO-NOMA based MEC in task offloading for delay minimization. *IEEE Access* **2023**, *11*, 162–170. [CrossRef]
27. Tuong, V.D.; Truong, T.P.; Nguyen, T.-V.; Noh, W.; Cho, S. Partial computation offloading in NOMA-assisted mobile-edge computing systems using deep reinforcement learning. *IEEE Internet Things J.* **2021**, *8*, 13196–13208. [CrossRef]
28. Feng, H.; Guo, S.; Yang, L.; Yang, Y. Collaborative data caching and computation offloading for multi-service mobile edge computing. *IEEE Trans. Veh. Technol.* **2021**, *70*, 9408–9422. [CrossRef]
29. Ren, D.; Gui, X.; Zhang, K. Adaptive request scheduling and service caching for MEC-assisted IoT networks: An online learning approach. *IEEE Internet Things J.* **2022**, *9*, 17372–17386. [CrossRef]
30. Yang, S.; Liu, J.; Zhang, F.; Li, F.; Chen, X.; Fu, X. Caching-enabled computation offloading in multi-region MEC network via deep reinforcement learning. *IEEE Internet Things J.* **2022**, *9*, 21086–21098. [CrossRef]
31. Zhou, T.; Yue, Y.; Qin, D.; Nie, X.; Li, X.; Li, C. Mobile device association and resource allocation in HCNs with mobile edge computing and caching. *IEEE Syst. J.* **2023**, *17*, 976–987. [CrossRef]
32. Wu, Q.; Zhao, Y.; Fan, Q.; Fan, P.; Wang, J.; Zhang, C. Mobility-aware cooperative caching in vehicular edge computing based on asynchronous federated and deep reinforcement learning. *IEEE J. Sel. Top. Signal Process.* **2023**, *17*, 66–81. [CrossRef]
33. Zhou, H.; Wang, Z.; Zheng, H.; He, S.; Dong, M. Cost minimization-oriented computation offloading and service caching in mobile cloud-edge computing: An A3C-based approach. *IEEE Trans. Netw. Sci. Eng.* **2023**, *10*, 1326–1338 [CrossRef]
34. Zhang, W.; Zeadally, S.; Zhou, H.; Zhang, H.; Wang, N.; Leung, V. Joint service quality control and resource allocation for service reliability maximization in edge computing. *IEEE Trans. Commun.* **2023**, *71*, 935–948. [CrossRef]
35. Li, X.; Zhang, H.; Zhou, H.; Wang, N.; Long, K.; Al-Rubaye, S.; Karagiannidis, G. Multi-agent drl for resource allocation and cache design in terrestrial-satellite networks. *IEEE Trans. Wirel. Commun.* **2023**, *22*, 5031–5042. [CrossRef]
36. Yang, Z.; Liu, Y.; Chen, Y.; Al-Dhahir, N. Cache-aided NOMA mobile edge computing: A reinforcement learning approach. *IEEE Trans. Wirel. Commun.* **2020**, *19*, 6899–6915. [CrossRef]
37. Li, S.; Li, B.; Zhao, W. Joint optimization of caching and computation in multi-server NOMA-MEC system via reinforcement learning. *IEEE Access* **2020**, *8*, 112762–112771. [CrossRef]
38. Huynh, L.N.T.; Pham, Q.-V.; Nguyen, T.D.T.; Hossain, M.D.; Shin, Y.-R.; Huh, E.-N. Joint Computational offloading and data-content caching in NOMA-MEC networks. *IEEE Access* **2021**, *9*, 12943–12954. [CrossRef]
39. Yang, Z.; Ding, Z.; Fan, P.; Al-Dhahir, N. A general power allocation scheme to guarantee quality of service in downlink and uplink NOMA systems. *IEEE Trans. Wirel. Commun.* **2016**, *15*, 7244–7257. [CrossRef]
40. Cheng, Y.; Liang, C.; Chen, Q.; Yu, F.R. Energy-efficient D2D-assisted computation offloading in NOMA-enabled cognitive networks. *IEEE Trans. Veh. Technol.* **2020**, *70*, 13441–13446. [CrossRef]
41. Li, C.; Wang, H.; Song, R. Mobility-aware offloading and resource allocation in NOMA-MEC systems via DC. *IEEE Commun. Lett.* **2022**, *26*, 1091–1095. [CrossRef]
42. Huang, H.; Ye, Q.; Zhou, Y. 6G-empowered offloading for realtime applications in multi-access edge computing. *IEEE Trans. Netw. Sci. Eng.* **2023**, *10*, 1311–1325. [CrossRef]
43. Chen, J.; Xing, H.; Xiao, Z.; Xu, L.; Tao, T. A DRL agent for jointly optimizing computation offloading and resource allocation in MEC. *IEEE Internet Things J.* **2021**, *8*, 17508–17524. [CrossRef]
44. Dai, Y.; Zhang, K.; Maharjan, S.; Zhang, Y. Edge intelligence for energy-efficient computation offloading and resource allocation in 5G beyond. *IEEE Trans. Veh. Technol.* **2020**, *69*, 12175–12186. [CrossRef]
45. Long, D.; Wu, Q.; Fan, Q.; Fan, P.; Li, Z.; Fan, J. A power allocation scheme for MIMO-NOMA and D2D vehicular edge computing based on decentralized DRL. *Sensors* **2023**, *23*, 3449. [CrossRef] [PubMed]

Disclaimer/Publisher's Note: The statements, opinions and data contained in all publications are solely those of the individual author(s) and contributor(s) and not of MDPI and/or the editor(s). MDPI and/or the editor(s) disclaim responsibility for any injury to people or property resulting from any ideas, methods, instructions or products referred to in the content.

Article

Capture-Aware Dense Tag Identification Using RFID Systems in Vehicular Networks

Weijian Xu [1], Zhongzhe Song [1], Yanglong Sun [2], Yang Wang [3] and Lianyou Lai [1,*]

[1] School of Ocean Information Engineering, Jimei University, Xiamen 361000, China; xwjxwj@jmu.edu.cn (W.X.); 202111810010@jmu.edu.cn (Z.S.)
[2] Navigation Institute, Jimei University, Xiamen 361000, China; syl@jmu.edu.cn
[3] School of Informatics, Xiamen University, Xiamen 361000, China; yangwang@stu.xmu.edu.cn
* Correspondence: kaikaixinxinlly@jmu.edu.cn; Tel.:+86-137-9977-0802

Abstract: Passive radio-frequency identification (RFID) systems have been widely applied in different fields, including vehicle access control, industrial production, and logistics tracking, due to their ability to improve work quality and management efficiency at a low cost. However, in an intersection situation where tags are densely distributed with vehicle gathering, the wireless channel becomes extremely complex, and the readers on the roadside may only decode the information from the strongest tag due to the capture effect, resulting in tag misses and considerably reducing the performance of tag identification. Therefore, it is crucial to design an efficient and reliable tag-identification algorithm in order to obtain information from vehicle and cargo tags under adverse traffic conditions, ensuring the successful application of RFID technology. In this paper, we first establish a Nakagami-m distributed channel capture model for RFID systems and provide an expression for the capture probability, where each channel is modeled as any relevant Nakagami-m distribution. Secondly, an advanced capture-aware tag-estimation scheme is proposed. Finally, extensive Monte Carlo simulations show that the proposed algorithm has strong adaptability to circumstances for capturing under-fading channels and outperforms the existing algorithms in terms of complexity and reliability of tag identification.

Keywords: vehicular networks; RFID; tag identification; capture effect

1. Introduction

The vehicle networking industry is booming due to the rapid development of emerging technologies such as edge computing, wireless communication, and artificial intelligence [1–7]. Compared to visual inspection technology, which relies on "seeing" assets, RFID solutions rely on "listening" by using ultra-high-frequency (UHF) signals to remotely interrogate RFID tags attached to or embedded in objects, which greatly improves the efficiency of traffic control systems, including electronic toll collection, cargo inspection, and vehicle entry control, among other applications [8,9].

A complete UHF RFID system consists of readers, UHF tags, and back-end servers [10]. Each tag has a globally unique electronic product code (EPC). However, collisions may occur when multiple vehicles or cargo tags use backward scattering modulation to respond to readers simultaneously on the same cascaded channel [11,12]. Therefore, the RFID system needs to effectively coordinate the network and employ anti-collision algorithms. Collisions between readers will also certainly occur in the application system for intensive readers. In particular, with the rise of reinforcement learning [13–16], it is necessary for RFID systems that a multi-reader anti-collision algorithm based on reinforcement learning is used. However, this paper focuses on the field of identifying tags for a single reader in densely distributed vehicular networks.

Anti-collision algorithms are divided into two categories: the tree-based algorithm [17–19] (including query tree and tree splitting) and the ALOHA-based algorithm [20–23]. The former requires high computing costs when solving the problem of tag starvation. In addition, when the length of the tag ID is long, the tree-based algorithm causes frequent time slot collision, resulting in a large identification delay [20]. In contrast, the ALOHA-based algorithm guarantees equal opportunity for tags in random access, especially in table-dense scenarios with better adaptability [21]. It is favored by most RFID vendors due to its compliance with EPC Class 1 Gen2 UHF standards [24]. The ALOHA-based tag-identification algorithm is designed for rapid vehicle identification on expressways [22,23]. However, the aforementioned works do not comprehensively take into account actual difficulties and simply concentrate on tag recognition in ideal circumstances, neglecting the capture effect.

In Figure 1, as roadside observers, readers can swiftly discern multiple densely clustered tags within each vehicle as they congregate at intersections and obtain reliable target information by using RFID technology [25]. However, in wireless communication, the capture effect is a ubiquitous phenomenon [26]. For instance, in scenarios of checkpoints and intersections, tag density increases significantly with vehicle aggregation, which causes a complex RFID channel condition and frequent timeslots collision. The impact of the capture effect on the system is amplified as the number of collision timeslots increases. Furthermore, for the densely distributed tags, identification delay is a crucial issue for vehicular networks. Thus, it is necessary to mitigate the capture effect under collision timeslots to reduce identification delay and improve system efficiency.

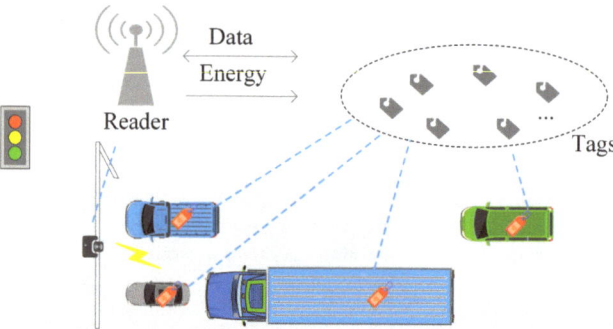

Figure 1. Vehicles with a high volume of cargo tags are statically detected at traffic intersections.

An analytic model for the capture effect categorizes tags based on signal strength [27]. Subsequently, to study the relationship between frame length and the number of tags that can be identified under unreliable channels, some work has focused on improving the efficiency of tag identification through capture-aware anti-collision schemes on the reader side [28–31]. The capture-aware backlog-estimation algorithm (CMEBE) estimates tag population and the probability of the capture effect with two-dimensional searches for minimum values [28]. However, it is difficult for the the CMEBE algorithm to guarantee identification performance under a large-scale tag environment. A capture-aware estimation (CAE) algorithm also simultaneously estimates the tag population and the probability of capture effect based on the number of idle slots in a frame [29]. Although it has low computational complexity, the algorithm fails in tag identification when the number of tags is much larger than the frame length. Minimum mean square error (MMSE) and Bayesian mean square estimation (CBMS) are proposed respectively in Ref. [30,31] for large-scale tag identification scenarios. However, the computational cost is increased because the algorithm uses a 2D search. Although researchers have begun to study the effect of capture on the system identification rate of RFID systems, this study is still in its infancy. They only give a fixed value of the capture probability, and the estimation performance and computational cost need to be improved.

Some researchers have explored the capture effect by establishing an RFID model through the stochastic distribution of tag positions, using a uniform or sigmoid-shaped distribution function for distance. In Ref. [32], the authors discussed an RFID model related to the capture effect under unreliable fading channel conditions. In Ref. [33,34], the authors provided a collision-avoidance algorithm for mobile tags under complex channel conditions. However, these studies fail to account for the impact of the capture effect on system identification efficiency in scenarios where tags are densely distributed within a specific area. In fact, due to the dense concentration of numerous tags in a particular region, the wireless channel becomes highly complex during tag identification [35,36], significantly affecting the identification efficiency of the system. Relying on path-loss analysis to determine capture probability is insufficient. In Ref. [37], the authors discussed the advantages of the RFID system in traffic recognition compared to visual detection under dense tags and proposed a traffic sign inventory-management system. However, it does not further study the reader's recognition performance in the context of dense tags. Therefore, it is unclear how the capture effect behaves when tags are located at the same distance or densely distributed. In this work, we construct a channel model to explore and analyze the capture probability of dense tag distributions and their influence on the performance of tag-identification algorithms. The contributions of this paper are as follows:

- A fading channel-capture model is established to analyze the capture probability in different channel environments, and the closed expression of the Nakagami-m fading channel capture probability is derived.
- We propose an advanced capture-aware estimation algorithm that quickly adjusts the initial frame length through the first few timeslots in a frame, reduces the delay caused by the lack of prior knowledge of the number of tags, and improves the estimation performance of both tags and capture probabilities.
- Considering the capture effect in fading channels and the duration of the slots, this paper dynamically adjusts the size of the next frame by combining the estimate of the number of tags and the capture probability, thus greatly improving the tag-identification rate. Compared with other excellent algorithms, the estimation method proposed in this paper shows better identification performance.

The remainder of the paper is organized as follows. Section 2 introduces the background, Section 3 establishes the system model, Section 4 shows the proposed estimation algorithm and the optimal frame length strategy. The numerical and analytical results are presented in Section 5, and Section 6 concludes the paper with an overview of some crucial points.

2. Background
2.1. Brief Analysis of ALOHA-Based Anti-Collision Algorithm

Each time slot in the UHF RFID system is expected to have three states: collision, successful, and idle. Based on statistics, the probability of collision timeslots, successful timeslots, and idle timeslots in the ALOHA algorithm follows a binomial distribution

$$p(r) = \binom{n}{r}\left(\frac{1}{l}\right)\left(1-\frac{1}{l}\right)^{n-r} \qquad (1)$$

where n is the number of tags in the system, r is the number of tags in the same timeslot, and l is the size of the frame length. Therefore, the number of idle timeslots is $N_i = lp(0)$, the number of success timeslots is $N_s = lp(1)$, and the number of collision timeslots can be expressed as $N_c = l - N_i - N_s$. Compared to the basic frame-slotted ALOHA (FSA) with a fixed frame length, the dynamic frame-length adjustment algorithm (DFSA) dynamically adjusts the size of the next frame according to the number of tags, which makes tag identification more stable [38]. Therefore, the next frame length needs to be determined by the backlog of the number of unidentified tags, which requires the system to estimate it in advance. At present, the anti-collision scheme in the two major standards, EPC C1 G2 standard and ISO18000-6C standard TypeA [39], is based on the DFSA.

2.2. The Entire System under the Capture Effect

In fact, due to the near–far effect and the random fluctuation of the received signal power caused by fading and shading, the probability of a packet being successfully received is significant even when multiple packets are transmitted at the same timeslot, which is called the capture effect [40]. An example of detection results from ALOHA, when the capture effect occurs, can be seen in Figure 2, where the collision state can transform into a successful state with a certain probability because of the capture effect. Therefore, the application of traditional tag estimation and its optimal frame configuration is challenging due to inaccurate backlog estimation, which in turn poses an obstacle to the identification performance of UHF RFID tags [34].

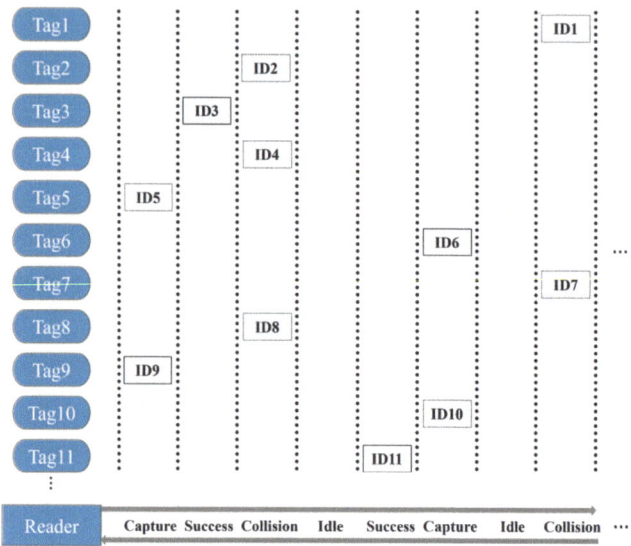

Figure 2. Detection results from ALOHA when the capture effect occurs.

Under the capture effect, if the capture probability is α, then the numbers of idle timeslots, successful timeslots, and collision timeslots are as follows

$$\begin{aligned} N_i^{cap} &= N_i \\ N_s^{cap} &= N_s + N_c \alpha \\ N_c^{cap} &= N_c - N_c \alpha \end{aligned} \quad (2)$$

If the duration of all three states is equal, the system throughput can be expressed as

$$\eta = \frac{N_s^{cap}}{N_i^{cap} + N_s^{cap} + N_c^{cap}} = \frac{n}{l}\left(1 - \frac{1}{l}\right)^{n-1} + \left(1 - \left[\left(1 - \frac{1}{l}\right)^n \left(1 + \frac{n}{l-1}\right)\right]\right)\alpha. \quad (3)$$

Figure 3 presents the expectation of system throughput under the non-capture effect and capture effect when a fixed frame length and the DFSA are adopted. The results show that the capture effect can enhance system throughput. This work has defaulted to the traditional frame-length adjustment scheme for DFSA, which sets the next frame length to an unrecognized number of tags. The first curve in the legend using DFSA converges to the optimal value of FSA under non-capture effects. The second curve cannot converge to the optimal value of FSA under the capture effect, which indicates that the next frame length cannot simply be set to an unidentified number of tags under the capture effect. Therefore, the adjustment of the next frame length under the capture effect needs further research, which will be discussed in Section 4.

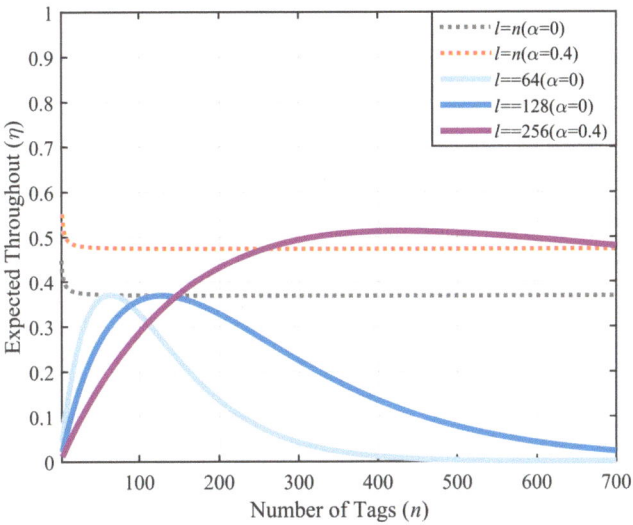

Figure 3. A comparison of FSA and DFSA with the capture effect.

3. Capture Model of High-Density Tag Distribution

To analyze the capture effect in densely distributed tags and its impact on anti-collision algorithms, this work models this complex channel and uses a fading channel to simulate interference and the attenuation of RF signals during transmission.

From Figure 4, the reader communicates with n tags densely stored in each container through one channel, which is characterized by its connection to two channels, namely the forward link and the reverse link. The forward channel coefficient $h_{f,D}$ describes the signal propagation from the reader to the tag, while the reverse channel coefficient $h_{b,D}$ describes the signal propagation from the tag to the reader after scattering.

In a fading channel, signal strength varies randomly due to multipath propagation, shadowing effects, and other factors. Thus, this work takes the strength of the signal received between the tags as a sample of the signal during that period.

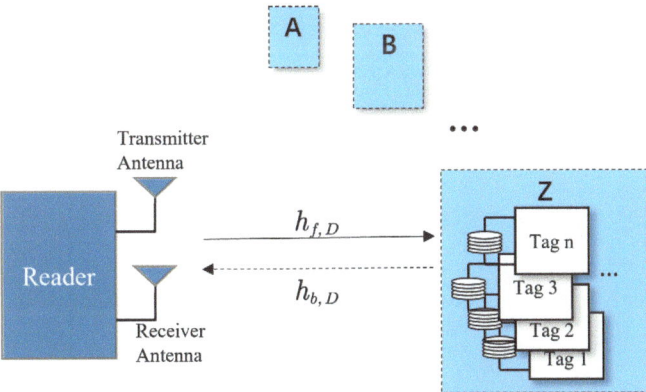

Figure 4. A monostatic system with transmit and receive antennas to be colocated at the reader (the reader device is positioned on the left side of the diagram, while the dotted line box on the right represents areas with high-density tag distribution).

In Section 2, this work concludes that the capture effect occurs during the collision timeslot. Therefore, this work assumes that there are s ($s \in [2, s_{max}]$) tags in a collision timeslot, where s_{max} is the maximum number of tags. The t-th tag is successfully captured, and the remaining $s-1$ tags are interference tags or noise for the RFID capture model, where tags are located at the same distance. Then, the probability of the t-th tag being captured can be defined as

$$M_s(q) = Pr\{P_{R_t} > qP_{R_{IS}}\} \tag{4}$$

where P_{R_t} is the received power from the t-th tag, $P_{R_{IS}} = \sum_{k=1}^{s-1} P_{Rd,rx,k}$ is the hidden power from other tags, and q represents the power ratio threshold, which represents the minimum carrier-to-Interference Ratio (CIR) required by the reader to receive signals successfully. This CIR is primarily determined by the modulation mode and coding scheme used by the reader. The capture probability can then be written as [41]

$$\alpha = s\int_0^\infty f_{p_t}(p_t)M_s(q)dp_t = s\int_0^\infty f_{p_t}(p_t)\left[\int_0^{p_t/q} f_{p_{s-1}}(p_{s-1})dp_{s-1}\right]dp_t \tag{5}$$

where $f_{p_{s-1}}(p_{s-1})$ is the pdf (for the power) that results from the convolution of $s-1$ non-central Chi-square pdf's.

According to the expression [34] for the received power of a reader in free space, assuming constant transmit power and carrier frequency of the reader, the received power is inversely proportional to distance d. However, when tags are densely distributed and located very close to each other, the distance between each tag and the reader is also close. Assuming equal reflected power from each interference tag, this work uses the default that the power of other interference signals received by the reader in each collision slot is approximately equal. In addition, these articles [42–44] assume that the wireless channel of the RFID system experiences Rayleigh fading and Rician fading. However, Nakagami-m is a multi-application model that is suitable for various channel conditions [45], and its distribution is closer to experimental data in various wireless communication environments [46]. Therefore, this work considers a Nakagami-m fading channel, ignoring the effect of propagation path loss and assuming that the mean received power of all interfering signals is the same. It is worth noting that the Nakagami-m fading channel model is only used to simulate the noise and attenuation effects on the signal during transmission, and it cannot accurately describe the mutual coupling effect between tags. And the instantaneous received power in Nakagami-m fading channel is then given by

$$f_p(p) = \frac{m^m p^{m-1}}{\bar{p}^m \Gamma(m)} e^{-\frac{mp}{\bar{p}}} (p \geq 0, m \geq 0.5) \tag{6}$$

where m is the channel fading parameter, $\bar{p}$ is the average power of the tag's signal, and $\Gamma(m)$ represents the Gamma function. This work was derived by adopting the Laplace and inverse Laplace transformation of the expressions for the PDF, but the details are omitted in this paper for brevity. Consequently, the compound PDF results from the convolution of $n-1$ PDFs of received power can be written as

$$f_{p_{s-1}}(p_{s-1}) = \frac{m^{m(s-1)}}{\Gamma(ms-m)\bar{p}^{m(s-1)}} p_{s-1}^{ms-m-1} e^{-\frac{mp_{s-1}}{\bar{p}}}. \tag{7}$$

Substituting (6) and (7) into (5), we can obtain [47]

$$\alpha^{Nak}(s,q,m) = \frac{s}{\Gamma(m)\Gamma(ms-m)} \sum_{k=0}^\infty \frac{(-1)^k \Gamma(ms+k)}{k!(ms-m+k)q^{ms-m+k}}. \tag{8}$$

It can be seen from the above formula that the capture effect for the RFID system is not only affected by the capture threshold q and the channel parameter m but also by the number of tags s in the collision timeslots.

4. Capture-Aware Algorithm for Large-Scale Tag Identification

4.1. Proposed Estimation Algorithm

The capture-aware algorithm can achieve parameter estimation by counting the number of different timeslot states observed after each frame and further adjusting the frame length to improve the efficiency of tag identification. However, due to the lack of prior knowledge of the number of tags, especially during large-scale intensive tag identification, it is difficult for the initial frame length to match the true number of tags, which negatively affects the system's efficiency of tag identification [29]. When the number of tags is much larger than the initial frame length, there are many collision timeslots in a frame, and tag identification efficiency is very low, and on the other hand, many idle timeslots can also affect the system's identification delay. Therefore, we propose a method of solving the problem of matching the initial frame length to an unknown number of tags by using the first several timeslots in a frame. In this work, we define the probability that there are no idle timeslots in the first w timeslots, which can be expressed as

$$P_w = 1 - \sum_{i=1}^{w} \binom{w}{i} p_I^i (1-p_I)^{w-i} \tag{9}$$

where the probability p_I of the idle timeslot can be expressed as a_I/l_+. When $l_+ = 128$, theoretical and simulation results P_w are given in Table 1, where $w = 1$ to 8 and the number of tags n varies from 50 to 1000.

Table 1. Probability of no idle slot in the first w slots when $l_+ = 128$.

n	$w = 1$	$w = 2$	$w = 4$	$w = 6$	$w = 8$
50	0.3244	0.1052	0.0110	0.0011	0.0001
	(0.3241)	(0.1055)	(0.0110)	(0.0010)	(0)
100	0.5436	0.2955	0.0873	0.0258	0.0076
	(0.5439)	(0.2956)	(0.0871)	(0.0257)	(0.0074)
400	0.9566	0.9151	0.8374	0.7663	0.7012
	(0.9557)	(0.9152)	(0.8374)	(0.7675)	(0.7010)
700	0.9959	0.9918	0.9836	0.9755	0.9675
	(0.9956)	(0.9905)	(0.9833)	(0.9750)	(0.9677)
1000	0.9996	0.9992	0.9984	0.9976	0.9969
	(1.0000)	(0.9995)	(0.9983)	(0.9977)	(0.9966)

where the results given in the brackets are computed from (9).

The results from Table 1 indicate that the probability P_w of having all non-empty timeslots in the first w timeslots increases with the number of tags regardless of the value of w. In the context of large-scale RFID tags, the initial frame length can be rapidly adjusted according to $l = \mu l_+$ when P_w is close to 1. It is worth noting that P_w being close to 1 is not an exact expression, and in experiments, we need to perform relevant debugging based on w and the parameter σ to determine the conditions for the rapid adjustment of the initial frame length, which will be introduced in Algorithm 1. Indeed, the estimation strategy of the proposed algorithm can be divided into two parts: the estimation of tags' numbers and the estimation of capture probability. Then, considering that the number of idle timeslots observed by the reader is not affected by the capture effect of Nakagami-m fading channels, the number of tags can be estimated using idle timeslots. Invert $N_i = lp(0)$ and replace N_i with the observed value c_i^{cap} and the proposed algorithm estimates the number of tags by

$$\hat{n} = \frac{\ln\left(c_i^{cap}/l\right)}{\ln(1-1/l)}, \quad c_i^{cap} \neq 0. \tag{10}$$

Algorithm 1: Pseudo-code operation of a reader.

```
1  begin
2    Initialization: l₊, w, κ, σ;
3    while unidentified tags ≠ 0 do
4      Interrogate(l₊)
5      for timeslot ← 1 to l₊ do
6        Update the statistics of Slot_diff(w)
7        if Slot_diff(w) > σ then
8          l = κl₊
9        else
10         l = l₊
11         break
12       end if
13     end for
14     while unidentified tags ≠ 0 do
15       Interrogate(l)
16       Receive tag responses and observe cᵢᶜᵃᵖ, cₛᶜᵃᵖ, c꜀ᶜᵃᵖ
17       Estimate n̂ according to cᵢᶜᵃᵖ and formula (10)
18       Estimate α̂ according to cₛᶜᵃᵖ c꜀ᶜᵃᵖ and formula (11)
19       l = AdjustFrame(n̂, α̂, k*)
         //Compute l_opt according to Equations (12)–(14)
20     end while
21   end while
22 end
```

In addition, the number of unidentified tags in the $i+1$-th frame is $backlog = n_{est} - c_s$ if the estimated number of tags is n_{est} after the i-th frame; then, the capture probability α can be shown to be

$$\hat{\alpha}_{cap} = \arg\min_{\alpha_{cap} \in P} E\left\{\left(c_s^{cap} - N_s'\right)^2 + \left(c_c^{cap} - N_c'\right)^2\right\} \tag{11}$$

based on the minimum mean square error. The proposed algorithm can accurately search the capture probability under the fading channel, where c_s^{cap} and c_c^{cap} denote the observed numbers of successful and collision timeslots, respectively; N_s' and N_c' denote their expectations, respectively; and P can be expressed as $P = \{\alpha_{cap} | 0 \leq \alpha_{cap} \leq 1\}$.

4.2. Frame Length Adjustment

In Section 2, it is observed that collision timeslots have the potential to convert into successful timeslots in the capture environment. However, setting a frame length that is too long may increase idle timeslots and ultimately lead to time slot waste. Conversely, reducing the frame length can increase the likelihood of capturing transmission slots but may also decrease system identification efficiency. Therefore, this paper proposes an optimal method for frame-length adjustment. Combining Equation (3) and capture-aware efficiency with the duration of different time slots can then be written as

$$\eta_{id} = \frac{N_s t_s + N_c t_c \alpha_{cap}}{N_i t_i + N_s t_s + N_c t_c} \tag{12}$$

where t_i, t_s and t_c can be expressed as the duration of the idle time slot, the successful time slot, and the collision time slot, respectively. Divide the numerator and denominator of the above equation by t_s. Let $u = t_i/t_s$, $v = t_c/t_s$ and take a linear model ($l = kn, k \in R^+$) into account [48]. Combining the $\lim_{n\to\infty}(1 - 1/kn)^n = e^{-1/k}$, we can obtain

$$\eta_{id} \approx \frac{\frac{1}{\left(ke^{\frac{1}{k}}+k-1\right)} + \alpha^{Nak}}{\frac{1+ku}{\left(ke^{\frac{1}{k}}-k-1\right)} + v + (1-v)\alpha^{Nak}}. \tag{13}$$

Then, the optimal frame length can be written by

$$l_{opt} = [k^*\hat{n}] \tag{14}$$

where $k^* = \underset{k \in \mathbb{k}}{argmax}(\eta_{id})$, $\mathbb{k}$ denotes the searching range of k and $[\cdot]$ is a floor integer function.

Table 2 shows the performance comparison between equal timeslots and unequal timeslots. According to the different frame size and system-throughput values corresponding to different channel shape parameters, we find that the method under an unequal time slot is better than the method under an equal time slot, which further verifies that the time slot setting in EPC C1 G2 protocol is still suitable under the influence of fading channel capture effect. Next, according to Equation (14), this work establishes the relationship between n and the optimal frame length l_{opt} for different channel parameters m adopting the EPC C1 G2 standard, as depicted in Figure 5. It is evident that when the number of tags is constant, the RFID system requires a greater optimal frame length as m increases.

Table 2. The values of l_{opt} and η_{id} when $n = 700$.

Fading Channels (m)	$u = v = 1$		$u = 0.125, v = 0.5$	
	l_{opt}	η_{id}	l_{opt}	η_{id}
0.5	393	0.6303	699	0.8427
1	466	0.5522	828	0.8116
1.5	508	0.5131	898	0.7938
2	535	0.4885	944	0.7823
2.5	555	0.4712	978	0.7738
3	571	0.4581	1004	0.7672
3.5	583	0.4478	1025	0.7618

where the results are computed from the Equation (12)–(14).

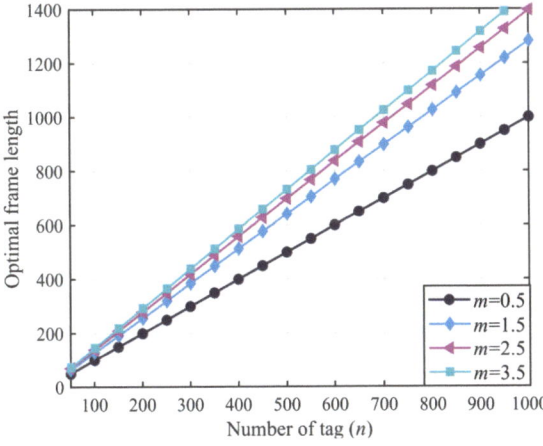

Figure 5. Optimal frame length l_{opt} as function of tags n ($q = 2, u = 0.125, v = 0.5$).

4.3. Description of the Algorithm

In the proposed algorithm, an identification cycle is depicted, which is implemented as shown in Algorithm 1.

To recognize densely distributed large-scale tags, the reader will send a query command with a short frame l_+ after initializing the parameters. Then, the tag will randomly select a time slot from the frame and respond to the query command. After that, the $Slot\_diff(w)$ and σ are compared, where $Slot\_diff(w)$ is used to record the calculation results of the first w slot states, and σ can control the pace of the frame length adjustment ($w = 4$, $\sigma = 1$ in this work). If the former is larger than the latter, a dynamic adjustment of the initial frame length l is performed according to the method of $l = \kappa l_+$. Otherwise, the adjustment is completed. After sending the query command with the frame length l, the reader will count the idle slots after each frame ends to estimate the number of tags. It will also count the number of successful slots and collision slots to estimate the capture probability. Finally, the optimal frame length is found according to the Formulas (12)–(14), and the reader will send a query command with the next frame length until all tags are recognized.

5. Numerical and Analytical Result

In this section, numerical results are presented to verify the performance of the proposed algorithm. The random data used in the experiments were generated using MATLAB's internal function (MATLAB Version: 9.14.0.2194193 (R2023a)). Firstly, the capture effect was simulated using the Monte Carlo method, and 10,000 independent experiments were performed to obtain the average results. Simulation results, including Figures 6 and 7, are presented to investigate the impact of various parameters on the capture performance of Nakagami-m fading channels in a dense tags environment. In the experiment, the numbers of collision slot tags were set to two, three, and four, and the theoretical value in the closed expression of capture probability in the Nakagami-m fading channel in Section 3 was compared with the actual value. Then, in order to more intuitively explore the influence of other factors on the capture effect, a 3D surface diagram of the capture probability under the Nakagami-m fading channel was presented in Figure 6. Additionally, in order to validate the performance and versatility of the proposed algorithm in estimating the capture probability, the capture probability under the actual Nakagami-m fading channel was not used. Instead, the search range for the capture probability was set to a larger range, $\alpha = \{0, 0.1, 0.2, \ldots, 1\}$, and the step size was 0.1. The prior distribution of the tag was set to be uniform over the range of μ, where μ varies from 50 to 1000. The search range for the tag of the CMEBE and MMSE, CBMS algorithms was set to $N = \left\{c_s^{cap} + 2c_c^{cap} \leq n \leq N_{\max} | n \in \mathbb{Z}\right\}$, where $N_{\max} = 1000$. In the experiments, the initial frame length of the proposed algorithm was set to 128. The parameters w, κ, and σ were set to 4, 2, and 0.85 in the proposed algorithm, respectively. Based on the analysis in Table 2, the setting of time slot duration was $t_i = 50$ μs, $t_s = 400$ μs, and $t_c = 200$ μs ($u = 0.125$, $v = 0.5$), which is consistent with the EPC C1 G2 Protocols, and the VOGT [49] algorithm used an equal duration for all timeslots. To evaluate the performance of an algorithm, the estimation error is defined as $error = \left|\frac{x-\hat{x}}{x}\right| \times 100\%$.

Figure 6 shows that when the capture threshold is constant, the capture effect is related to these two factors, which are the shape parameter m of the fading channel and the number of tags s in the collision timeslots. Among them, the value of the capture probability decreases with the increase in the shape parameter m of the channel; in other words, a larger m will make the capture effect more difficult to occur. The capture probability also decreases as the number of tags in the collision timeslots increases.

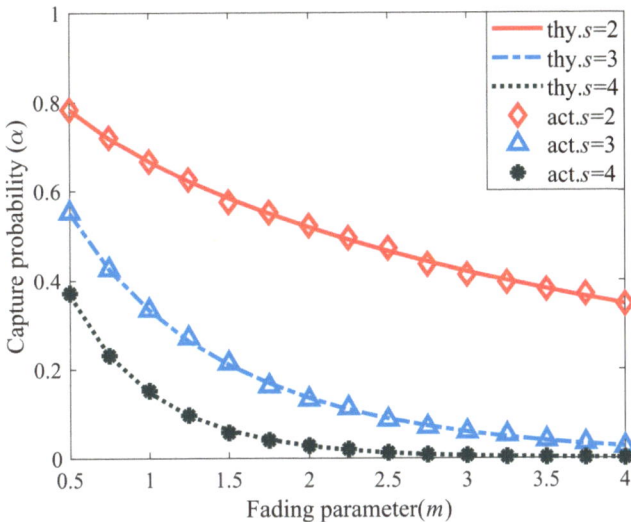

Figure 6. Capture probability on the Nakagami-m fading channel ($q = 2$).

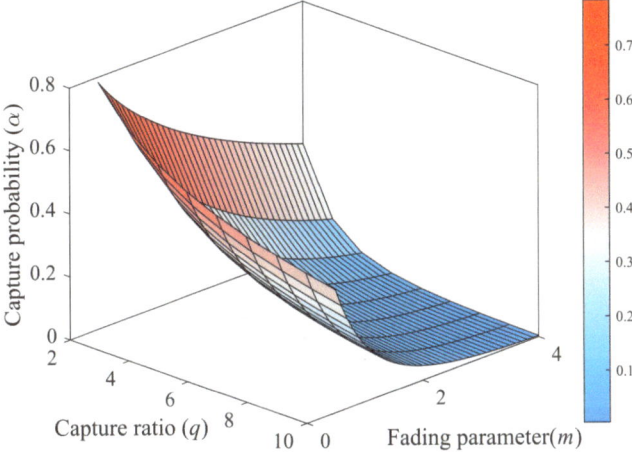

Figure 7. 3D Capture probability on Nakagami-m fading ($s = 2$).

Figure 7 presents the same trend as Figure 6 and also shows that the capture probability is closely related to the coding and modulation capability of the reader. When the capture threshold is set to a large size, the conditions for satisfying the capture will be more stringent, such as $q = 7$, $m = 2$, and the capture probability is only 0.08 ($s = 2$).

Afterward, we conducted simulations to evaluate the estimation performance of the proposed algorithm. The simulation results are shown in Figures 8 and 9. In Figure 8, we compare the estimation error of the capture probability of the proposed algorithm with that of other algorithms. The results show that the estimation error of the capture probability of all four algorithms decreases as the capture probability increases. Compared with the other algorithms, the estimation algorithm for capture probability proposed in this paper demonstrates good estimation performance in the range of capture probabilities from 0 to 0.5. For example, when the capture probability is 0.2, the estimation error of the proposed algorithm is only 3.46%, much lower than that of the other three algorithms. Additionally, Figure 7 shows that the capture probability is usually small under a fading channel, which further demonstrates the superiority of the proposed algorithm.

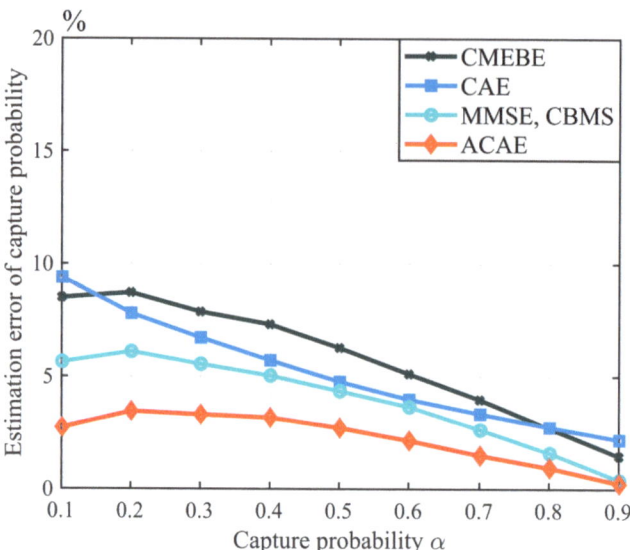

Figure 8. Capture probability estimation of the proposed algorithm ($l = 128$, $n = 700$).

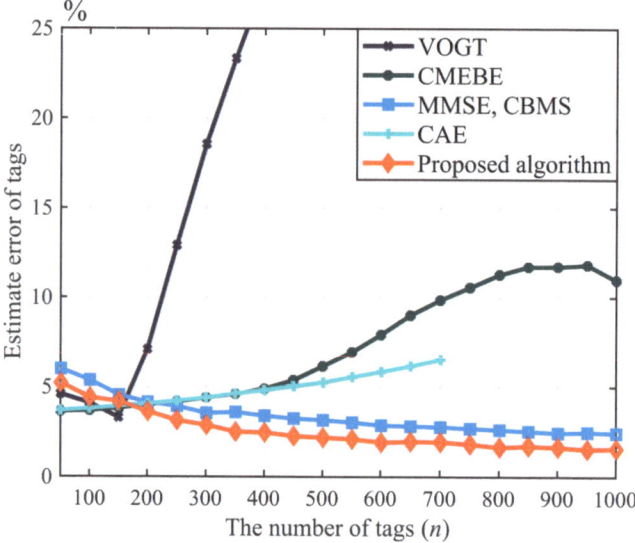

Figure 9. Tag estimation of proposed algorithm under fading environment ($m = 1.5$, $q = 2$).

Figure 9 illustrates the tag estimation performance of various algorithms in the presence of the capture effect. Notably, CMEBE and CAE exhibit suboptimal tag-estimation accuracy when the number of tags is large. Specifically, when there are more than 700 tags, CAE's estimation algorithm becomes invalid due to a lack of free timeslots caused by mismatched initial frame length during tag estimation, which significantly impairs its performance. Because the VOGT algorithm does not have the capture sensing ability, it has a large estimation error when the number of tags exceeds 200. In contrast, both MMSE and CBMS and our proposed algorithm employ an identical initial frame-length-adjustment mechanism, resulting in similar performance trends for tag estimation. Simulation results indicate that with 1000 tags, the proposed algorithm achieves a tag estimation error of only 1.94%, which is a reduction of 29.96% compared to the MMSE, CBMS algorithm and 70.29% compared to the CAE algorithm. Therefore, this paper's proposed algorithm exhibits superior tag estimation performance in the presence of a capture effect under fading channels.

The capture probability of a Nakagami-m fading channel is not only related to the encoding and modulation method of the reader but also depends on the degree of fading and the number of tags in the collision epoch. Therefore, compared to a fixed capture effect, we need to deeply explore the impact of different estimation algorithms on the stability of tag identification under the parameters of channel fading.

To simplify the analysis of the role of the capture effect in collision-avoidance strategies, we assume that the reader can count the number of tags in the collision epoch and use the formula, combined with the setting of the capture threshold, to obtain the probability of the occurrence of capture effect in different collision epochs under the fading channel. In this experiment, the number of tags is $n = 700$, and the capture threshold is set to $q = 2$.

When the capture threshold is fixed, the probability of the occurrence of the capture effect is mainly affected by the shape parameters of the fading channel. Therefore, different shape parameters will affect the efficiency of tag identification in the system. However, by combining with the channel environment, we can set the optimal next frame length to achieve the optimal overall efficiency of tag identification. To determine the next frame length in the VOGT algorithm, this article employs the method $l_{next} = \hat{n} - c_s^{cap}$. In contrast, for the CAE and CMEBE algorithms, method $l_{next} = \hat{\alpha} + (1 - \hat{\alpha})\left(\hat{n} - c_s^{cap}\right)$ is utilized to determine the subsequent frame length. As for both MMSE and CBMS and our proposed algorithm, we adjust the next frame length using the formulas (12)–(14).

Figure 10 shows the influence of shape parameters of the Nakagami-m fading channel on the tag-identification efficiency under different algorithms, in which channel shape parameters reflect the fading degree of the channel. When $m = 1$, the fading channel follows Rayleigh distribution. It can be seen that the identification efficiency of the five algorithms decreases with the increase in the channel-shape parameter m, among which the VOGT algorithm has the worst tag-identification performance. When the shape parameter $m = 0.5$, the proposed algorithm is very close to the tag-identification efficiency of CMEBE, CAE, MMSE, and CBMS, all reaching 83%, mainly because their estimation performance is not different when the capture probability is large. However, with the increase in the channel shape parameters, the capture effect becomes smaller. When $m = 3$, the identification efficiency of the proposed algorithm is 76.83%, only declining by about six percentage points, indicating that the proposed algorithm exhibits strong stability compared to other algorithms.

Next, we employ the Big O notation to obtain the computational complexity of the proposed algorithm with that of pre-existing algorithms. Here, we specify that the set of estimation tags n and capture effect α of the above algorithm are N and P, respectively. Assuming that $\varepsilon = |P|$, $\varphi = |N|$, where $|\cdot|$ represents the cardinality of the set, the VOGT method needs to gradually search for an extreme value within a certain number of tags, and its computational complexity can be approximated as $O\left(\sum_{i=1}^{\varphi} \tilde{n}_i\right)$. CMEBE, MMSE, and CBMS all require a two-dimensional search for the number of tags and probability of capture effect. The computational complexity of CMEBE, MMSE, and CBMS can be approximated as $O\left(\varepsilon \sum_{i=1}^{\varphi} \tilde{n}_i\right)$, where $\tilde{n}_i$ represents the i search value of the number of tags, and CAE and the algorithms proposed in this paper involve one-dimensional searches, with the CAE algorithm having a complexity approximation of $O(1)$. The proposed algorithm has an estimated complexity of $O(\varepsilon)$.

Thus, based on the proposed model, a conclusion can be drawn from the experiment that the estimation performance and tag-identification efficiency of the proposed algorithm are better than those of other algorithms. Among them, Table 3 indicates that the proposed scheme possesses the features of low complexity and high identification efficiency in vehicular network tag identification.

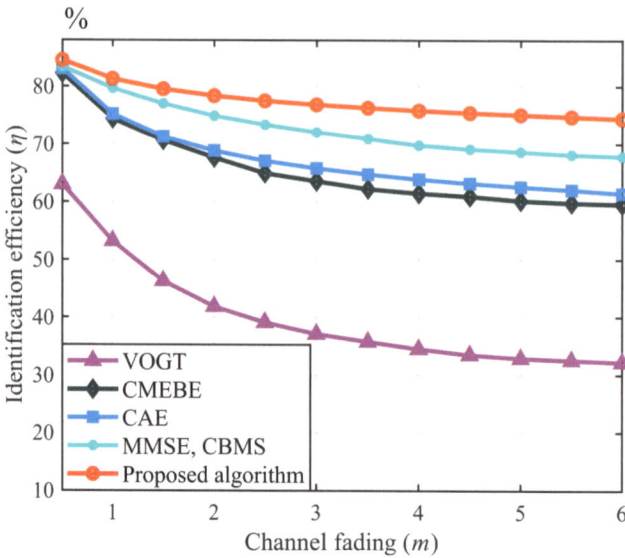

Figure 10. Comparison of the identification efficiency of five algorithms.

Table 3. Comparison of estimation methods.

Estimation Method	Estimated Object	Number of Identification Tags	Estimation Error of Capture Probability (%)	Estimation Error of Tags (%)	Identification Efficiency (%)	Computational Complexity	Run Time (s)
VOGT	$\hat{n}$	Small	-	46.74	54.21	$O(\sum_{i=1}^{\varphi} \tilde{n}_i)$	5.7462
CMEBE	$\langle \hat{n}, \hat{\alpha} \rangle$	Small	8.50	9.84	72.86	$O(\varepsilon \sum_{i=1}^{\varphi} \tilde{n}_i)$	14.3671
CAE	$\langle \hat{n}, \hat{\alpha} \rangle$	Small	9.38	6.53	73.21	$O(1)$	1.15
MMSE, CBMS	$\langle \hat{n}, \hat{\alpha} \rangle$	Large	5.65	2.77	79.66	$O(\varepsilon \sum_{i=1}^{\varphi} \tilde{n}_i)$	32.61/33.13
Proposed algorithm	$\langle \hat{n}, \hat{\alpha} \rangle$	Large	2.73	1.94	81.31	$O(\varepsilon)$	15.42

where the estimation error of capture probability when $\alpha = 0.2$ and the estimation error of tags and identification efficiency when $n = 700$.

6. Conclusions

In this paper, a Nakagami-m fading-channel-capture model is established based on vehicle cargo tags in a large-scale environment with dense tags, and an effective capture-aware estimation algorithm is proposed. Both theoretical analysis and simulation results have shown that the proposed capture-aware algorithm is superior to the reference algorithms. In addition, we derive the optimal frame length for different durations of slots, further confirming that the duration settings of slots in the EPC C1 G2 standard are also applicable in the channel capture model. Next, we will develop a reader equipped with capture-aware technology that is in line with the current level of hardware development and conduct real-world testing on dense vehicle tags or cargo tags to further enhance its performance.

Author Contributions: Conceptualization, W.X. and Z.S.; methodology, W.X. and Y.S.; software, Z.S.; validation, L.L.; formal analysis, W.X.; investigation, Y.W.; resources, W.X.; data curation, L.L.; writing—original draft preparation, W.X. and Z.S.; writing—review and editing, Z.S., Y.S. and Y.W.; visualization, W.X. All authors have read and agreed to the published version of the manuscript.

Funding: This work was supported by the Guidance Projects of FuJian Science and Technology Agency: 2022H0022 and the Project of the Xiamen Science and Technology Bureau: 2022CXY0317.

Institutional Review Board Statement: Not applicable.

Informed Consent Statement: Not applicable.

Data Availability Statement: The data that support the findings of this study are available from the corresponding authors upon reasonable request.

Conflicts of Interest: The authors declare no conflict of interest.

References

1. Wu, Q.; Nie, S.; Fan, P.; Liu, H.; Qiang, F.; Li, Z. A swarming approach to optimize the one-hop delay in smart driving inter-platoon communications. *Sensors* **2018**, *18*, 3307. [CrossRef] [PubMed]
2. Wu, Q.; Zheng, J. Performance modeling and analysis of the ADHOC MAC protocol for VANETs. In Proceedings of the 2015 IEEE International Conference on c Communications (ICC), London, UK, 8–12 June 2015; pp. 3646–3652.
3. Wu, Q.; Zhao, Y.; Fan, Q.; Fan, P.; Wang, J.; Zhang, C. Mobility-aware cooperative caching in vehicular edge computing based on asynchronous federated and deep reinforcement learning. *IEEE J. Sel. Top. Signal Process.* **2022**, *17*, 66–81. [CrossRef]
4. Wu, Q.; Shi, S.; Wan, Z.; Fan, Q.; Fan, P.; Zhang, C. Towards V2I age-aware fairness access: A dqn based intelligent vehicular node training and test method. *arXiv* **2022**, arXiv:2208.01283.
5. Fan, J.; Yin, S.; Wu, Q.; Gao, F. Study on refined deployment of wireless mesh sensor network. In Proceedings of the 2010 6th International Conference on Wireless Communications Networking and Mobile Computing (WiCOM), Chengdu, China, 23–25 September 2010; pp. 1–5.
6. Wu, Q.; Zheng, J. Performance modeling and analysis of the ADHOC MAC protocol for vehicular networks. *Wirel. Netw.* **2016**, *22*, 799–812. [CrossRef]
7. Wu, Q.; Xia, S.; Fan, P.; Fan, Q.; Li, Z. Velocity-adaptive V2I fair-access scheme based on IEEE 802.11 DCF for platooning vehicles. *Sensors* **2018**, *18*, 4198. [CrossRef]
8. Olaby, O.; Hamadache, M.; Soper, D.; Winship, P.; Dixon, R. Development of a novel railway positioning system using RFID technology. *Sensors* **2022**, *22*, 2401. [CrossRef]
9. Shariq, M.; Singh, K.; Maurya, P.K.; Ahmadian, A.; Taniar, D. AnonSURP: An anonymous and secure ultralightweight RFID protocol for deployment in internet of vehicles systems. *J. Supercomput.* **2022**, *78*, 8577–8602. [CrossRef]
10. Park, S.; Lee, H. Self-recognition of vehicle position using UHF passive RFID tags. *IEEE Trans. Ind. Electron.* **2012**, *60*, 226–234. [CrossRef]
11. Qin, H.; Chen, W.; Chen, W.; Li, N.; Zeng, M.; Peng, Y. A collision-aware mobile tag reading algorithm for RFID-based vehicle localization. *Comput. Netw.* **2021**, *199*, 108422. [CrossRef]
12. Wang, H.; Wang, S.; Yao, J.; Pan, R.; Huang, Q.; Zhang, H.; Yang, J. Effective anti-collision algorithms for RFID robots system. *Assem. Autom.* **2019**, *40*, 55–64. [CrossRef]
13. Wu, G.; Xu, Z.; Zhang, H.; Shen, S.; Yu, S. Multi-agent DRL for joint completion delay and energy consumption with queuing theory in MEC-based IIoT. *J. Parallel Distrib. Comput.* **2023**, *176*, 80–94. [CrossRef]
14. Zhang, P.; Chen, N.; Shen, S.; Yu, S.; Kumar, N.; Hsu, C.H. AI-Enabled Space-Air-Ground Integrated Networks: Management and Optimization. *IEEE Netw.* **2023**.. [CrossRef]
15. Shen, S.; Wu, X.; Sun, P.; Zhou, H.; Wu, Z.; Yu, S. Optimal privacy preservation strategies with signaling Q-learning for edge-computing-based IoT resource grant systems. *Expert Syst. Appl.* **2023**, *225*, 120192. [CrossRef]
16. Wang, C.; Jiang, C.; Wang, J.; Shen, S.; Guo, S.; Zhang, P. Blockchain-Aided Network Resource Orchestration in Intelligent Internet of Things. *IEEE Internet Things J.* **2022**, *10*, 6151–6163. [CrossRef]
17. Li, G.; Sun, H.; Li, Z.; Wu, P.; Inserra, D.; Su, J.; Fang, X.; Wen, G. A Dynamic Multi-ary Query Tree Protocol for Passive RFID Anti-collision. *CMC Comput. Mater. Contin.* **2022**, *72*, 4931–4944. [CrossRef]
18. Zhang, H.; Gao, L.; Luo, H.g.; Zhai, Y. Research on the RFID anticollision strategy based on decision tree. *Wirel. Commun. Mob. Comput.* **2022**, *2022*, 2913157. [CrossRef]
19. Yang, X.; Wu, B.; Wu, S.; Liu, X.; Zhao, W.W. Time Slot Detection-Based M-ary Tree Anticollision Identification Protocol for RFID Tags in the Internet of Things. *Wirel. Commun. Mob. Comput.* **2021**, *2021*, 6638936. [CrossRef]
20. Zhang, G.; Tao, S.; Cai, Q.; Gao, W.; Jia, J.; Wen, J. A fast and universal RFID tag anti-collision algorithm for the Internet of Things. *IEEE Access* **2019**, *7*, 92365–92377. [CrossRef]
21. Zhong, D. An ALOHA-Based Algorithm Based on Grouping of Tag Prefixes for Industrial Internet of Things. *Secur. Commun. Netw.* **2022**, *2022*, 1812670. [CrossRef]
22. Nafar, F.; Shamsi, H. Design and implementation of an RFID-GSM-based vehicle identification system on highways. *IEEE Sens. J.* **2018**, *18*, 7281–7293. [CrossRef]
23. Qu, Z.; Sun, X.; Chen, X.; Yuan, S. A novel RFID multi-tag anti-collision protocol for dynamic vehicle identification. *PLoS ONE* **2019**, *14*, e0219344. [CrossRef]
24. Global, E. EPC radio-frequency identity protocols class-1 generation-2 UHF RFID protocol for communications at 860 MHz–960 MHz. *Version* **2008**, *1*, 23.
25. Menon, V.G.; Jacob, S.; Joseph, S.; Sehdev, P.; Khosravi, M.R.; Al-Turjman, F. An IoT-enabled intelligent automobile system for smart cities. *Internet Things* **2022**, *18*, 100213. [CrossRef]
26. Salah, H.; Ahmed, H.A.; Robert, J.; Heuberger, A. A time and capture probability aware closed form frame slotted ALOHA frame length optimization. *IEEE Commun. Lett.* **2015**, *19*, 2009–2012. [CrossRef]

27. Shin, W.J.; Kim, J.G. A capture-aware access control method for enhanced RFID anti-collision performance. *IEEE Commun. Lett.* **2009**, *13*, 354–356. [CrossRef]
28. Li, B.; Wang, J. Efficient anti-collision algorithm utilizing the capture effect for ISO 18000-6C RFID protocol. *IEEE Commun. Lett.* **2011**, *15*, 352–354. [CrossRef]
29. Yang, X.; Wu, H.; Zeng, Y.; Gao, F. Capture-aware estimation for the number of RFID tags with lower complexity. *IEEE Commun. Lett.* **2013**, *17*, 1873–1876. [CrossRef]
30. Wang, Y.; Wu, H.; Zeng, Y. Capture-aware estimation for large-scale RFID tags identification. *IEEE Signal Process. Lett.* **2015**, *22*, 1274–1277. [CrossRef]
31. Wu, H.; Wang, Y.; Zeng, Y. Capture-aware Bayesian RFID tag estimate for large-scale identification. *IEEE/CAA J. Autom. Sin.* **2017**, *5*, 119–127. [CrossRef]
32. Chen, Y.; Feng, Q.; Senior Member, I.; Jia, X.; Chen, H. Modeling and analyzing RFID generation-2 under unreliable channels. *J. Netw. Comput. Appl.* **2021**, *178*, 102937. [CrossRef]
33. Wang, H.; Yu, R.; Pan, R.; Liu, M.; Huang, Q.; Yang, J. Fast tag identification for mobile RFID robots in manufacturing environments. *Assem. Autom.* **2021**, *41*, 292–301. [CrossRef]
34. Su, J.; Sheng, Z.; Liu, A.X.; Han, Y.; Chen, Y. Capture-aware identification of mobile RFID tags with unreliable channels. *IEEE Trans. Mob. Comput.* **2020**, *21*, 1182–1195. [CrossRef]
35. Katagi, T.; Ohmine, H.; Miyashita, H.; Nishimoto, K. Analysis of mutual coupling between dipole antennas using simultaneous integral equations with exact kernels and finite gap feeds. *IEEE Trans. Antennas Propag.* **2016**, *64*, 1979–1984. [CrossRef]
36. Sarkar, B.; Takeyeva, D.; Guchhait, R.; Sarkar, M. Optimized radio-frequency identification system for different warehouse shapes. *Knowl.-Based Syst.* **2022**, *258*, 109811. [CrossRef]
37. Chen, W.; Childs, J.; Ray, S.; Lee, B.S.; Xia, T. RFID Technology Study for Traffic Signage Inventory Management Application. *IEEE Trans. Intell. Transp. Syst.* **2022**, *23*, 17809–17818. [CrossRef]
38. Schoute, F. Dynamic frame length ALOHA. *IEEE Trans. Commun.* **1983**, *31*, 565–568. [CrossRef]
39. Mokhtari, A.; Atani, R.E.; Maghsoodi, M. Using capture effect in DFSA anti-collision protocol in RFID systems according to ISO18000-6C standard. *Majlesi J. Mechatron. Syst.* **2012**, *1*, 26.
40. Wang, J.; Shi, Y.; Yang, C.; Ji, S.; Su, H. Research on fading characteristics of ultrahigh frequency signals in Karst landform around radio quiet zone of FAST. *Radio Sci.* **2020**, *55*, 1–10. [CrossRef]
41. Wang, Y.; Shi, J.; Chen, L.; Lu, B.; Yang, Q. A novel capture-aware TDMA-based MAC protocol for safety messages broadcast in vehicular ad hoc networks. *IEEE Access* **2019**, *7*, 116542–116554. [CrossRef]
42. Sabesan, S.; Crisp, M.J.; Penty, R.V.; White, I.H. Wide area passive UHF RFID system using antenna diversity combined with phase and frequency hopping. *IEEE Trans. Antennas Propag.* **2013**, *62*, 878–888. [CrossRef]
43. Gao, Y.; Chen, Y.; Bekkali, A. Performance of passive UHF RFID in cascaded correlated generalized Rician fading. *IEEE Commun. Lett.* **2016**, *20*, 660–663. [CrossRef]
44. Zhang, Y.; Gao, F.; Fan, L.; Lei, X.; Karagiannidis, G.K. Secure communications for multi-tag backscatter systems. *IEEE Wirel. Commun. Lett.* **2019**, *8*, 1146–1149. [CrossRef]
45. Jameel, F.; Haider, M.A.A.; Butt, A.A. Performance analysis of VANETs under Rayleigh, Rician, Nakagami-m and Weibull fading. In Proceedings of the 2017 International Conference on Communication, Computing and Digital Systems (C-CODE), Islamabad, Pakistan, 8–9 March 2017; pp. 127–132.
46. Zhang, Y.; Gao, F.; Fan, L.; Lei, X.; Karagiannidis, G.K. Backscatter communications over correlated Nakagami-m fading channels. *IEEE Trans. Commun.* **2018**, *67*, 1693–1704. [CrossRef]
47. Wang, Y.; Shi, J.; Chen, L. Capture Effect in the FSA-Based Networks under Rayleigh, Rician and Nakagami-m Fading Channels. *Appl. Sci.* **2018**, *8*, 414. [CrossRef]
48. Khandelwal, G.; Lee, K.; Yener, A.; Serbetli, S. ASAP: A MAC protocol for dense and time-constrained RFID systems. *EURASIP J. Wirel. Commun. Netw.* **2007**, *2007*, 18730. [CrossRef]
49. Vogt, H. Efficient object identification with passive RFID tags. In *Pervasive Computing, Proceedings of the First International Conference, Pervasive 2002 Zurich, Switzerland, 26–28 August 2002. Proceedings 1*; Springer: Berlin/Heidelberg, Germany, 2002; pp. 98–113.

Disclaimer/Publisher's Note: The statements, opinions and data contained in all publications are solely those of the individual author(s) and contributor(s) and not of MDPI and/or the editor(s). MDPI and/or the editor(s) disclaim responsibility for any injury to people or property resulting from any ideas, methods, instructions or products referred to in the content.

Communication

Mutual Coupling Reduction of a Multiple-Input Multiple-Output Antenna Using an Absorber Wall and a Combline Filter for V2X Communication

Yuanxu Fu *, Tao Shen and Jiangling Dou

Faculty of Information Engineering and Automation, Kunming University of Science and Technology, Kunming 650032, China; shentao@kust.edu.cn (T.S.); Jianglingdou@kust.edu.cn (J.D.)
* Correspondence: 20171104001@stu.kust.edu.cn ; Tel.: +86-150-871-56156

Abstract: This paper presents an MIMO antenna for vehicle-to-everything (V2X) communication, which adopts two ways of combline filters and absorption wall decoupling. A combline filter and an absorption wall are used, respectively, for internal and external decoupling. The combline filter is incorporated between the ground of the two adjacent antennas, which reduces the mutual coupling between them. Additionally, the mutual coupling of radiation between adjacent antennas is significantly reduced by the absorber wall. These combline filters and absorber walls use the method of electromagnetic field distribution to explain the reduction in the mutual coupling between the adjacent antennas. The transmission coefficient and surface current distribution explain the effectiveness of the decoupling structure. When the frequency is between 3.8 and 4.8 GHz, the simulation and measurement results show that S_{11} is less than -10 dB, the bandwidth is 25% and the peak gain is 7.8 dBi. In addition, the proposed MIMO antenna has a high isolation between antenna units (>37 dB), and the envelop correlation coefficient (ECC) is less than 0.005.

Keywords: MIMO; V2X; combline filter; absorber wall

Citation: Fu, Y.; Shen, T.; Dou, J. Mutual Coupling Reduction of a Multiple-Input Multiple-Output Antenna Using an Absorber Wall and a Combline Filter for V2X Communication. *Sensors* **2023**, *23*, 6355. https://doi.org/10.3390/s23146355

Academic Editors: Pingyi Fan and Qiong Wu

Received: 2 June 2023
Revised: 5 July 2023
Accepted: 10 July 2023
Published: 13 July 2023

Copyright: © 2023 by the authors. Licensee MDPI, Basel, Switzerland. This article is an open access article distributed under the terms and conditions of the Creative Commons Attribution (CC BY) license (https://creativecommons.org/licenses/by/4.0/).

1. Introduction

The fifth generation of mobile communications (5G) will meet the growing demand for high-speed data transmission, large capacity and ubiquitous connectivity, as the global positioning system (GPS) [1,2], satellite digital audio radio service (SDARS) [3,4], cellular communication [5–7] and vehicle-to-everything (V2X) communication [8,9] meet the needs of navigation, communication and entertainment. As the amount of information sources increases, V2X communication will become more crowded and there will be a faster fading issue, which leads to a decline in efficiency. As a result, multiple-input multiple-output (MIMO) antennas have been proposed to improve the channel-capacity-limited spectrum resources [10–13]. However, the coupling interference of MIMO antenna systems causes the radiation performance of a single antenna to decrease. Recently, several low coupling design techniques have been proposed to reduce the mutual coupling of adjacent antennas. The details are as follows: defective ground structure (DGS) [14], parasitic element [15], electromagnetic band gap (EBG) [16], neutralization line [17] and decoupling network [18] methods. Some of these methods require the antenna spatial distance, and others degrade the radiation of MIMO antennas or require complex construction MIMO antenna designs. To solve the these problems, the coupling of antenna elements is reduced by metamaterials. In [19], a Taichi-Bagua-like metamaterial with a single layer substrate has achieved high isolation. In [20], ENG metamaterial structures have achieved isolation enhancement. In [21], the isolation of MIMO antennas was improved by a complementary split ring resonator CSRR metamaterial.

To improve the isolation of MIMO antennas, an absorbing metamaterial is proposed as a new platform to reduce the coupling of antenna elements. Currently, there are two ways to isolate MIMO unit antennas. One method uses planar isolation to reduce the coupling of surface currents and thus reduce the coupling between antennas [22,23]. The other method uses a three-dimensional absorption wall, which mainly isolates the radiation coupling of the antenna [24,25]. However, such absorbent materials do not meet the requirements of miniaturization, and few multiple technologies methods are used to reduce the coupling of MIMO antennas.

In this paper, the absorption unit size of the absorber wall is only 13×13 mm^2, and the coupling of the MIMO antenna is reduced by a combline filter and absorption wall. These combline filters and absorber walls use the method of electromagnetic field distribution to explain the reduction in the mutual coupling between adjacent antennas. The transmission coefficient and surface current distribution explain the effectiveness of the decoupling structure. The rest of this paper is as follows. Section 2 illustrates the MIMO antenna configuration and design method. Additionally, the design procedure of the proposed combline filter reduction techniques is explained in detail in Section 3. Section 4 presents a concrete analysis of the absorption wall decoupling principle. Section 5 mainly describes the S parameter, MIMO isolation, antenna gain, antenna radiation efficiency and envelope correlation coefficient (ECC) measurement and simulation results. The conclusions are provided in Section 6.

2. Configuration

Figure 1 represents the geometry of the MIMO antenna. The antenna structure comes from reference [26], which has a low profile, high efficiency and high gain. It consists of a two-layer substrate connected by a coaxial feed. Substrate 1 and 2 adopt the Ro4003 material ($\epsilon_r = 3.55$ and $\tan \delta = 0.0027$) with a thickness of h = 1.524 mm and the FR-4 material ($\epsilon_r = 4.4$ and $\tan \delta = 0.02$) with a thickness of h = 1.5 mm, respectively. A 50 ohm coaxial cable is used to feed. The antenna has a favorable radiation efficiency and gain. The isolation can reach 20 dB. The antenna parameters are as follows: W = 100 mm, L = 80 mm, T_1 = 22 mm, H_1 = 5 mm and L_m = 33.5 mm. The diameter of the isolation hole between the feed and the ground is R_2 = 4 mm. Figure 2 shows the mutual coupling reduction of the MIMO antenna using the absorber wall and the combline filter, which is based on adding combline filtering and a metamaterial absorption wall to the structure in Figure 1. The combline filter belongs to the internal decoupling structure and the absorber wall is the external decoupling structure. The combline filter has a simple structure and cannot affect the radiation and efficiency of the antenna. Figure 2a–c shows the physical components of the MIMO antenna with a combline filter and absorber wall as follows: (a) is the radiation layer; (b) is the combline filter; and (c) is the ground plane and feed port. The combline filter is composed of seven microstrip lines, which are added between adjacent antennas to achieve adjacent antenna decoupling. The absorption wall substrate 3 adopts an FR-4 substrate with a thickness of 5 mm, which includes four absorption units arranged on both sides to absorb radiation to improve the isolation of the adjacent antenna. The following is the specific analysis process of decoupling between the two methods.

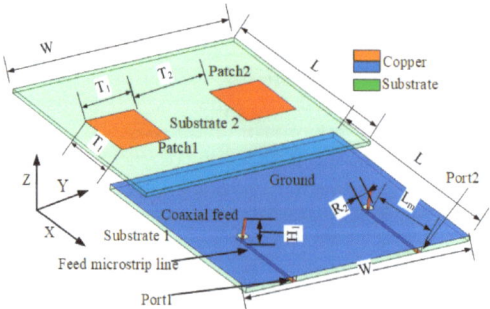

Figure 1. Geometry of the MIMO antenna.

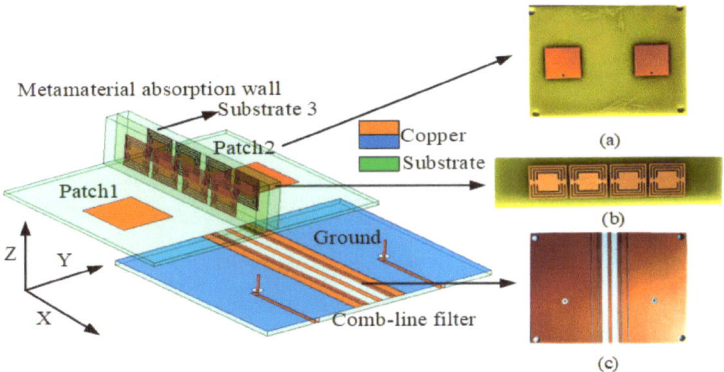

Figure 2. Design and fabrication of an MIMO antenna with a combline filter and absorber wall. (**a**) Radiation layer fabrication. (**b**) Absorber wall fabrication. (**c**) Ground plane and feed port fabrication.

3. Ground Adopts a Combline Filter to Reduce Antenna Element Coupling

We analyze the effect of the combline filter to reduce the coupling of the MIMO antenna from three aspects. First, the decoupling and bandwidth of the original MIMO unit are compared with those after combline filtering. Second, a theoretical model and equivalent circuit are used to calculate the specific parameter values of the MIMO antenna of the cofilter. Finally, an electromagnetic simulation verifies the decoupling capability of the combline filter.

As shown in Figure 3, the decoupled MIMO antenna is added to a combline filter at the ground position to form an isolation band. Figure 4 shows the change in the S parameter after adding the combline filter: S_{11} represents the reflection coefficient and S_{12} represents isolation. It is determined that the unit antenna bandwidth of the MIMO antenna increases by 20%, and the coupling decreases by 5 dB. The advantages of this filter are seven microstrip lines and eight microstrip coupling distances, whose structure is simple and does not change the original structure of the antenna, so it minimally affects the antenna radiation and gain. The S parameter is easy to analyze for the symmetrical coupling structure of the combline filter. Figure 5 shows the structure of the combline filter in the ground position of the MIMO antenna, which demonstrates the symmetry of the structure and the parameters to be analyzed. According to the filtering method and principle, we analyze the equivalent circuit diagram formed by microstrip lines in detail in Figure 6.

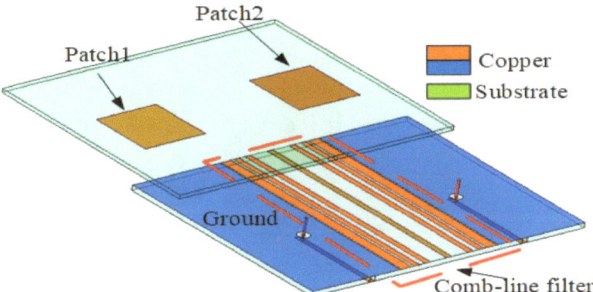

Figure 3. The MIMO antenna with a combline filter.

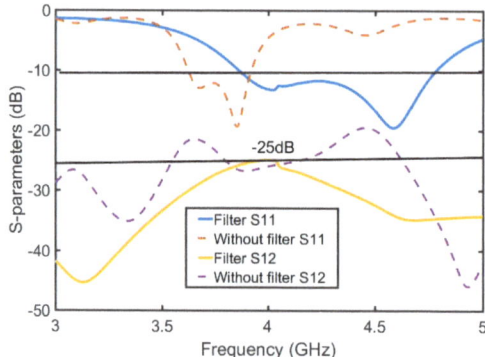

Figure 4. Comparison of S parameters with and without a combline filter.

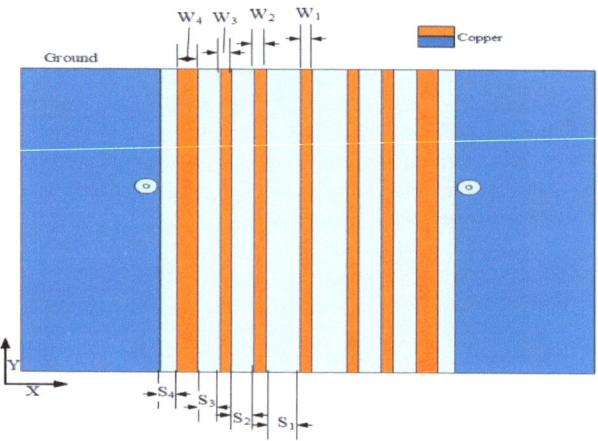

Figure 5. Comber-line filter structure diagram.

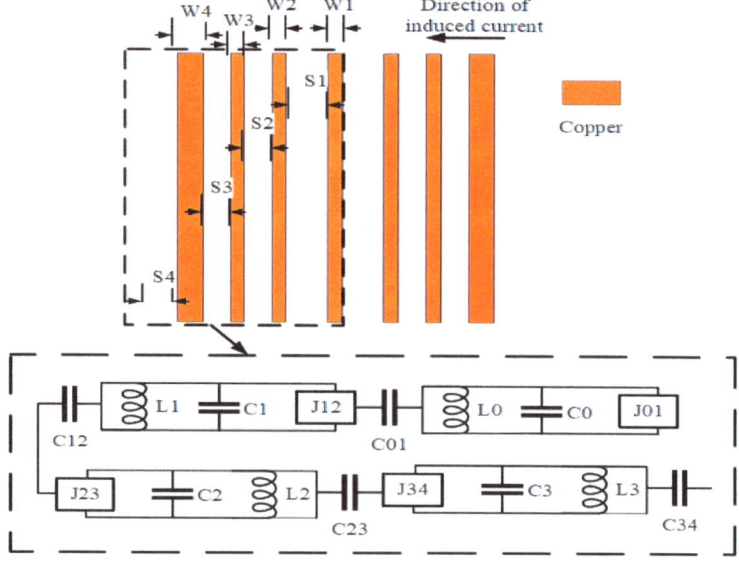

Figure 6. Combline filter equivalent circuit.

The microstrip line is composed of an LC circuit and a J converter, and the gap is replaced by $C_{i,i+1}$ for mutual coupling. The specific calculation steps are as follows:

(1) According to Equation (1), the comb filter is a combination of a multiorder filter and a microstrip coupled filter. We set the lower cutoff frequency of the combline filter as $f_1 = 3$ GHz ($S_{11} < -10$ dB), the upper cutoff frequency of the combline filter as $f_2 = 5$ GHz ($S_{11} < -10$ dB), the center frequency of the combline filter as $f_0 = 4$ GHz (minimum of S_{11}), the roll down coefficient as $a = -25$ dB and the dielectric constant as $\epsilon = 3.55$. The frequency range of the filter and reflection coefficients is from 3 to 5 GHz. We obtain n as approximately 7 by calculation.

(2) The electrical length of open-ended microstrip stubs θ_i is calculated through ADS, which is substituted into Equation (3) to obtain the propagation constant β_i value ($i = 0, 1, 2, 3$).

(3) The microstrip capacitance C_i ($i = 0, 1, 2, 3$) and the microstrip coupled capacitance $C_{i,i+1}$ ($i = 0, 1, 2, 3$) in the equivalent circuit are calculated by Equations (4)–(6).

(4) According to Equations (7) and (8), the width of the microstrip W_i ($i = 1, 2, 3, 4$) and the coupling distance between microstrip lines S_i ($i = 1, 2, 3, 4$) are calculated by the microstrip capacitance C_i and the mutual coupling $C_{i,i+1}$. The above calculation roughly describes the combline parameters, and further optimization is executed in Ansys HFSS 2021 software.

$$n = 1 + \frac{\ln(2a/\epsilon) - \mathrm{arcosh}(\tan(\pi f_1/2f_0)/\tan(\pi f_2/2f_0))}{\mathrm{arcosh}\left(\sqrt{1+\tan^2(\pi f_1/2f_0)}/\sqrt{1+\tan^2(\pi f_2/2f_0)}\right)} \quad (1)$$

$$\mathrm{FBW} = \frac{f_2 - f_1}{f_0} \quad (2)$$

$$\beta_i|_{i=0\sim3} = Z_0\left(\frac{\cos\theta_i + \theta_i \csc^2\theta_i}{2}\right) \quad (3)$$

$$J_{i,i+1}|_{i=0\sim3} = \sqrt{\frac{\beta_i Z_0}{g_i g_{i+1}}} \quad (4)$$

$$C_i|_{i=0\sim3} = \frac{376.7}{\sqrt{\epsilon}}\left(1 - \frac{J_{i,i+1}}{Z_0}\right) \quad (5)$$

$$C_{i,i+1}|_{i=0\sim3} = 376.7\sqrt{\epsilon} - C_i \quad (6)$$

$$W_i|_{i=1\sim4} = \frac{\mathrm{FBW}}{4\epsilon}(C_{i-1} - C_{i-1,i}) \quad (7)$$

$$S_i|_{i=1\sim4} = \frac{\mathrm{FBW}}{\pi}\left(\frac{C_{i-1,i}+1}{C_{i-1}+1}\right) \quad (8)$$

where n is the filtering order (number of microstrip), f_1 is the lower cutoff frequency of the combline filter, f_2 is the upper cutoff frequency of the combline filter, f_0 is the center frequency of the combline filter, θ_i is the electrical length of the open-ended microstrip stubs, β_i is the propagation constant, $g_i g_{i+1}$ is the value of the low-pass filter (LPF) prototype element, C_i is the microstrip capacitance, $C_{i,i+1}$ is the microstrip coupled capacitance, $J_{i,i+1}$ is the converter, W_i is the width of the microstrip and S_i is the coupling distance between microstrip lines [27,28].

Table 1 shows the results of further optimization by HFSS software. The overall isolation (decoupling) is below -25 dB after optimizing S_4 in Figure 7. To verify that the combline filter has a decoupling effect, HFSS is used to simulate the current intensity distribution of 4.5 GHz and 4 GHz at the combline filter microstrip line in Figure 8. Figure 8a,b show the current intensity displayed at 4.5 GHz and 4 GHz, respectively. At 4.5 GHz, most of the current is retained in the first half of the filter, and almost no current passes through the second half of the filter. Therefore, when $S_{12} = -33$ dB, the combline filter has a strong isolation state (decoupling optimal value) and exhibits no blocking current function at 4 GHz when the isolation $S_{12} = -27$ dB.

Table 1. Microstrip line width and coupling distance of the combline filter.

Parameter	Value (mm)	Parameter	Value (mm)
S_1	4	W_1	2
S_2	0.24	W_2	1
S_3	0.5	W_3	0.75
S_4	0.5	W_4	3.6

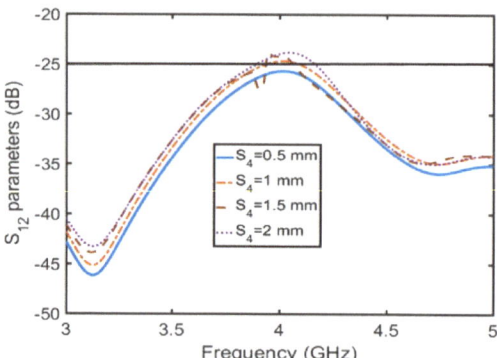

Figure 7. Coupling distance S_4 slightly adjusts the isolation S_{12}.

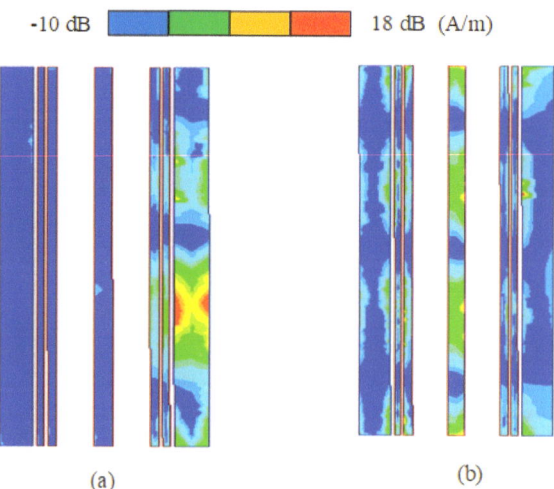

Figure 8. Current intensity distribution of the combline filter at (**a**) 4.5 GHz and (**b**) 4 GHz.

4. The Absorber Wall Is Used to Reduce the Radiation Coupling of Antenna Elements

First, the absorption function of the absorption unit is analyzed, then the relationship between the absorption function of the absorption array and the isolation degree is analyzed and the arrangement distance of the absorption unit is optimized. Finally, the effect of the absorption array is verified by simulation. As shown in Figure 9, the absorption unit is an annular metal embedded with microstrip lines to form an absorption structure. The ring size in the structure is as follows: $L_a = 13$ mm, $L_b = 11$ mm, $L_c = 9$ mm and $L_d = 7$ mm. The embedded microstrip lines L_{a1}, L_{a2} and L_{a3} are adjusted to the absorption frequency and absorption rate of the structure. Figure 10a shows that the adjustment of L_{a2} adjusts the absorption frequency, which decreases as the length of L_{a2} increases. Figure 10b shows that L_{a3} is a bidirectional regulation of frequency and absorption. L_{a1} is the length directly affecting the absorption rate in Figure 11. After optimizing the three parameters, when $L_{a1} = 2$ mm, $L_{a2} = 1$ mm and $L_{a3} = 2$ mm, the absorption rate at 4.6 GHz reaches 97%.

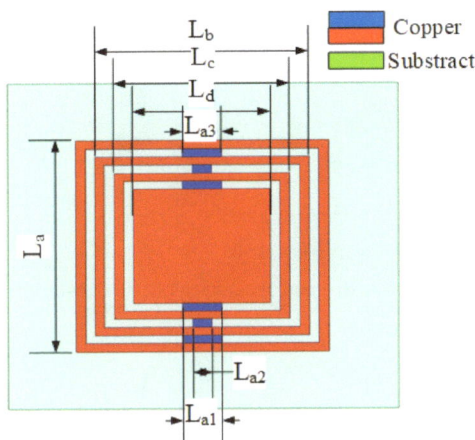

Figure 9. Absorber material unit.

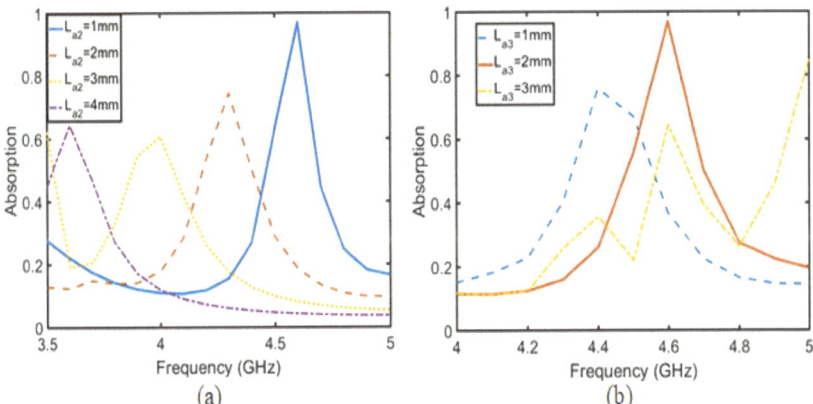

Figure 10. Optimization diagram of unit absorption rate. (**a**) Optimization diagram of parameter L_{a2} and absorptivity. (**b**) Optimization diagram of parameter L_{a3} and absorptivity.

To further improve the absorption bandwidth, the absorption wall is designed as a 1×4 horizontal array. As shown in Figure 12, the distance between parameters C_1 and C_2 is determined to optimize the absorption bandwidth. Figures 13 and 14 show the absorption diagram of the array and the isolation diagram of the MIMO antenna unit, respectively. When $C_1 = 2$ mm, the absorption rate is proportional to the isolation degree, which indicates that the greater the absorption bandwidth and absorption rate in Figure 13a, the greater the isolation in Figure 13b. The same principle applies to Figure 14. When $C_2 = 2$ mm, the absorption rate is proportional to the isolation degree. The final optimization parameter values are shown by HFSS software in Table 2. The optimized MIMO antenna with an absorber wall and combline filter was simulated, as shown in

Figure 15a. When the 4.5 GHz isolation degree $S_{12} = -44$ dB, the overall absorption current distribution is concentrated in the last three absorption units, while in Figure 15b, the current distribution is concentrated in the two units at 4 GHz, so the isolation degree $S_{12} = -39$ dB. In this scenario, the absorption rate is proportional to isolation.

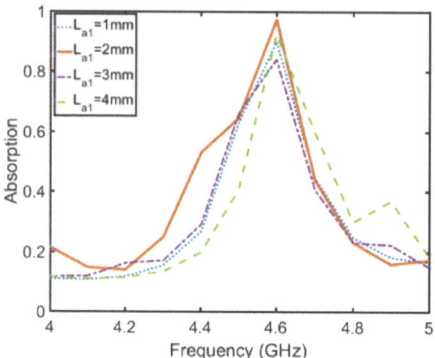

Figure 11. Optimization diagram of parameter L_{a1} and absorptivity.

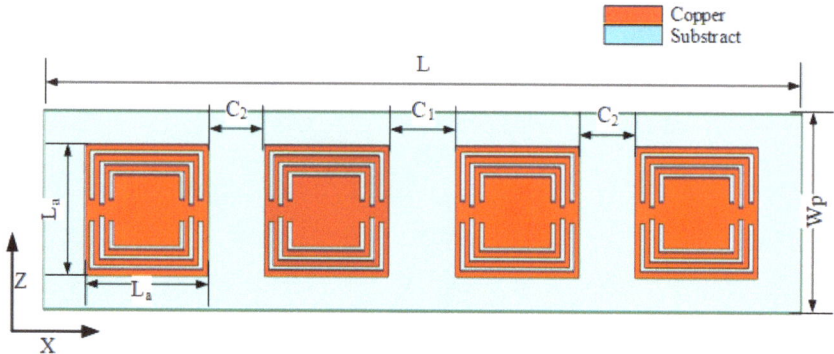

Figure 12. The 1×4 absorption array.

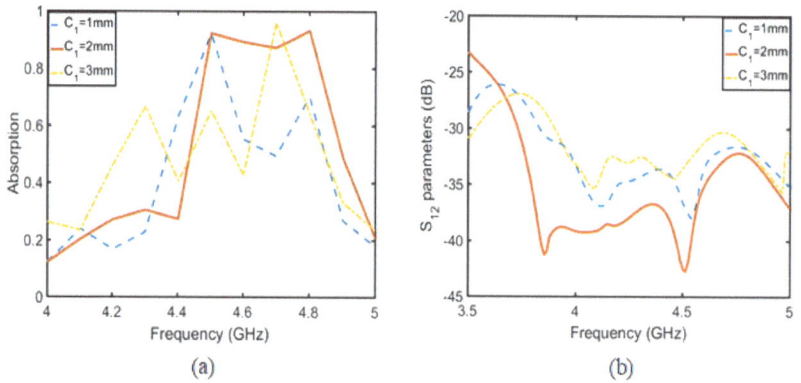

Figure 13. Optimization diagram C_1 regarding the absorber wall unit distance. (**a**) Relationship between C_1 and absorption rate. (**b**) Relationship between C_1 and isolation degree S_{12}.

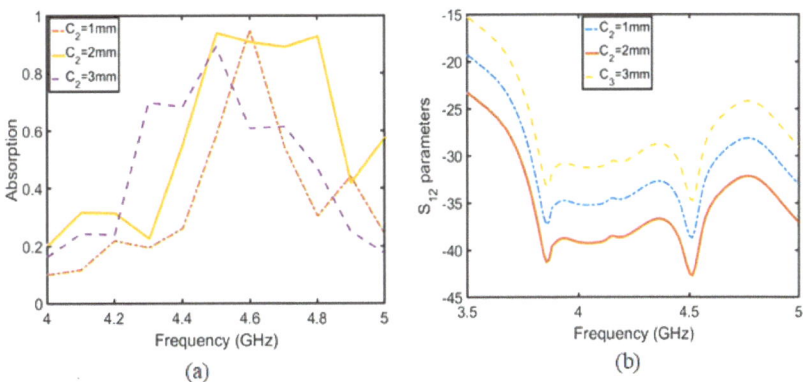

Figure 14. Optimization diagram C_2 of absorber wall unit distance. (**a**) Relationship between C_2 and absorption rate. (**b**) Relationship between C_2 and isolation degree S_{12}.

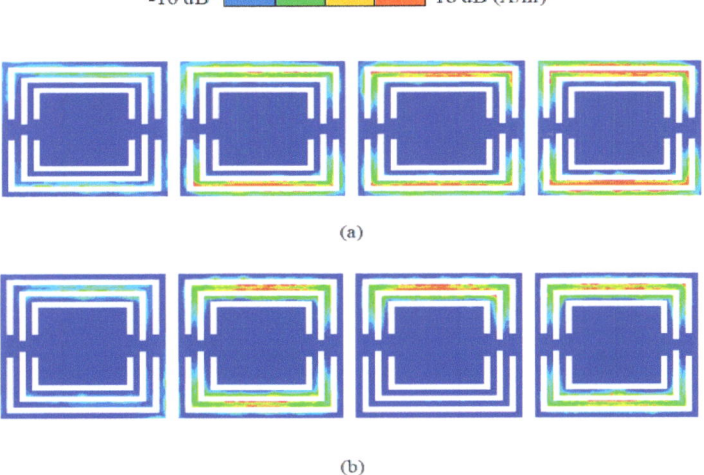

Figure 15. Absorption array current distribution at (**a**) 4.5 GHz and (**b**) 4 GHz.

Table 2. Dimensions of the proposed absorber wall.

Parameter	Value (mm)	Parameter	Value (mm)
L_a	13	L_{a1}	2
L_b	11	L_{a2}	1
L_c	9	L_{a3}	2
L_d	7	C_1	2
W_p	20	C_2	2

5. Measured and Simulated Results

To verify the validity of our design concept, a prototype of the proposed MIMO antenna was fabricated and measured. As shown in Figure 16a,b, an R&S ZNB20 vector network analyzer is used to measure the reflection coefficient and mutual coupling of the two MIMO antennas, for which the insertion loss is less than 0.05 dB. The experimental setup in the anechoic chamber and the two MIMO antennas are shown in Figure 16c,d.

The port of the antenna is welded to a 50 ohm SMA. HFSS software was used to simulate the MIMO antenna.

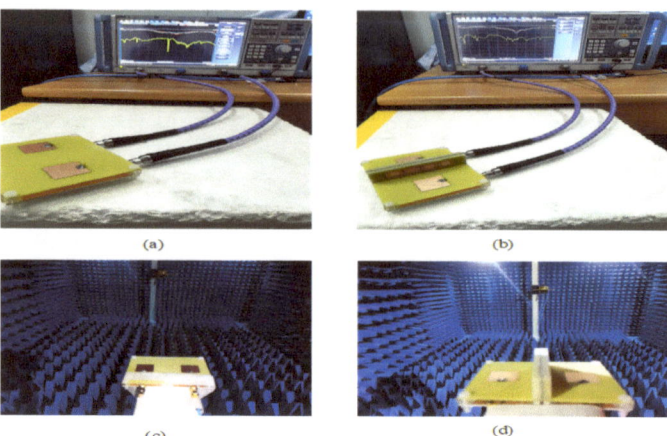

Figure 16. The two MIMO antennas in the S-parameter analysis: (**a**) only the filter structure and (**b**) the filter and absorber structure. The two MIMO antennas in the radiation analysis: (**c**) only the filter structure and (**d**) the filter and absorber structure.

5.1. S-Parameter Analysis

Figure 17a shows that the reflection coefficient simulation and measurement results essentially agree, with less than −10 dB in the frequency range of 3.8–4.8 GHz. When the operating frequency is 3.8–4.8 GHz, as shown in Figure 17b, the mutual coupling of the MIMO antennas with the two absorber and combline filter structures is much lower than that of MIMO antennas with only the combline filter structure. In the absorption band of 4.5–4.8 GHz, the mutual coupling is less than −40 dB.

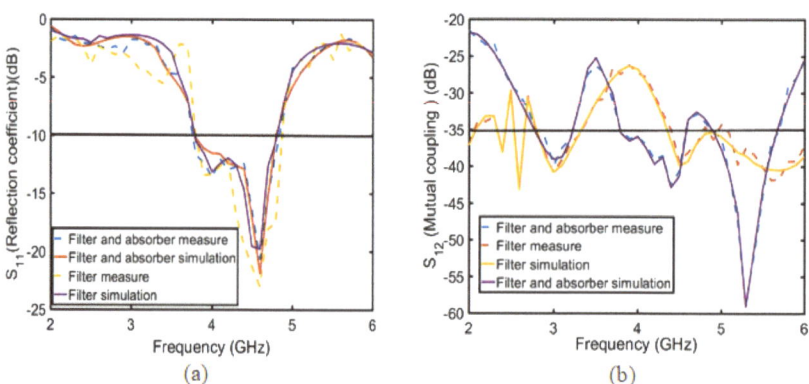

Figure 17. Only the filter structure and the filter and absorption structure S parameters of the simulation and measurement analysis: (**a**) S_{11} and (**b**) S_{12}.

5.2. Antenna Radiation Analysis

The effect of array absorption on the antenna performance is analyzed from three perspectives: gain, radiation efficiency and polarization radiation. As shown in Figure 18a, the antenna gain decreases with the absorption array compared with that without the absorber wall, indicating that the absorber wall absorbs the radiated power of the antenna, resulting in a sharp reduction in the gain. In the absorption band of 4.5–4.8 GHz, the gain is only 7 dB, which is 2 dB lower than that of the MIMO antenna without the absorber wall. Therefore, the radiation efficiency in Figure 18b also decreases. The absorption also has an impact on the antenna radiation polarization. As shown in Figure 19a,b, when the absorber wall is not used for only the combline filter at 4 GHz and 4.5 GHz, both the radiation gain and the radiation direction reach the maximum radiation value at the position of radiation of 0 degrees. After the absorber wall is adopted, the maximum value of the co-polarization radiation is tilted at 15 degrees at 4 GHz, and the cross-polarization angle is almost 90 degrees, as shown in Figure 19c. Similarly, in Figure 19d, the co-polarization is 30 degrees with the plan at 4.5 GHz, and the cross-polarization is also 90 degrees.

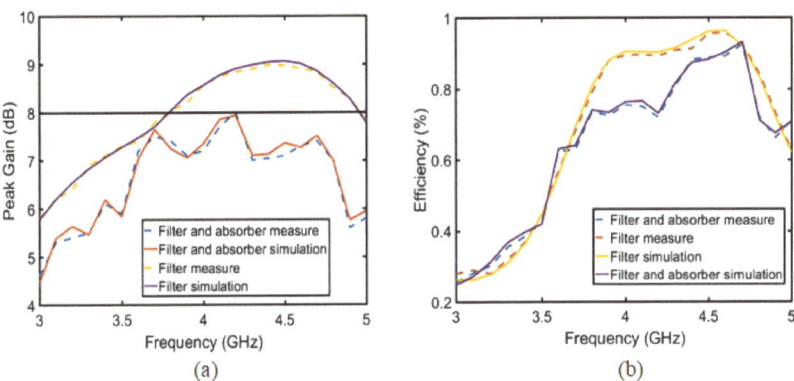

Figure 18. Only the filter structure and the filter and absorption structure radiation of the antenna simulation and measurement analysis: (**a**) peak gain and (**b**) efficiency.

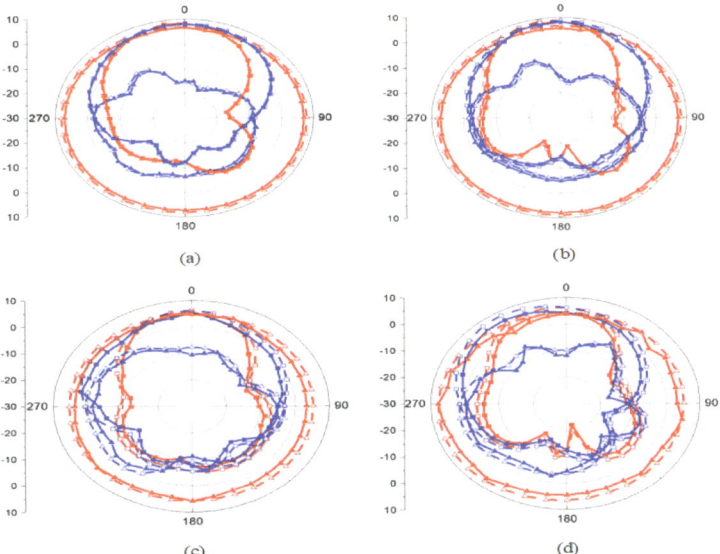

Figure 19. Only the filter structure of the antenna radiation: (**a**) 4 GHz and (**b**) 4.5 GHz. The filter and absorber structure of antenna radiation: (**c**) 4 GHz and (**d**) 4.5 GHz (□ is the co-polarization simulation XOZ, ■ is the co-polarization measure XOZ, □ is the co-polarization simulation YOZ, ■ is the co-polarization measure YOZ, △ is the cross-polarization simulation XOZ, ▲ is the cross-polarization measure XOZ, △ is the cross-polarization simulation YOZ and ▲ is the co-polarization measure YOZ).

5.3. Diversity Performance Analysis

The envelope correlation coefficient (ECC) is calculated for validating the diversity performance of the antenna, which can be expressed based on the radiation patterns of the antenna in an isotropic propagation environment. The method of ECC calculation is based on the scattering parameters, which can be expressed as

$$\rho_e = \frac{|S_{ii}^* S_{ij} + S_{ji}^* S_{jj}|^2}{(1-|S_{ii}|^2-|S_{ji}|^2)(1-|S_{jj}|^2-|S_{ij}|^2)} \quad (9)$$

where $S_{i(j),i(j)}$ ($i \neq j$ and i, j = 1, 2, 3) represents the complex S-parameters for Port i and/or Port j and the symbol "*" denotes the Hermitian product [29]. Figure 20 shows the ECC curves calculated by the simulated and measured complex 3D radiation patterns and scattering parameters for adjacent and opposite ports, i.e., ports 1 and 2. It can be seen that the ECC value calculated by the simulated and measured radiation patterns and scattering parameters agree well with each other within the entire band, which are all below 0.005 from 3.8 to 4.8 GHz. In short, the simulated and measured results demonstrate its low correlation for ensuring good channel characteristics.

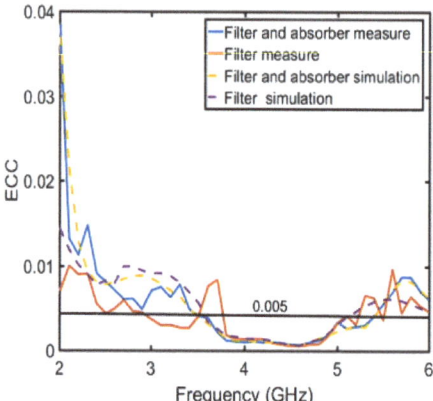

Figure 20. Only the filter structure and the filter and absorption structure ECC of the simulation and measurement analysis.

6. Conclusions

In summary, a two-port multiple-input multiple-output (MIMO) antenna from 3.8 to 4.8 GHz is presented for 5G vehicle communication. As shown in Table 3, compared with other MIMO antennas, ours has a moderate overall size, great advantages in bandwidth and gain and improves the isolation degree of adjacent antennas with a smaller ECC. The simulated and measured results show that the proposed MIMO antenna provides an overlapping S_{11} (−10 dB) and a bandwidth of 25% (3.8–4.8 GHz) with a peak gain of 7.8 dBi. Moreover, the fabricated MIMO antenna offers excellent diversity performance, the isolation between antenna elements is very high (>37 dB) and the envelop correlation coefficient (ECC) is lower than 0.005. The demonstrated antenna is a promising candidate for 5G vehicle communication systems for a wide variety of platforms.

Table 3. Performance comparison with the literature.

Reference	$\lambda_0 \times \lambda_0 \times h$ (mm^2)	Operating Frequency (GHz)	Maximum Bandwidth (%)	Peak Gain (dBi)	Average Efficiency (%)	Isolation (dB)	ECC	Number of Decoupling Methods
[16]	0.32 × 0.32 × 0.1	2.2–2.5	2.3	3.71	70	>−30	<0.05	1
[17]	1.2 × 1.2 × 0.15	2.4, 3.4	54	6.5	70	>−25	<0.07	1
[18]	0.5 × 0.5 × 1	1.8–2.2	10	3.8	77	>−20	<0.1	1
[19]	1.5 × 1.5 × 0.2	4.7–5.2	10	4	75	>−30	<0.025	1
[20]	2.0 × 2.0 × 0.5	3.6–5.2	20	4	70	>−25	<0.02	1
[21]	2.1 × 2.4 × 0.7	3.8, 5.2	10	4.5	70	>−20	<0.02	1
[22]	2.4 × 2.4 × 0.5	5.2–5.8	12	7	80	>−25	<0.04	2
[23]	0.5 × 0.5 × 0.2	5.5–5.8	6	3.8	72	>−25	<0.03	1
[24]	2.4 × 2.0 × 0.8	3.5, 5.2	6	5.2	68	>−35	<0.05	2
[25]	0.5 × 0.5 × 1.2	2.2–2.8	15	6	78	>−25	<0.05	1
This work	1.1 × 1.2 × 0.3	3.8–4.8	25	7.8	78	>−37	<0.005	2

Author Contributions: Conceptualization, Y.F.; methodology, Y.F.; investigation, T.S.; writing—original draft preparation, Y.F. and J.D. All authors have read and agreed to the published version of the manuscript.

Funding: This work was supported by the Natural Science Foundation of China (61971208); College personnel training project; Yunnan Young Top Talents of Ten Thousands Plan (Shen Tao, Yunnan Social Development No. 2018 73); Yunnan Key Laboratory of Computer Technologies Application Open Fund; Yunnan Fundamental Research project Of fund grant number 202101AU070164; Yunnan Fundamental Research Projects, number: 202301AV07003; the National Natural Science Foundation of China, number: 61971208; the Yunnan Fundamental Research Projects under Grant 202301AV070003 and 202101BE070001-008 and the Major Science and Technology Projects in Yunnan Province under Grant 202302AG050009 and 202202AD080013.

Institutional Review Board Statement: Not applicable.

Informed Consent Statement: Not applicable.

Data Availability Statement: Not applicable.

Conflicts of Interest: The authors declare no conflict of interest.

References

1. Sanchana, S.; Manjari, A.P.; Jyothi, B.N. Application Development for Real-Time Location Tracking for Underwater Vehicles Using Low-Cost GPS with GSM. In Proceedings of the 2021 Sixth International Conference on Wireless Communications, Signal Processing and Networking (WiSPNET), Chennai, India, 25–27 March 2021; pp. 108–112
2. Zhang, Q.; Chen, X.; Zhang, S.; Jin, Z. Design and Implementation of Remote Upgrade System for Vehicle Terminal Based on GPRS. In Proceedings of the 2021 IEEE Asia-Pacific Conference on Image Processing, Electronics and Computers (IPEC), Dalian, China, 14–16 April 2021; pp. 636–640.
3. Bilgic, M.M.; Yegin, K. Modified Annular Ring Antenna for GPS and SDARS Automotive Applications. *IEEE Antennas Wirel. Propag. Lett.* **2016**, *15*, 1442–1445. [CrossRef]
4. Kammerer, J.; Lindenmeier, S. Invisible antenna combination embedded in the roof of a car with high efficiency for reception of SDARS - and GPS - signals. In Proceedings of the 2013 IEEE Antennas and Propagation Society International Symposium (APSURSI), Orlando, FL, USA, 7–13 July 2013; pp. 2075–2076.

5. Wu, Q.; Nie, S.; Fan, P.; Liu, H.; Qiang, F.; Li, Z. A Swarming Approach to Optimize the One-hop Delay in Smart Driving Inter-platoon Communications. *Sensors* **2018**, *18*, 3307. [CrossRef] [PubMed]
6. Wu, Q.; Zheng, J. Performance modeling and analysis of the ADHOC MAC protocol for VANETs. In Proceedings of the 2015 IEEE International Conference on Communications (ICC), London, UK, 8–12 June 2015; pp. 3646–3652
7. Wu , Q.; Zheng, J. Performance Modeling and Analysis of the ADHOC MAC Protocol for Vehicular Networks. *Wirel. Netw.* **2016**, *22*, 799–812. [CrossRef]
8. Sehla, K.; Nguyen, T.M.T.; Pujolle, G.; Velloso, P.B. Resource Allocation Modes in C-V2X: From LTE-V2X to 5G-V2X. *IEEE Internet Things J.* **2022**, *9*, 8291–8314. [CrossRef]
9. Maglogiannis, V.; Naudts, D.; Hadiwardoyo, S.; van den Akker, D.; Marquez-Barja, J.; Moerman, I. Experimental V2X Evaluation for C-V2X and ITS-G5 Technologies in a Real-Life Highway Environment. *IEEE Trans. Netw. Serv. Manag.* **2022**, *19*, 1521–1538. [CrossRef]
10. Che, J.K.; Chen, C.C.; Locke, J.F. A Compact 4-Channel MIMO 5G Sub-6 GHz/LTE/WLAN/V2X Antenna Design for Modern Vehicles. *IEEE Trans. Antennas Propag.* **2021**, *69*, 7290–7297. [CrossRef]
11. Liu, Q.; Zhu, L. A Low-Profile Dual-Band Filtering Hybrid Antenna with Broadside Radiation Based on Patch and SIW Resonators. *IEEE Open J. Antennas Propag.* **2021**, *2*, 1132–1142 . [CrossRef]
12. Ishfaq, M.K.; Babale, S.A.; Chattha, H.T.; Himdi, M.; Raza, A.; Younas, M.; Rahman, T.A.; Rahim, S.K.A.; Khawaja, B.A. Compact Wide-Angle Scanning Multibeam Antenna Array for V2X Communications. *IEEE Antennas Wirel. Propag. Lett.* **2021**, *20*, 2141–2145. [CrossRef]
13. Zhou, Z.; Ge, Y.; Yuan, J.; Xu, Z.; Chen, Z.D. Wideband MIMO Antennas With Enhanced Isolation Using Coupled CPW Transmission Lines. *IEEE Trans. Antennas Propag.* **2023**, *71*, 1414–1423. [CrossRef]
14. Sandi, E.; Diamah, A.; Permata Putri, R.A. Combination of EBG and DGS to Improve MIMO Antenna Isolation. In Proceedings of the 2022 IEEE International Conference on Aerospace Electronics and Remote Sensing Technology (ICARES), Yogyakarta, Indonesia, 24–25 November 2022; pp. 1–6
15. Sokunbi, O.; Attia, H.; Hamza, A.; Shamim, A.; Yu, Y.; Kishk, A.A. New Self-Isolated Wideband MIMO Antenna System for 5G mm-Wave Applications Using Slot Characteristics. *IEEE Open J. Antennas Propag.* **2023**, *4*, 156–158. [CrossRef]
16. Ismail, M.F.; Rahim, M.K.A.; Samsuri, N.A.; Murad, N.A.; Pramudita, A.A. High Isolation MIMO Antenna using Electromagnetic Band Gap—EBG Structure. In Proceedings of the 2021 International Symposium on Antennas and Propagation (ISAP), Taipei, Taiwan, 19–22 October 2021; pp. 1–2.
17. Feng, B.; Chung, K.L.; Lai, J.; Zeng, Q. A Conformal Magneto-Electric Dipole Antenna With Wide H-Plane and Band-Notch Radiation Characteristics for Sub-6-GHz 5G Base-Station. *IEEE Access* **2021**, *7*, 17469–17479. [CrossRef]
18. Lau, C.Y.; Cheng, K.K.M. Pattern-Reconfigurable 2-element MIMO Antenna Design by using Novel Switchable Decoupling and Matching Network. In Proceedings of the 2021 IEEE Asia-Pacific Microwave Conference (APMC), Brisbane, Australia, 28 November–1 December 2021; pp. 247–249.
19. Zhou, W.; Li, Y. A Taichi-Bagua-like Metamaterial Covering High Isolation MIMO for 5G Application. In Proceedings of the 2021 IEEE 4th International Conference on Electronic Information and Communication Technology (ICEICT), Xi'an, China, 18–20 August 2021; pp. 948–949.
20. Supreeyatitikul, N.; Phungasem, A.; Aeimopas, P. Design of Wideband Sub-6 GHz 5G MIMO Antenna with Isolation Enhancement Using an MTM-Inspired Resonators. In Proceedings of the 2021 Joint International Conference on Digital Arts, Media and Technology with ECTI Northern Section Conference on Electrical, Electronics, Computer and Telecommunication Engineering, Cha-am, Thailand, 3–6 March 2021; pp. 206–209.
21. Sharma, M.K.; Tripathi, K.; Sharma, A.; Jha, M. Orthogonal Elements CSRR-based UWB-MIMO Antenna with Improved Isolation and Multiple Band Rejection Function. In Proceedings of the 2021 7th International Conference on Advanced Computing and Communication Systems (ICACCS), Coimbatore, India, 19–20 March 2021; pp. 84–89.
22. Garg, P.; Jain, P. Isolation Improvement of MIMO Antenna Using a Novel Flower Shaped Metamaterial Absorber at 5.5 GHz WiMAX Band. *IEEE Trans. Circuits Syst. II Express Briefs* **2020**, *67*, 675–679. [CrossRef]
23. Kaur, M.; Singh, H.S. Isolation Improvement of the MIMO PIFA using Metamaterial Absorber Array. In Proceedings of the 2020 International Symposium on Antennas & Propagation (APSYM), Cochin, India, 14–16 December 2020; pp. 29–31.
24. Deng, S.; Zhang, X.; Chen, C.; Ding, J.; Zhang, H.; Chen, W. Mutual Coupling Reduction of Dual-band MIMO Antenna Array Using Near-Perfect Metamaterial Absorber Wall. In Proceedings of the 2022 IEEE International Symposium on Antennas and Propagation and USNC-URSI Radio Science Meeting (AP-S/URSI), Denver, CO, USA, 10–15 July 2022; pp. 1782–1783.
25. Hoang, D.H.T.; Nguyen, V.X.T.; Do, T.T.; Nguyen, D.N.; Vu, T.A.; Tran, Q.N. Mutual Coupling Reduction of Mimo Antenna Using Metamaterial Absorber for 5G Base Stations. In Proceedings of the 2022 Asia-Pacific Microwave Conference (APMC), Yokohama, Japan, 29 November–2 December 2022; pp. 767–769.
26. Kim, Y.-B.; Kim, J.-W.; Lee, H.L. Low Profile Multi-Slot Loaded Antenna With Enhanced Gain-to-Volume Ratio and Efficiency. *IEEE Access* **2020**, *8*, 225407–225415. [CrossRef]
27. Zhang, N.; Li, N.; Yang, R.-Q. Design and implementation of comb line filter based on strip line. *Magn. Mater. Devices* **2021**, *1*, 38–41.

28. Wu, C.-H.; Lin, Y.-S.; Wang, C.-H.; Chen, C.H. Novel microstrip coupled-line bandpass filters with shortened coupled sections for stopband extension. *IEEE Trans. Microw. Theory Tech.* **2006**, *54*, 540–546.
29. Zhang, K.; Jiang, Z.H.; Zhou, W.L.; Peng, P.; Hong, W. A Compact, Band-Notched Ultra-Wideband Fully-Recessed Antenna with Pattern Diversity for V2X Communications. *IEEE Open J. Antennas Propag.* **2022**, *3*, 1302–1312. [CrossRef]

Disclaimer/Publisher's Note: The statements, opinions and data contained in all publications are solely those of the individual author(s) and contributor(s) and not of MDPI and/or the editor(s). MDPI and/or the editor(s) disclaim responsibility for any injury to people or property resulting from any ideas, methods, instructions or products referred to in the content.

Article

Next-Hop Relay Selection for Ad Hoc Network-Assisted Train-to-Train Communications in the CBTC System

Sixing Ma, Meng Li *, Ruizhe Yang, Yang Sun , Zhuwei Wang and Pengbo Si

Faculty of Information Technology, Beijing University of Technology, Beijing 100124, China; masixing958@emails.bjut.edu.cn (S.M.); yangruizhe@bjut.edu.cn (R.Y.); sunyang@bjut.edu.cn (Y.S.); wangzhuwei@bjut.edu.cn (Z.W.); sipengbo@bjut.edu.cn (P.S.)
* Correspondence: limeng720@bjut.edu.cn

Abstract: In the communication-based train control (CBTC) system, traditional modes such as LTE or WLAN in train-to-train (T2T) communication face the problem of a complex and costly deployment of base stations and ground core networks. Therefore, the multi-hop ad hoc network, which has the characteristics of being relatively flexible and cheap, is considered for CBTC. However, because of the high mobility of the train, it is likely to move out of the communication range of wayside nodes. Moreover, some wayside nodes are heavily congested, resulting in long packet queuing delays that cannot meet the transmission requirements. To solve these problems, in this paper, we investigate the next-hop relay selection problem in multi-hop ad hoc networks to minimize transmission time, enhance the network throughput, and ensure the channel quality. In addition, we propose a multiagent dueling deep Q learning (DQN) algorithm to optimize the delay and throughput of the entire link by selecting the next-hop relay node. The simulation results show that, compared with the existing routing algorithms, it has obvious improvement in the aspects of delay, throughput, and packet loss rate.

Keywords: communication-based train control (CBTC); train-to-train(T2T) communication; relay selection; multiagent dueling DQN

Citation: Ma, S.; Li, M.; Yang, R.; Sun, Y.; Wang, Z.; Si, P. Next-Hop Relay Selection for Ad Hoc Network-Assisted Train-to-Train Communications in the CBTC System. *Sensors* **2023**, *23*, 5883. https://doi.org/10.3390/s23135883

Academic Editors: Qiong Wu and Pingyi Fan

Received: 15 May 2023
Revised: 21 June 2023
Accepted: 23 June 2023
Published: 25 June 2023

Copyright: © 2023 by the authors. Licensee MDPI, Basel, Switzerland. This article is an open access article distributed under the terms and conditions of the Creative Commons Attribution (CC BY) license (https://creativecommons.org/licenses/by/4.0/).

1. Introduction

Recently, with the rapid development of urbanization, urban rail transit has become one of the main transportations. With the development of technology, the communication-based train control (CBTC) system plays an important role in urban rail transit to guarantee the safe operation of rail trains [1]. To ensure their safety and reliability, CBTC systems have strict requirements on transmission delay and channel quality [2]. Long communication delay and link interruptions may lead to emergency brakes or collisions [3]. Therefore, it is crucial to design a CBTC communication system with low latency and high channel quality.

In traditional CBTC systems, long-term evolution for metro (LTE-M) and wireless local area networks (WLANs) are more widely used in train-to-wayside communication [4]. The train information is first transmitted to the ground-zone controller (G-ZC), which is used to generate control commands for all trains in its management area [5]. After obtaining the commands, the wayside node sends the commands back to trains. However, due to the huge computational burden of the G-ZC and non-direct transmission link [6], the transmission delay of important control commands is excessively large. Therefore, the T2T direct transmission approach was proposed [4], while the G-ZC was also changed to an onboard-zone controller (On-ZC). Unlike the G-ZC, the On-ZC only needs to generate its own commands, which greatly reduces computation latency.

Although the direct T2T transmission greatly reduces latency [7], if we continue to use WLAN or LTE, interruptions and delays caused by hard handoff at the base station boundary are still unavoidable. In addition, the deployment of the terrestrial core network and base station are complex, which makes network construction and maintenance cost high. Therefore, new technologies such as reconfigurable intelligent surfaces [8,9] and wireless ad hoc networks have been proposed to improve T2T communication.

The wireless ad hoc network is a novel approach to improve the performance of CBTC systems. In the wireless ad hoc network, the packages are sent to the wayside node from the on-board node. Similar to the vehicular ad hoc network (VANET), the role of wayside nodes is to assist packet transmission [10]. Therefore, the packages are transmitted through the transmission network formed by the wayside nodes hop-by-hop and finally to the running train, so that the transmission link is more stable. Furthermore, the deployment of wireless ad hoc network nodes is less costly, and it does not have a fixed topology or require any fixed infrastructure to communicate, allowing it to be deployed more flexibly and configured quickly.

The relay selection strategy plays an important role in wireless ad hoc networks to reduce transmission delay, improve throughput, and decrease packet loss. However, since wireless ad hoc networks are rarely used in CBTC systems, no suitable routing strategy has been proposed. Therefore, the existing strategies from VANETs and mobile ad hoc networks (MANETs) should be considered and improved to adapt to the CBTC system.

In VANET and MANET relay selection strategies, the routing algorithms are generally divided into two types: proactive routing and reactive routing [11]. Traditional proactive routing approaches, such as optimized link-state routing (OLSR) [12] and destination sequenced distance vector (DSDV) [13], require significant overhead for node exploration and maintenance of routing tables, which is not feasible for CBTC systems with strict requirements for low latency. At the same time, although the traditional reactive routing greedy perimeter stateless routing (GPSR) can directly select the next-hop relay, it cannot fully consider various factors such as node congestion and channel quality. In the CBTC system, the transmission channel can be affected by the high mobility of the train, causing shadow fading and multipath fading, which leads to a sudden change in transmission link. Therefore, the relay selection strategy must be able to make decisions according to the varying channel state.

On the basis of past training experience, the learning-based routing algorithm can make real-time decisions well by observing current channel state. Meanwhile, in deep reinforcement learning (DRL), the optimization problem of multiple factors can be transformed to maximize cumulative rewards [14]. We can combine diverse factors to design rewards, in order to achieve optimization of these indicators. Therefore, learning-based routing algorithms are more suitable for CBTC networks. Existing DRL-based algorithms are often used in relay selection for VANET and MANET [15,16]; these algorithms consider more focus on single-hop delay, outage probability, and power consumption as the criteria for next-hop selection. However, they ignore the whole-link delay and throughput.

In this paper, our objective is to design a low-latency and high-throughput routing method in an ad hoc network for a CBTC system. However, packet transmission still faces challenges in multi-hop relay selection. For example, due to the high-speed train, the transmission distance between the train and wayside node is limited. In order to decrease the outage probability, we set strict distance limitations when transmitting with trains. In addition, we comprehensively consider the transmission delay, queuing delay, and channel quality, aiming to optimize the overall performance of the link by selecting the next-hop node. Moreover, the process of selecting the next-hop node can be formulated as a Markov decision process (MDP) [17,18]. We propose a multi-agent DRL method to solve the problem. The main contributions of this paper are summarized as follows:

1. We formulate the next-hop relay selection problem in a CBTC system. The goal is to select relay nodes with low transmission delay and high throughput in both the train and the wayside node communication range. Meanwhile, in order to balance the single-hop transmission delay and the whole-link hop count, we propose the concept of "hop tradeoff" to minimize the entire link latency.
2. To handle the time-varying channel state and node congestion, we propose a DRL algorithm to optimize the long-term system reward. Using a multiagent approach [14],

all nodes are trained centrally with dueling DQN [19], and then each node makes the next-hop decision individually, in order to avoid nodes with a long queuing delay and poor channel quality.
3. Lastly, we conduct simulations with a different number of nodes between two trains and different buffer sizes. Meanwhile, the proposed algorithm is compared with several existing algorithms in terms of whole-link delay, packet loss rate, and throughput. The simulation results indicate that the proposed scheme works well against congested networks. In particular, it also significantly superior to other routing algorithms in the aspects of whole-link delay, throughput, and packet loss rate.

The remainder of this paper is organized as follows: in Section 2, some related work about routing selection in ad hoc networks is introduced; in Section 3, we present a multi-hop relay selection model for ad hoc networks in CBTC systems; the joint optimization problem of channel throughput and total-link delay is formulated in Section 4; then, we introduce the multiagent deep reinforcement method to solve the formulated problem in Section 5; some simulation results and analyses are presented in Section 6; in Section 7, we conclude the paper and propose some future work.

2. Related Work
2.1. Traditional Communication Method in CBTC System

In the traditional CBTC system, WLAN is widely used in the communication between trains and the wayside base station. Zhu et al. [20] proposed a WLAN-based redundant connect scheme for train–ground communication. The train connects the backup link and active link simultaneously to deal with the interruption at the coverage boundaries of the two access nodes. However, many WLAN standards based on IEEE802.11 are not suitable for high-speed mobile environments [5]. Meanwhile, WLAN is in the open frequency band, which can easily be interfered with by other devices [21]. LTE has strong anti-interference ability and is more stable in switching between access nodes; thus, LTE-based approaches have been proposed in the CBTC system. In [6], a sensing-based semi-persistent scheduling method for LTE-based T2T communication was proposed, which greatly improved the transmission delay of system safety information. However, both LTE and WLAN have the problem of packet loss and delay due to the switch between the access nodes, in addition to the high cost of the base stations and ground core network. Since wireless ad hoc networks do not require a fixed infrastructure, their deployment is more flexible and cheaper. Therefore, ad hoc networks are also a better choice for T2T communication.

2.2. Traditional Ad Hoc Network Route Selection

In ad hoc network application, the packet routing is critical for optimizing transmission delay, throughput, and packet loss. There are two types of routing methods commonly used in ad hoc networks: proactive routing and reactive routing. In proactive routing, the OLSR protocol [12] is to store the information of each relay node into the routing table by sending HELLO packets in advance, and then select the shortest path from routing table. The DSDV protocol [13] uses the Bellman–Ford algorithm to select relay nodes in the routing table. Although these approaches allow the optimal route to be selected, they require a large amount of information to be exchanged among all nodes. Especially when the nodes are dynamic, this leads to a rapid increase in the amount of information exchanged. Nevertheless, in the CBTC system, the trains move rapidly, and transmission latency is strict; thus, it is more suitable to use reactive routing.

In reactive routing, nodes cannot know global information of the whole network and can only make decisions for selecting the next-hop node. In [22], the GPSR protocol was derived, and the node which is closest to the destination within the communication range was selected as next-hop node. This minimizes the number of hops in the entire link, thus reducing latency. In order to solve the high outage probability caused by the high-speed movement of vehicles, the authors of [23] proposed a hybrid relay node selection strategy. The relay with the best channel quality and the relay closest to the destination were simultaneously selected as the next-hop node. Although this method improved the transmission successful rate, the overhead of transmitting both packets simultaneously was huge.

2.3. The Relay Selection of Ad Hoc Network Enabled by DRL

The deep learning-based relay selection method can make a better choice for next-hop relay where transmission rate and node congestion change in real time, and it can also synthesize multiple performance factors to make comprehensive decisions. Therefore, the routing method based on learning-based algorithms performs better in mobile communication network.

Learning-based routing methods are widely used in MANETs and VANETs [24]. As the mobility of nodes may lead to constant variations in the network topology [25], the learning-based approach can make decisions according to real-time changes and is, thus, more suitable for both scenarios. In [26], Wang et al. proposed a multiagent method, wherein all agents share the same training experience based on DQN, but make the next-hop decision individually. The aim of this method is to select the optimal route in MANET, which minimizes transmission delay and queuing delay. In order to minimize delay and make a reasonable power allocation in a vehicular-to-vehicular (V2V) communication network, Zhang et al. [27] proposed a deep reinforcement learning method to choose the optimal relay according to the velocity, location, and packet number of each vehicle. In [16], He et al. proposed a Q-learning algorithm to find optimal UAV relays to assist V2V communication. Their proposed method improved the delivery ratio and delivery latency with comprehensive consideration of the state transition probability of communication interruption, delay consumption, and energy consumption.

However, these methods only consider the delay of a single hop and ignore the total delay of the whole link. In some cases, if we only consider the shortest delay of a single hop, it is likely to choose the closest next-hop relay. This causes the hop number of the whole link to increase, along with the total link latency. Meanwhile, most methods use an infinitely long cache during the computation of system throughput. However, in real life, the buffer size is limited. Therefore, the impact of packet loss on throughput must be considered.

3. System Model

As shown in Figure 1, we consider a T2T communication over a multi-hop wireless ad hoc network. In this scenario, since the coverage of one hop is very limited, the train needs the assistance of wayside relays for multi-hop transmission. There are multiple relays within the communication range of each train and wayside node; hence, they need to select the most suitable next-hop wayside relays among these candidate nodes. For example, R_2 may communicate with R_3, R_4, and R_5, but R_3 is chosen as the next hop node by considering factors such as channel quality and delay.

Therefore, for N trains running on the rail, denoted as $T = \{T_1, T_2, \ldots, T_n, \ldots T_N\}$, there are M wayside relays distributed beside the rail, denoted as $R = \{R_1, R_2, \ldots, R_m, \ldots R_M\}$. The train has high mobility; in order to ensure the quality of the T2W transmission, we assume that there are two orthogonal frequency bands available: band 1 for train-to-wayside (T2W) transmission [28] and band 2 for wireless wayside-to-wayside (W2W) transmission. Since the transmission is on two orthogonal channels, there is no interference between T2W and W2W transmission, while multiple W2W transmissions at the same time will cause interference.

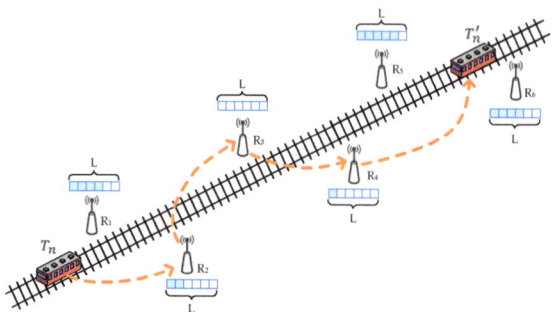

Figure 1. Next-hop relay selection in T2T communication.

In multi-hop transmission, all relays follow the decode-and-forward (DF) principle. Furthermore, we assume that the whole system is stationary in the time slot t, and that the transmit power of the nodes does not change. All channels follow quasi-static Rayleigh fading, such that the channel gain between node a and node b can be represented as follows:

$$h_{a,b} = X_{a,b}\, d_{a,b}^{-\beta/2}, \tag{1}$$

where $h_{a,b}$ is the instantaneous channel gain of link $a \to b$, $X_{a,b}$ is the fading coefficient, and $d_{a,b}$ and β indicate the distance between two nodes and the path-loss exponent [29].

3.1. Communication Model

3.1.1. Train-to-Wayside (T2W) Link

The transmission of T2W link is on an independent channel; thus, there is no other link interference, and the signal-to-noise ratio (SNR) of the T2W transmission is

$$\gamma_{T_n,R_m} = \frac{p_{T_n}\left|h_{T_n,R_m}\right|^2}{N_0}, \tag{2}$$

where p_{T_n} is the transmission power of the train T_n, h_{T_n,R_m} is the channel gain between the train T_n and the wayside relay R_m, and N_0 is the noise power. Hence, the channel throughput [15] between train T_n and the wayside relay R_m is

$$C_{T_n,R_m} = B\log_2(1 + \gamma_{T_n,R_m}). \tag{3}$$

As for the final hop, wayside node R_m transmits to destination train T'_n, which is also calculated in the same way as above, except that the transmission direction is different. The throughput between R_m and destination train T'_n can be expressed as

$$C_{R_m,T'_n} = B\log_2(1 + \gamma_{R_m,T'_n}). \tag{4}$$

3.1.2. Wayside-to-Wayside (W2W) Link

At time slot t, it is possible that more than one wireless W2W link transmits information simultaneously; thus, there is interference between W2W links [16]. During the packet transmission, the one-hop link from wayside relay i to wayside relay j at time slot t is denoted as $l_{i,j}(t)$, where $i,j \in \{R_1, R_2, \ldots, R_m, \ldots R_M\}$. Moreover, $\rho_{i,j}(t)$ represents the $i \to j$ link transmission status. $\rho_{i,j}(t) = 1$ denotes that wayside node i is transmitting with node j. Therefore, the SNR for the transmission of wayside node i and wayside node j can be represented as

$$\gamma_{i,j} = \frac{P_i\left|h_{i,j}\right|^2}{N_0 + \sum\limits_{l_{i,j}(t) \neq l_{i',j'}(t)} \rho_{i',j'} P_{i'}\left|h_{i',j'}\right|^2}, \tag{5}$$

where $l_{i',j'}(t)$ is the interference link during the time slot t.

The transmission throughput between wayside relay i and wayside relay j is

$$C_{i,j} = B \log_2(1 + \gamma_{i,j}). \tag{6}$$

3.1.3. Outage Analysis

In wireless networks, outage events occur when the actual mutual information is less than the required data rate [30]. To ensure the reliability of information transmission, the SNR of the channel must be greater than the SNR threshold value γ_{th} to transmit. At time slot t, the transmission condition for train T_n with wayside node R_m is $\gamma_{T_n,R_m} > \gamma_{th}$, and the transmission condition for wayside node i and wayside node j is $\gamma_{i,j} > \gamma_{th}$. In particular, the maximum transmission distance $R_{\max}$ between train and wayside relay while the train T_n is moving can be calculated as

$$\frac{p_{T_n}\left|X_{T_n,R_m}d_{T_n,R_m}^{-\beta/2}\right|^2}{N_0} \geq \gamma_{th}. \tag{7}$$

We can obtain the maximum distance $R_{\max}$ as

$$R_{\max} = d_{T_n,R_m} = \left(\frac{N_0}{p_{T_n}X_{T_n,R_m}^2}\gamma_{th}\right)^{-\frac{1}{\beta}}. \tag{8}$$

3.2. Optimal Relay Selection

3.2.1. Mobile Reliable Model

Due to the mobility of trains, the train T_n may move out of the communication range of the wayside relay R_m, resulting in an outage event; hence, the distance between the train and wayside nodes should be limited. The candidate wayside node locations are denoted as $(x_{R_1}, y_{R_1}), (x_{R_2}, y_{R_2}), (x_{R_m}, y_{R_m}) \ldots (x_{R_M}, y_{R_M})$. In the transmission delay T_c between train and wayside node, the channel SNR must satisfy the SNR threshold condition, whether the train is in the initial position (x_{R_1}, y_{R_1}) or at the end of transmission position (x'_{T_1}, y'_{T_1}). Meanwhile, the speed of packet transmission in the channel is much faster than the speed of the train; thus, we assume that the train drives at initial speed $v(t)$ during the transmission. In the transmission time delay T_c, the distance of the train moving in the x- and y-directions can be calculated as follows [15]:

$$S_x = v(t) \times T_c \times e_x(t), \tag{9}$$

$$S_y = v(t) \times T_c \times e_y(t), \tag{10}$$

where $e_x(t)$ and $e_y(t)$ are the x- and y-directions of the train. Therefore, the location where the train ends its transmission is $(x'_{T_1}, y'_{T_1}) = (x_{T_1} + S_x, y_{T_1} + S_y)$. The conditions that candidate wayside nodes need to satisfy are

$$d_{T1,Rm} = \sqrt{(x_{T1} - x_{Rm})^2 + (y_{T1} - y_{Rm})^2} < R_{\max}, \tag{11}$$

$$d'_{T1,Rm} = \sqrt{(x'_{T1} - x_{Rm})^2 + (y'_{T1} - y_{Rm})^2} < R_{\max}. \tag{12}$$

A node can only be a candidate transmission node for a train if its location is within the transmission range of the train at the beginning and end of the transmission.

3.2.2. Delay Model

During the packet transmission, a time delay is generated. In this section, we build a delay model to calculate the transmission delay between each node. We define the total delay $D_{i,j}$ of packet transmission between wayside nodes i, j into two main components, which are the transmission delay T_n caused by the node sending the packet and the queuing delay T_q due to node congestion. In this subsection, the time delay calculation is the same for both T2T transmission and T2W transmission; thus, both are expressed as the transmission between nodes a and b.

When the node sends data packets, the transmission delay can be represented as

$$T_c = \frac{L}{C_{a,b}}, \tag{13}$$

where L is the number of bits in the packet, and $C_{a,b}$ is the transmission rate of the channel between node a and node b.

Queuing delay [27] is unavoidable in the transmission of large amounts of data. Therefore, it is crucial to build a node queuing model. The queue follows the first-in first-out (FIFO) rule. When the CBTC system is stable, we assume that each wayside node can receive multiple data streams simultaneously to eliminate any scheduling effects, and that the average arriving rate and queuing situation are basically fixed. In order to calculate the queuing delay, we use Little's formula.

In Little's law, the average waiting time of a queue can be calculated as the queue length divided by the effective throughput. Since the buffer length of our designed model is limited, we calculate the effective throughput by considering the packet loss rate and the packet error rate. According to Little's law [31,32], the packet delay at next-hop node b can be expressed as

$$T_q = \frac{Q_b}{Th_b}, \tag{14}$$

where Q_b is the average number of packets queued at node b, and Th_b is the effective throughput of node b.

The effective throughput of node b is indicated as

$$Th_b = \lambda_b \Delta t \times (1 - p_f), \tag{15}$$

where λ_b is the average arriving rate of node b, $\lambda_b \Delta t$ is the total number of packets arriving in a time slot Δt, and p_f is the link $a \rightarrow b$ unsuccessful transmission rate. There are many factors that affect the transmission unsuccessful rate, such as the packet error rate p_{fe} and the packet loss rate p_{fl}. If a packet is lost or transmission error occurs, this will cause this packet to be unusable; hence, the probability of unsuccessful transmission rate can be expressed as

$$p_f = 1 - (1 - p_{fl})(1 - p_{fe}). \tag{16}$$

As for the calculation of the packet error rate, we can assume that the channel is modulated using the quadrature phase shift keying (QPSK) coding method. Therefore, the bit error rate (BER) [2] is

$$p_{be} = \frac{1}{2}\left(1 - \sqrt{\frac{\gamma_{a,b}}{1 + \gamma_{a,b}}}\right), \tag{17}$$

where $\gamma_{a,b}$ is the SNR between the node a and node b; when the SNR between two nodes is large enough, $p_{be} \cong \frac{1}{4 \cdot \gamma_{a,b}}$. Meanwhile, as the length of the packet is L, the BER of the whole packet can be represented as

$$p_{fe} = 1 - (1 - p_{be})^L. \tag{18}$$

For packet loss rate p_{fl}, we build a node queuing model to solve this problem [33,34]. Packet loss is due to the limited buffer length of the node. Therefore, if the total packet length exceeds the buffer length, the packet will not be received by the node, causing an increment in packet loss rate. We define M as the maximum number of packets that the buffer can hold, while Q_{t-1} is denoted as the number of packets left in the previous time slot, and A_t is the average number of packets arriving at the node in time slot t. We can derive the average number of arriving packets per time slot as $A_t = \lambda_b \Delta t$, where λ_b is the arrival rate of node b, and L is the length of the packet. The level of packet loss is F_t, which can be expressed as

$$F_t = \max[0, (Q_{t-1} + A_t) - M]. \tag{19}$$

When the train or wayside relay steadily sends packets to the next hop, A_t and F_t remain constant during transmission. Therefore, $\lim_{t \to \infty} A_t = A$ and $\lim_{t \to \infty} F_t = F$. In time slot t, the packet loss rate of this node can be calculated by

$$p_{fl} = \lim_{t \to \infty} \frac{\sum_{t=1}^{T} F_t}{\sum_{t=1}^{T} A_t} = \frac{E\{F\}}{E\{A\}} = \frac{E\{F\}}{\lambda_b \Delta t}, \tag{20}$$

where $E\{x\}$ is the mathematical expectation. If there is no packet overflow, then the packet loss rate is zero. Otherwise, the packet loss rate is the number of packet losses divided by the number of node arrivals. The packet loss rate and the packet error rate are calculated in Equations (18) and (20), respectively, and then introduced into Equations (14) and (15) to calculate the queuing delay. Therefore, the total delay of the k hop from node a to node b can be expressed as

$$D_{a,b}^k = T_p + T_c. \tag{21}$$

3.2.3. Hop Tradeoff

In the next-hop selection, if we only pursue large throughput and small delay for one hop, this will result in an increase in the hop count of the entire T2T link. Therefore, we design a "hop tradeoff" indicator to optimize the number of hops on the entire link. The initial train T_n and the destination train T'_n need to transmit information, and the distance of the whole T2T link is d_{T_n, T'_n}. During the transmission process, the distance between node a and node b for one-hop distance is $S_{a,b}$. We calculate the number of hops $k_{a,b}$ required to complete the entire T2T link for the one-hop distance $S_{a,b}$, which can be represented as

$$k_{a,b} = \frac{d_{T_n, T'_n}}{S_{a,b}}. \tag{22}$$

4. Problem Formulation

In the CBTC scenario, there are different numbers of wayside nodes to assist information transmission depending on the distance between two trains; thus, the link selection for transmission is particularly important. To solve the problem of multi-hop relay selection in wireless ad hoc networks, we propose an optimal transmission model based on a discrete Markov process. The aim is to design a relay node selection decision that satisfies low latency and high throughput, so that information can be transmitted quickly and accurately between trains to each other.

Since the next-hop selection depends only on the current state and the next state changes with the current action selection, next hop selection can be considered as an MDP. The transfer probability between states in the MDP is unknown; thus, we can use the DRL approach to better solve our proposed problem. In DRL, the agent finds the optimal policy that maximizes the long-term reward value according to the channel state information (CSI) and the congestion level of the node. In this paper, we use multiagent DRL (MADRL), in which each node acts as an agent. As agents need to make decisions in a shorter period of time, the agent's network is trained offline centrally by collecting information between nodes, and then porting it to each agent for online decision making. Therefore, each agent needs to select the next-hop node independently according to the current state, without additional communication for further training. In DRL, there are several key components, as described below.

(1) State Space

In each time slot, the agent updates and learns the policy by observing the variation of the state. In particular, the state contains two components: the number of packages queued in each node and the channel throughput of the links between the two nodes. In time slot t, the state space is defined as

$$s_t = \{Q(t), C(t), V(t)\}, \qquad (23)$$

where $Q(t)$ indicates the queue length of each node. $C(t)$ is the throughput between the transmission node and other nodes. $V(t) = \{0,1\}$; when $V(t) = 0$, at time slot t, the wayside node sends packets, whereas $V(t) = 1$ means that the train sends the packet.

(2) Action Space

According to the channel state and queue state of each candidate node, the optimal next-hop node $A(t)$ is selected. The action space can be given by

$$a_t = \{A(t)\}, \qquad (24)$$

where $A(t) = \{0, 1 \ldots m \ldots M\}$. If $A(t) = m$, then the next hop is node m.

(3) Reward Function

In the selection of the next-hop node, the optimization objective is to minimize the delay and maximum the throughput of the entire T2T link while ensuring that the next-hop SNR is greater than threshold value. The packet is successfully transmitted from the initial train to the destination train after $k \in \{1, 2, \ldots, K\}$ hops through wayside nodes $i, j \in \{R_1, R_2, \ldots, R_m, \ldots R_M\}$. Furthermore, T_n is the packet source train, and T'_n is the packet destination train.

- Overall Transmission Time of Whole Link

The total transmission time for packet transmission between source train T_n and destination train T'_n is

$$\tau_{T_n, T'_n} = D^1_{T_n, i} + \sum_{k=2}^{K-1} D^k_{i,j} + D^K_{j', T'_n}. \qquad (25)$$

- Throughput of Whole Link

In order to better measure the quality of each hop in the transmission link and considering the packet loss and packet error rate, the throughput of the entire link is defined as follows [35,36]:

$$C_{T_n, T'_n} = (1 - p'_f) \cdot \min\{E[C_{T_n, i}], E[C_{i,j}] \ldots E[C_{j', T'_n}]\}, \qquad (26)$$

where p'_f is the unsuccessful transmission rate of whole link; p_f for the one-hop unsuccessful rate is calculated using Equation (16). $C_{i,j}$ denotes the throughput between waysides node i and j.

- The Optimization Goal

The optimization goal for the proposed MADDQN is to reduce the latency and improve throughput for the whole link; hence, the proposed optimization goal is defined as

$$\max \omega_1 \frac{1}{\tau_{T_n,T'_n}} + \omega_2 C_{T_n,T'_n} \tag{27}$$

$$s.t: C1: \gamma_{i,j}^k > \gamma_{th},$$

$$C2: D_{i,j}^k > D_{th},$$

$$C3: d_{T_n,i} < R_{\max}.$$

Here, τ_{T_n,T'_n} and C_{T_n,T'_n} are the whole-link latency and throughput between train T_n and train T'_n, respectively. ω_1 and ω_2 are the weight factors of the delay and throughput ($\omega_1 + \omega_2 = 1$). C1 is to ensure that the SNR of the channel is greater than threshold value. C2 indicates that each hop delay needs to be less than the target transmission time. If the one-hop delay takes too much time, the transmission is considered to fail. C3 indicates that, when the train is transmitted with the wayside node i, the distance should be less than the maximum transmission distance.

The defined reward function comprehensively considers the throughput and delay of each hop, as well as adds the indicator $k_{i,j}$ in Section 3.2.3. Therefore, the reward function is defined as

$$F_t = \begin{cases} \omega_1 \frac{1}{k_{i,j} D_{i,j}^k} + \omega_2 C_{i,j}^k + r_s + r_c, & \text{if } C1\text{--}C3 \text{ are satisfied} \\ \omega_1 \frac{1}{k_{i,j} D_{i,j}^k} + \omega_2 C_{i,j}^k + r_s, & \text{otherwise} \end{cases}, \tag{28}$$

where r_s is an additional reward for the final hop directly to the train. This reward is established to prevent other wayside nodes close to the train from being selected, which may lead to an increase in hops and delay. r_c is the outage penalty caused by the next-hop node out of the communication range and a long single-hop delay under C1–C3.

5. Problem Solution

Since value-based functions are suitable for solving discrete space problems, and next-hop relay selection is a discrete action, we choose a value-based reinforcement learning approach for policy optimization. Our proposed scheme has a large number of channel states and queuing states, which leads to a high dimension of the Q-table, and makes the Q-learning [37] algorithm difficult to coverage during training. However, the DQN algorithm can solve this problem, featuring a combination of a deep neural network (DNN) and Q-value. DQN does not directly select the action with the highest Q-value in the Q-table, but fits the $Q_\pi(s_t, a_t, \theta)$ through the neural network [38]. Compared to recording the Q-value for an all-action-state situation, DQN just needs to store the weights of each neuron to calculate the Q-values for all policies $\pi(s_t, a_t)$, which greatly reduces the storage space and makes the algorithm converge faster [39,40].

Since each node needs to make the decision to select the next-hop node, we use the multiagent dueling DQN (MADDQN) approach. MADDQN treats each node as an independent agent, and all the agents are trained centrally. When making the next-hop selection, the trained network parameters are shared with all nodes, and each node selects the next-hop node individually. The specific process of MADDQN is shown in Figure 2.

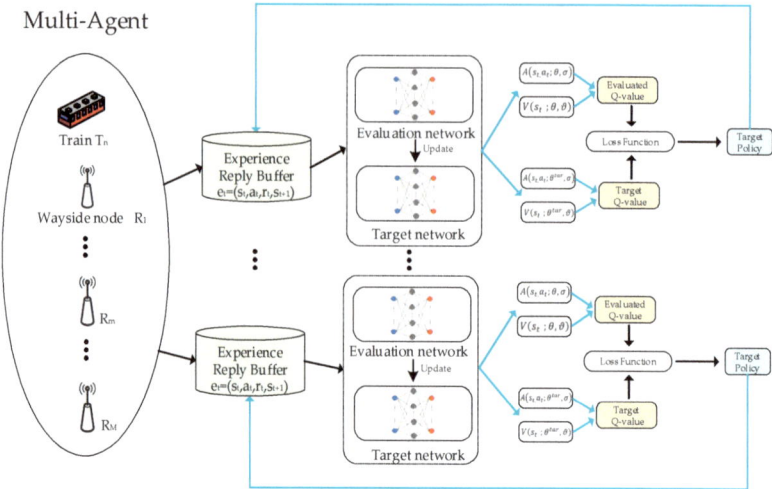

Figure 2. Framework of MADDQN.

5.1. DQN

In each time slot t, each node acts as an agent to observe the current state s_t of the system, including the congestion level and channel state information of all nodes. Then, the agent chooses a suitable action a_t to select the next-hop node. After selecting an action a_t, a new state s_{t+1} is obtained, and the reward r_t corresponding to this action is also computed. The goal of the agent is to find a policy $\pi(s_t, a_t)$, which maximizes the expected discounted cumulative reward $E\left[\sum_{i=t}^{K-1} \gamma^{i-t} r_{i+1}\right]$ [9]. Therefore, the action-state value function is used to calculate the expected discounted cumulative reward of each relay selection policy and then select the policy π with the largest reward. The state-action value function is defined as

$$Q_\pi(s_t, a_t) \triangleq E\left[\sum_{i=t}^{K-1} \gamma^{i-t} r_{i+1} \Big| s_t, a_t, \pi\right], \quad (29)$$

where $\gamma \in (0, 1)$ is the discount factor, which represents the ratio between immediate and long-term reward. r_{i+1} is the immediate reward at time slot $i + 1$.

In the next time slot of action selection, not only is the next-hop relay selected with maximum $Q_\pi(s_t, a_t)$, but the $\varepsilon - greedy$ algorithm is also added to explore extra actions. In order to try more possible actions and avoid falling into the local maximum, the agent has ε possibility to choose an action randomly. The $\varepsilon - greedy$ algorithm is denoted as

$$a_t = \begin{cases} \text{argmax} Q_\pi(s_t, a_t), & \text{with probability } 1 - \varepsilon \\ \text{random}, & \text{with probability } \varepsilon \end{cases}. \quad (30)$$

Due to the different channel states and node congestion levels, a large number of states are formed; thus, it is impossible to calculate the Q-value for each action and state. In DQN, the convolutional neural network (CNN) is trained to get the weight θ of each neuron. After inputting the current channel state and the next-hop relay selection action, the neural network can fit a state-action value $Q(s_t, a_t, \theta)$. To make the network converge faster, DQN has two networks: the target network and the evaluation network. During the training process, the weights of the evaluation network θ are continuously updated, and the weights of the evaluation network are assigned to the target network θ^{tar} at certain time intervals. Then, the weight θ is updated by the stochastic gradient descent method to minimize the result of loss function between the target network and the evaluation network. The loss function between the target network and the evaluation network is defined as

$$L(\theta) = E_{(s_t,a_t,r_t,a_{t+1})}(Q_{target} - Q(s_t,a_t,\theta))^2, \tag{31}$$

where the target value for each iteration of the network is represented as

$$Q_{target} = r_t + \gamma \max_{a_{t+1}} Q(s_t, a_t, \theta^{tar}). \tag{32}$$

To focus more on historical experiences and disrupt the correlation between experiences, DQN also uses the mechanism of experience replay. At each time slot, when nodes are trained centrally, the node acts as an agent, storing the training experience $e_t = (s_t, a_t, r_t, s_{t+1})$ into the experience pool, and then forming the sequence $D = \{e_1, e_2 \ldots e_N\}$. For each training, a small number of samples are randomly selected from the experience pool as a batch for network training, which makes the network converge better.

5.2. Dueling DQN

The dueling DQN network [19] makes further improvements on the DQN network structure. In DQN, the network directly outputs the state-action values $Q(s_t, a_t, \theta)$ corresponding to each relay selection policy. However, in dueling DQN, the output Q-value is split into two branches: the state value $V(s_t)$ indicating the value of the current channel and queuing state, and the action advantage value $A(s_t, a_t)$ representing the value brought by the relay selection action. Finally, the output values of the two branches are combined to make the estimation of Q more accurate. The combination of the two branches can be written as

$$Q(s_t, a_t; \theta, \sigma, \vartheta) = V(s_t; \theta, \vartheta) + A(s_t, a_t; \theta, \sigma), \tag{33}$$

where θ, σ, and ϑ are the coefficients of the neural network. In order to prevent multiple sets of state value $V(s_t)$ and action advantage value $A(s_t, a_t)$ with the same state-action value $Q(s_t, a_t; \theta, \sigma, \vartheta)$, and to make the algorithm more stable [38], Equation (30) is replaced by

$$Q(s_t, a_t; \theta, \sigma, \vartheta) = V(s_t; \theta, \vartheta) + \left(A(s_t, a_t; \theta, \sigma) - \frac{1}{|A|}\sum_{a_{t+1}} A(s_t, a_{t+1}; \theta, \sigma)\right). \tag{34}$$

The proposed multiagent dueling DQN is shown in Algorithm 1.

Algorithm 1 Dueling-DQN

1: Initialization:
Initialize the maximum buffer capacity M and packet length L;
Initialize the number of nodes along the rail N;
Initialize network memory size J, batch size B, greedy coefficient ε, and learning rate φ.
2: **for** episode in range K do:
3:   Reset channel quality C and the queue length Q of each node as initial state $S_{initial}$
4:   **While** $a(t)! = $ Destination Node **do**
5:     Choose action: with probability ε to choose next hop node in random.
6:     Otherwise, choose action a_t with $\text{argmax} Q_\pi(s_t, a_t, \theta)$.
7:     From current state s_t and action a_t of this hop, obtain the reward r_t for this action a_t and the next state s_{t+1}.
8:     Store $\langle s_t, a_t, r_t, s_{t+1}\rangle$ into experience reply to memory.
9:     Randomly take minibatch of $\langle s_t, a_t, r_t, s_{t+1}\rangle$ from experience reply to memory.
10:    Combine two branches $V(s_t; \theta, \vartheta)$ and $A(s_t, a_t; \theta, \sigma)$ into $Q(s_t, a_t; \theta, \sigma, \vartheta)$
11:    Calculate target Q-value

$$Q_{target} = \begin{cases} r_t, & \text{if } a_t \text{ is the destination node} \\ r_t + \gamma \max_{a_{t+1}} Q(s_{t+1}, a_{t+1}; \theta^{tar}, \sigma, \vartheta), & \text{otherwise}. \end{cases}$$

12:    Minimize loss function $L(\theta)$ using Equation (30)
13:    Update the target network after several steps using the parameters of the evaluation network
14: **end while**
15: **end for**

6. Simulation Results

In this section, we verify the effectiveness of the proposed deep learning-based relay selection algorithm by conducting simulation experiments in CBTC system.

6.1. Simulation Settings

In the simulation, TensorFlow 1.13.1 was imported in Python 3.6 as the simulation environment.

In order to simplify the system model, we performed a simulation of relay selection between two adjacent trains. If the SNR between the current node and the next-hop wayside relay is greater than the SNR threshold, then the communication between the two nodes is possible. Since each node has the same transmission power, the distance between nodes mainly determines the channel throughput; thus, the next-hop node which is closer to the current node has higher one-hop channel throughput. In the train system, packets transmitted by the train must pass through wayside nodes and cannot be delivered directly to the forward train. In addition, wayside relays are uniformly distributed on both sides of the track, as the distance between trains become longer, the number of hops required for transmission also increases.

Furthermore, in the process of training, each agent has its own training parameters; we set the batch size $B = 256$, greedy coefficient $\varepsilon = 0.1$, learning rate $\varphi = 0.001$, and memory size $J = 1024$. Some other main parameters of the communication system are shown in Table 1.

Table 1. Communication system parameters.

Parameters	Value
Number of trains	2
Number of wayside nodes	7
Bandwidth	10 MHz
Max buffer size	250 kb
Average packet size	25 kb
SNR threshold	31
Gaussian noise power spectral density	−174 dBm/Hz
Weight of latency, ω_1	0.5
Weight of channel throughput, ω_2	0.5

6.2. Performance Analysis

We compare the proposed MADDQN algorithm with two existing algorithms:

1. GPSR [22] (greedy perimeter stateless routing) is often used in the transmission of ad hoc networks, which collects the geographic location information of neighboring nodes and finds the next-hop node with the nearest geographic location to the destination through a greedy algorithm.
2. The random selection scheme randomly selects the next-hop node within the communication range without any optimization strategy.

6.2.1. Performance Comparison of Convergence

Firstly, in order to find the learning rate that makes the proposed model converge best, we conducted experiments at three different learning rates. As shown in Figure 3, when the learning rate was equal to 10^{-4}, the convergence rate of the agent was slow, and the total reward value did not reach the optimal value. To make the convergence speed up, we increased the learning rate to 10^{-3}, which made the convergence faster and the total reward higher. When the learning rate increased to 10^{-2}, it was easy for the agent to converge to a local optimum, resulting in poor convergence results. Therefore, in the training of the agents, the learning rate was set to 10^{-3}.

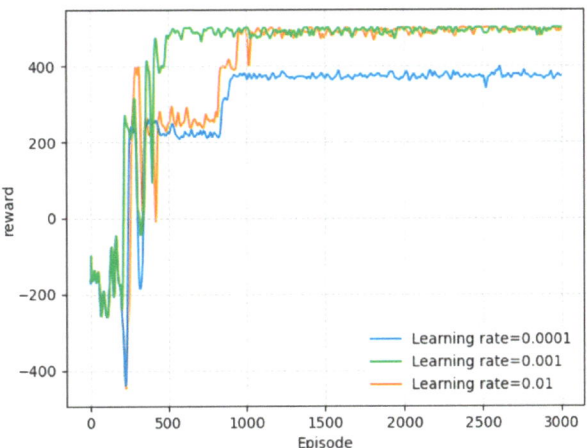

Figure 3. Total reward with different learning rates.

The goal of agent training is to better avoid outage events and reduce packet loss rate. As shown in Figure 4, the probability of outage events decreased dramatically in the first 1000 episodes, indicating that the agent learned to select next-hop relays within communication range. At the same time, the probability of network congestion gradually decreased during the training process, which illustrates that the agent successfully avoided congested nodes. The simulation results show that the MADDQN algorithm could effectively avoid outage and congestion events, ensuring the quality of transmission.

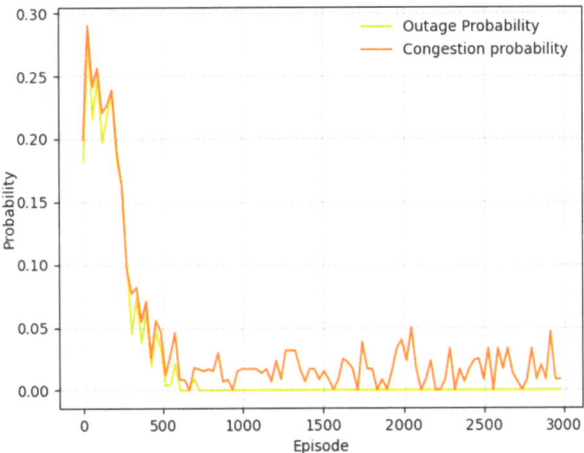

Figure 4. Outage probability and congestion probability of the proposed algorithm.

6.2.2. Performance Comparison of Different Aspects

The distance between two adjacent trains is different and the wayside nodes are uniformly distributed. Thus, when the distance between two trains become larger, the number of trackside nodes between them increases and the topology of the network changes. Figure 5 depicts the curves of the variation of the total delay as the number of nodes increase. It can be observed that the whole-link delay increased from 4.31 ms to 8.43 ms under the MADDQN algorithm. The main reason is that, with the increment in the number of relays, the number of hops required for the whole link increased; thus, the total delay increased. Compared with the random selection scheme and GPSR algorithm, the transmission delay of the MADDQN algorithm was reduced by an average of 2 ms and 0.5 ms, respectively. Although the traditional GPSR algorithm requires a small number of hops, it cannot avoid congested nodes, resulting in a large packet loss rate. In addition, the random scheme can neither select nodes with small queued tasks nor optimize the hop count; hence, the delay is longer than the previous two algorithms.

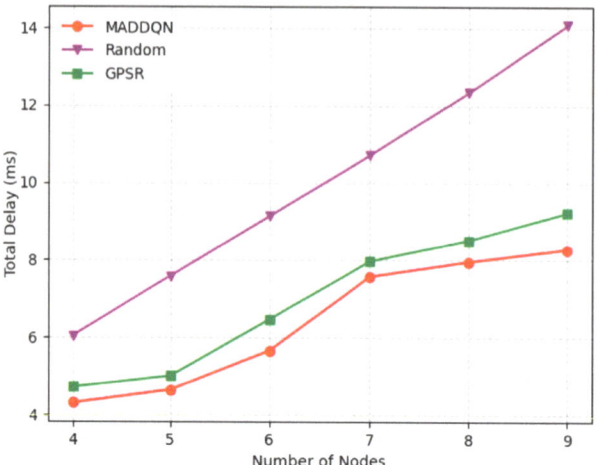

Figure 5. Total delay versus the number of nodes.

The effect of buffer size on total delay is investigated in Figure 6. When the buffer size was less than 300 Kb, the total latency increased significantly as the buffer size became larger. When the buffer size was small, the packet queue became shorter, resulting in lower packet delay. Moreover, when the buffer size reached 300 kb, the total delay gradually flattened out. The reason is that nodes were no longer dropping packets, and the queue length of each node tended to be stable. In addition, MADDQN improved by 0.5 ms compared to the GPSR algorithm and 3 ms compared to the random selection scheme, which illustrates that the proposed MADDQN algorithm could select nodes with shorter queues for transmission.

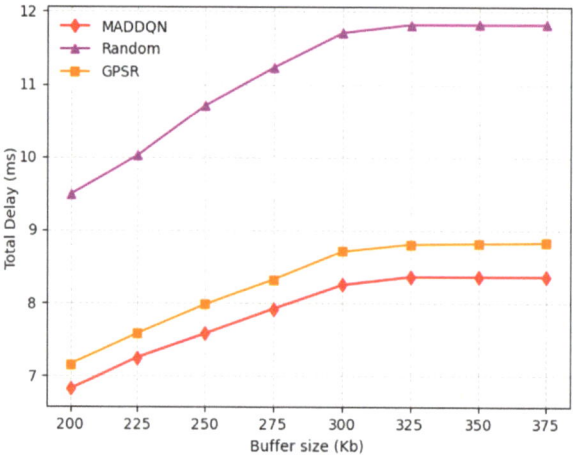

Figure 6. Total delay versus buffer size.

Figure 7 presents the relationship between the number of nodes and the average loss rate under different schemes. As the number of nodes increased from four to nine, the packet loss rate of MADDQN increased to 0.12. The reason is that, as the number of hops increased, the total packet loss rate also increased. Compared with the other two schemes, the proposed MADDQN could select the next-hop relay with a shorter queuing number for transmission, which greatly reduced the packet loss probability. While the GPSR algorithm could not avoid congested nodes, it required fewer hops for transmission; hence, the packet loss rate was also lower than the random scheme.

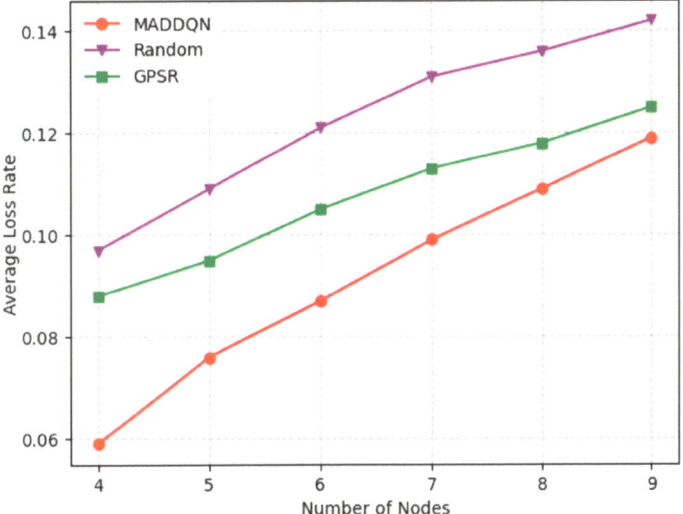

Figure 7. Average loss rate versus the number of nodes.

Figure 8 depicts the change in the average loss rate with the buffer size. The average packet loss rate rapidly decreased to zero when the buffer size was 300 kb. Due to the small buffer size, packets could easily overflow. Hence, the packet loss rate continued to decrease until the buffer was large enough. The average loss rate of the GPSR and random scheme before 300 kb was higher than MADDQN, which illustrates that our proposed method had a significant effect in avoiding congested nodes and reducing the number of hops, such that the packet loss rate was the lowest.

Figure 9 presents how the number of nodes affects the average throughput. It can be observed that the average throughput of the entire link decreased monotonically with the increment in the number of nodes. This is because, as shown in Figure 7, the packet loss rate increased with the number of nodes, it led to a reduction in the overall throughput. The throughput of MADDQN was greater than the other two methods, indicating that, when selecting the next hop node, MADDQN chose the node with relatively large channel throughput and fewer queued packets, ensuring channel quality.

Figure 10 shows the impact of buffer size on the throughput. It can be observed that the throughput rapidly increased before 300 kb and then reached a stable state with increasing buffer size. This is because as the buffer size gradually increased, it caused a decrease in packet loss rate; therefore, the system throughput increased. When the system had no packet loss, the throughput tended to stable. Meanwhile, the proposed algorithm had the highest throughput compared with the other two schemes; hence, it can be proven that MADDQN was effective in selecting the routing with a large throughput.

Figure 11 illustrates the relationship of the number of nodes and optimization goal under different weights of latency and channel throughput. The simulation results show that, with the rise in ω_1, the optimization goal was much larger. This is because, although the optimization goal optimized both delay and throughput, when the weight of delay was high, latency was optimized more, while the optimization of throughput was relatively weak. Moreover, when the node number was between four and six, delay accounted for a large proportion of the optimization objective. Thus, when ω_1 = 0.9, the total optimization objective was the highest. However, in order to optimize both delay and throughput to a better level, we chose the case ω_1 = 0.5, ω_2 = 0.5. In the application, different parameters can also be chosen according to different needs. For example, the weight of ω_1 can be increased for safety information with higher requirements on latency. For systems where throughput is more important, the proportion of ω_2 can be increased appropriately, but the sum of ω_1 and ω_2 must be equal to 1.

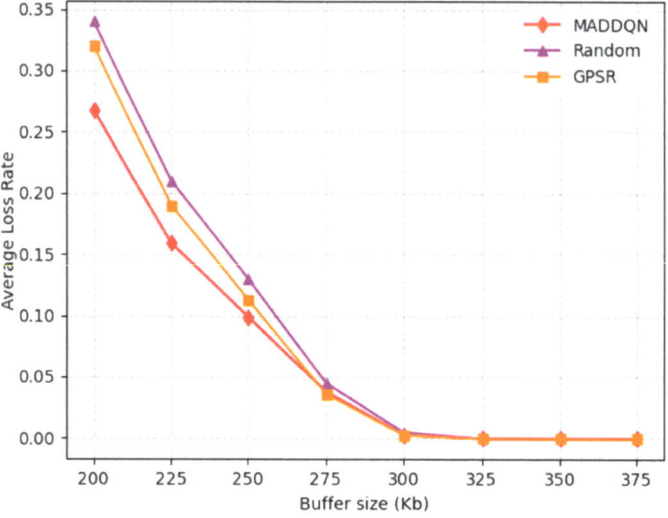

Figure 8. Average loss rate versus buffer size.

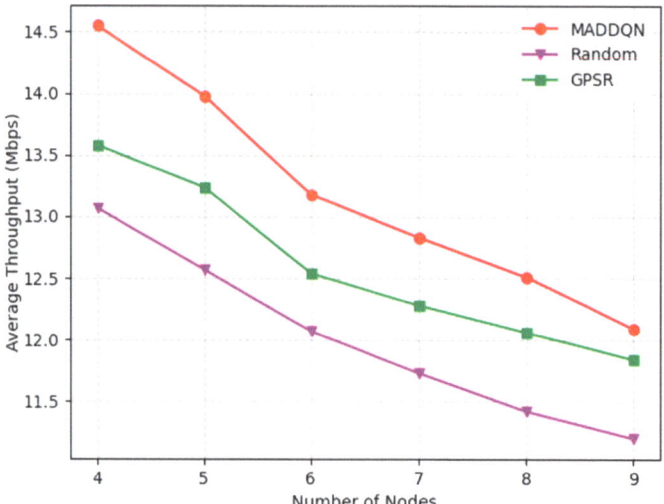

Figure 9. Average throughput versus the number of nodes.

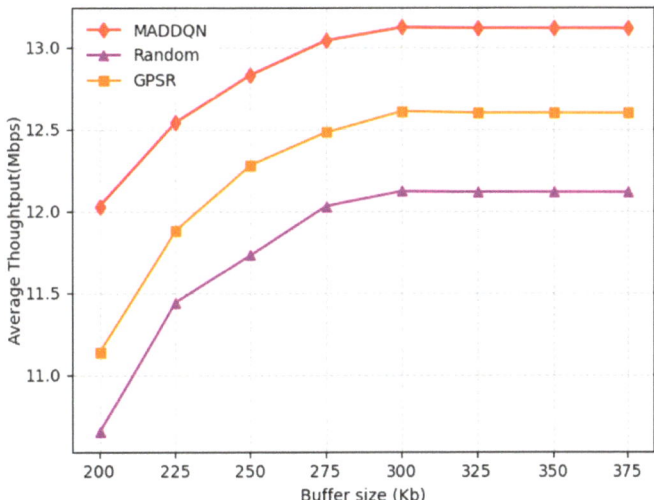

Figure 10. Average throughput versus buffer size.

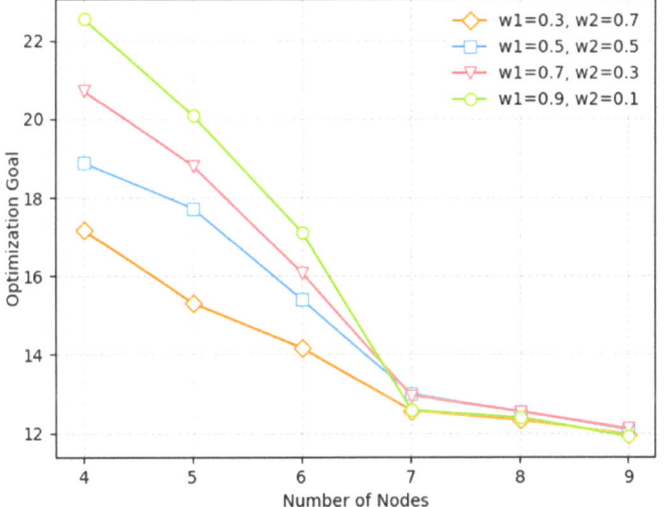

Figure 11. Relationship of optimization goal and number of nodes under different weights.

The effect of the number of nodes and the buffer size on the optimization goal is investigated in Figures 12 and 13. The optimization goal was derived from Equation (27), which is a comprehensive indicator of channel throughput and transmission delay. In Figure 12, the delay increased and the throughput decreased as the number of nodes grew; thus, the optimization objective was gradually reduced. This shows that, as the number of nodes increased, the overall performance of the system worsened. In Figure 13, both latency and throughput tended to rise as the buffer size increased, but latency rose faster, having a greater impact on the optimization objective. Therefore, the optimization goal showed a slight decrease after combining these two indicators. Moreover, we can observe that our proposed algorithm always outperformed the existing algorithms, indicating that the MADDQN algorithm could better tradeoff the channel throughput and transmission delay under any topological condition and buffer size to achieve the optimization goal.

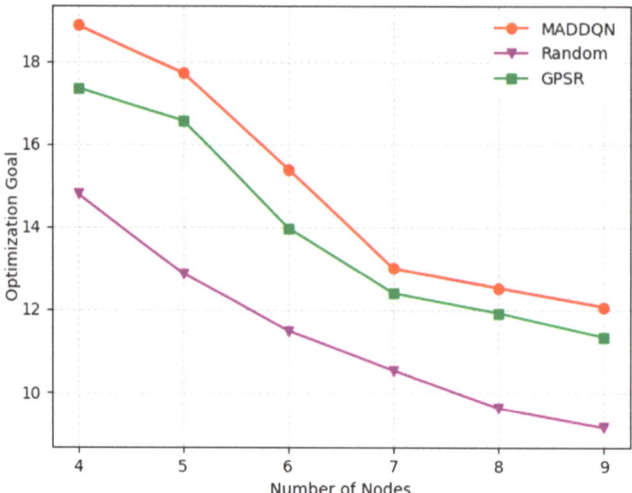

Figure 12. Relationship of optimization goal and number of nodes under different schemes.

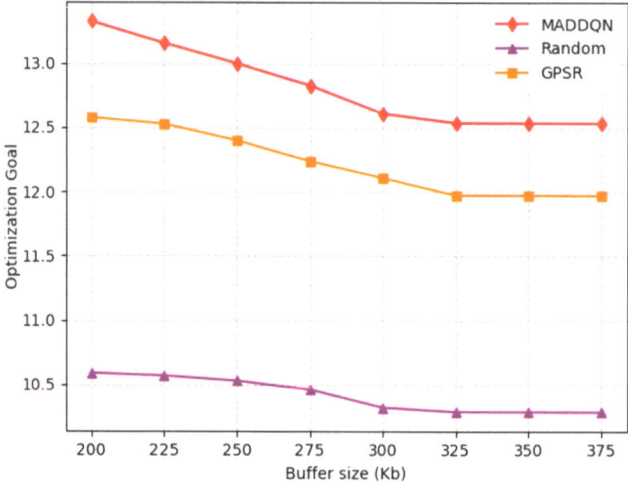

Figure 13. Relationship of optimization goal and buffer size under different schemes.

7. Conclusions and Future Work

In this paper, we designed a multi-hop relay selection strategy based on wireless ad hoc networks to assist T2T communication. The optimization goal of our proposed algorithm is to reduce the T2T transmission delay and increase the throughput of the entire link in a congested network. Since the channel status changes in real time, an MADDQN approach was proposed to better solve the problem. Simulation results showed that our proposed algorithm could effectively avoid congested nodes and reduce the number of hops for the whole-link transmission, thereby better achieving the optimization goal compared with existing routing algorithms. In future work, the energy consumption of the nodes and the problem of retransmission after packet loss should be considered. Moreover, some secure and energy-efficient technologies, such as reconfigurable intelligent surfaces (RIS), will be applied to the CBTC system to better assist in signal transmission.

Author Contributions: Conceptualization, S.M. and M.L.; methodology, S.M. and R.Y.; software, S.M.; validation, S.M., R.Y., and Y.S.; formal analysis, S.M. and M.L.; investigation, S.M. and Z.W.; resources, S.M., Z.W., and P.S.; data curation, S.M. and M.L.; writing—original draft preparation, S.M.; writing—review and editing, M.L. and P.S.; visualization, Y.S. and Z.W.; supervision, M.L. and P.S.; project administration, M.L., R.Y., and Z.W. All authors have read and agreed to the published version of the manuscript.

Funding: This work was partially supported through the Beijing Natural Science Foundation under Grants L211002, 4222002, and L202016, and the Foundation of the Beijing Municipal Commission of Education under Grants KM 202010005017 and KM202110005021.

Institutional Review Board Statement: Not applicable.

Informed Consent Statement: Not applicable.

Data Availability Statement: Not applicable.

Conflicts of Interest: The authors declare no conflict of interest.

References

1. Wang, X.; Liu, L.; Tang, T.; Sun, W. Enhancing Communication-Based Train Control Systems Through Train-to-Train Communications. *IEEE Trans. Intell. Transp. Syst.* **2019**, *20*, 1544–1561. [CrossRef]
2. Zhu, L.; Yu, F.R.; Ning, B.; Tang, T. Communication-Based Train Control (CBTC) Systems with Cooperative Relaying: Design and Performance Analysis. *IEEE Trans. Veh. Technol.* **2014**, *63*, 2162–2172. [CrossRef]
3. Liu, L.; Parag, P.; Tang, J.; Chen, W.Y.; Chamberland, J.F. Resource Allocation and Quality of Service Evaluation for Wireless Communication Systems Using Fluid Models. *IEEE Trans. Inf. Theory.* **2007**, *53*, 1767–1777. [CrossRef]
4. Wang, X.; Liu, L.; Tang, T. Train-Centric CBTC Meets Age of Information in Train-to-Train Communications. *IEEE Trans. Intell. Transp. Syst.* **2020**, *21*, 4072–4085. [CrossRef]
5. Sun, W.; Yu, F.R.; Tang, T.; Bu, B. Energy-Efficient Communication-Based Train Control Systems with Packet Delay and Loss. *IEEE Trans. Intell. Transp. Syst.* **2016**, *17*, 452–468. [CrossRef]
6. Wang, X.; Liu, L.; Tang, T.; Zhu, L. Next Generation Train-Centric Communication-Based Train Control System with Train-to-Train (T2T) Communications. In Proceedings of the 2018 International Conference on Intelligent Rail Transportation (ICIRT), Singapore, 12–14 December 2018.
7. Li, Y.; Zhu, L. Collaborative Cloud and Edge Computing in 5G based Train Control Systems. In Proceedings of the 2022 IEEE Global Communications Conference, Rio de Janeiro, Brazil, 4–8 December 2022.
8. Gong, S.; Lu, X.; Hoang, D.T.; Niyato, D.; Shu, D.; Shu, L.; Kim, D.I.; Liang, Y.C. Toward Smart Wireless Communications via Intelligent Reflecting Surfaces: A Contemporary Survey. *IEEE Commun. Surv. Tutor.* **2020**, *22*, 2283–2314. [CrossRef]
9. Ahmed, M.; Wahid, A.; Laique, S.S.; Khan, W.U.; Ihsan, A.; Xu, F.; Chatzinotas, S.; Han, Z. A Survey on STAR-RIS: Use Cases, Recent Advances, and Future Research Challenges. *IEEE Internet Things J.* **2023**. [CrossRef]
10. Ahmed, M.; Mirza, M.A.; Raza, S.; Ahmad, H.; Xu, F.; Khan, W.U.; Lin, Q.; Han, Z. Vehicular Communication Network Enabled CAV Data Offloading: A Review. *IEEE Trans. Intell. Transp. Syst.* **2023**. [CrossRef]
11. Gupta, L.; Jain, R.; Vaszkun, G. Survey of Important Issues in UAV Communication Networks. *IEEE Commun. Surv. Tutor.* **2016**, *18*, 1123–1152. [CrossRef]
12. Jacquet, P.; Muhlethaler, P.; Clausen, T.; Laouiti, A.; Qayyum, A.; Viennot, L. Optimized Link State Routing Protocol for Ad Hoc Networks. In Proceedings of the 2001 IEEE International Multi Topic Conference, Lahore, Pakistan, 30 December 2001.
13. Bai, F.; Sadagopan, N.; Helmy, A. IMPORTANT: A framework to systematically analyze the Impact of Mobility on Performance of Routing Protocols for Adhoc Networks. In Proceedings of the 2003 Twenty-Second Annual Joint Conference of the IEEE Computer and Communications Societies, San Francisco, CA, USA, 30 March–3 April 2003.
14. Ding, R.; Xu, Y.; Gao, F.; Shen, X. Trajectory Design and Access Control for Air–Ground Coordinated Communications System with Multiagent Deep Reinforcement Learning. *IEEE Internet Things J.* **2022**, *9*, 5785–5798. [CrossRef]
15. Ding, R.; Chen, J.; Wu, W.; Liu, J.; Gao, F.; Shen, X. Packet Routing in Dynamic Multi-Hop UAV Relay Network: A Multi-Agent Learning Approach. *IEEE Trans. Veh. Technol.* **2022**, *71*, 10059–10072. [CrossRef]
16. He, Y.; Zhai, D.; Jiang, Y.; Zhang, R. Relay Selection for UAV-Assisted Urban Vehicular Ad Hoc Networks. *IEEE Wirel. Commun. Lett.* **2020**, *9*, 1379–1383. [CrossRef]
17. Wu, Q.; Zheng, J. Performance Modeling and Analysis of the ADHOC MAC Protocol for VANETs. In Proceedings of the IEEE International Conference on Communication (ICC'15), London, UK, 8–12 June 2015.
18. Wu, Q.; Zheng, J. Performance Modeling and Analysis of the ADHOC MAC Protocol for Vehicular Networks. *Wirel. Netw.* **2016**, *22*, 799–812. [CrossRef]
19. Wu, Q.; Zhao, Y.; Fan, Q.; Fan, P.; Wang, J.; Zhang, C. Mobility-Aware Cooperative Caching in Vehicular Edge Computing Based on Asynchronous Federated and Deep Reinforcement Learning. *IEEE J. Sel. Top. Signal Process.* **2023**, *17*, 66–81. [CrossRef]

20. Zhu, L.; Yu, F.R.; Ning, B. Availability Improvement for WLAN-Based Train-Ground Communication Systems in Communication-Based Train Control (CBTC). In Proceedings of the 2010 IEEE 72nd Vehicular Technology Conference, Ottawa, ON, Canada, 6–9 September 2010.
21. Wang, Y.; Zhu, L.; Zhao, H. Handover Performance Test and Analysis in TD-LTE based CBTC Train Ground Communication Systems. In Proceedings of the 2017 Chinese Automation Congress, Jinan, China, 20–22 October 2017.
22. Karp, B.; Kung, H.-T. Gpsr: Greedy Perimeter Stateless Routing for Wireless Networks. In Proceedings of the 6th Annual International Conference on Mobile Computing and Networking, Boston, MA, USA, 6–11 August 2000.
23. Liu, K.; Niu, K. A Hybrid Relay Node Selection Strategy for VANET Routing. In Proceedings of the 2017 IEEE/CIC International Conference on Communications in China (ICCC), Qingdao, China, 22–24 October 2017.
24. Li, F.; Wang, Y. Routing in Vehicular Ad Hoc Networks: A Survey. *IEEE Veh. Technol. Mag.* **2007**, *2*, 12–22. [CrossRef]
25. Toor, Y.; Muhlethaler, P.; Laouiti, A.; La Fortelle, A.D. Vehicle Ad Hoc Networks: Applications and Related Technical Issues. *IEEE Commun. Surv. Tutor.* **2008**, *10*, 74–88. [CrossRef]
26. Wang, Z.; Han, R.; Li, H.; Knoblock, E.J.; Apaza, R.D.; Gasper, M.R. Deep Reinforcement Learning Based Routing in an Air-to-Air Ad-hoc Network. In Proceedings of the 2022 IEEE/AIAA 41st Digital Avionics Systems Conference (DASC), Portsmouth, VA, USA, 18–22 September 2022.
27. Zhang, H.; Chong, S.; Zhang, X.; Lin, N. A Deep Reinforcement Learning Based D2D Relay Selection and Power Level Allocation in mmWave Vehicular Networks. *IEEE Wirel. Commun. Lett.* **2020**, *9*, 416–419. [CrossRef]
28. Wu, Q.; Zheng, J. Performance Modeling of IEEE 802.11 DCF Based Fair Channel Access for Vehicular-to-Roadside Communication in a Non-Saturated State. In Proceedings of the IEEE International Conference on Communication (ICC'14), Sydney, NSW, Australia, 10–14 June 2014.
29. Chen, Y.; Feng, Z.; Xu, D.; Liu, Y. Optimal Power Allocation and Relay Selection in Dual-Hop and Multi-Hop Cognitive Networks. In Proceedings of the 2012 IEEE International Conference on Communications (ICC), Ottawa, ON, Canada, 10–15 June 2012.
30. Wang, Y.; Feng, Z.; Chen, X.; Li, R.; Zhang, P. Outage Constrained Power Allocation and Relay Selection for Multi-Hop Cognitive Network. In Proceedings of the 2012 IEEE Vehicular Technology Conference (VTC Fall), Quebec City, QC, Canada, 3–6 September 2012.
31. Kleinrock, L. *Queueing Systems, Volume I: Theory*; Wiley-Interscience: New York, NY, USA, 1975.
32. Liu, W.; Zhou, S.; Giannakis, G.B. Queuing with Adaptive Modulation and Coding over Wireless Links: Cross-Layer Analysis and Design. *IEEE Trans. Wirel. Commun.* **2005**, *4*, 1142–1153.
33. Ma, R.; Chang, Y.-J.; Chen, H.-H.; Chiu, C.-Y. On Relay Selection Schemes for Relay-Assisted D2D Communications in LTE-A Systems. *IEEE Trans. Veh. Technol.* **2017**, *66*, 8303–8314. [CrossRef]
34. Chen, Z.; Smith, D. MmWave M2M Networks: Improving Delay Performance of Relaying. *IEEE Trans. Wirel. Commun.* **2021**, *20*, 577–589. [CrossRef]
35. Xia, B.; Fan, Y.; Thompson, J.; Poor, H.V. Buffering in a Three-Node Relay Network. *IEEE Trans. Wirel. Commun.* **2008**, *7*, 4492–4496. [CrossRef]
36. Gui, J.; Deng, J. Multi-Hop Relay-Aided Underlay D2D Communications for Improving Cellular Coverage Quality. *IEEE Access* **2018**, *6*, 14318–14338. [CrossRef]
37. Liu, M.; Yu, F.R.; Teng, Y.; Leung, V.C.M.; Song, M. Performance Optimization for Blockchain-Enabled Industrial Internet of Things (IIoT) Systems: A Deep Reinforcement Learning Approach. *IEEE Trans. Ind. Inf.* **2019**, *15*, 3559–3570. [CrossRef]
38. Wang, Z.; Freitas, N.D.; Lanctot, M. Dueling Network Architectures for Deep Reinforcement Learning. *PMLR* **2016**, *48*, 1995–2003.
39. Du, J.; Cheng, W.; Lu, G.; Cao, H.; Chu, X.; Zhang, Z.; Wang, J. Resource Pricing and Allocation in MEC Enabled Blockchain Systems: An A3C Deep Reinforcement Learning Approach. *IEEE Trans. Netw. Sci. Eng.* **2022**, *9*, 33–44. [CrossRef]
40. Du, J.; Yu, F.R.; Lu, G.; Wang, J.; Jiang, J.; Chu, X. MEC-Assisted Immersive VR Video Streaming over Terahertz Wireless Networks: A Deep Reinforcement Learning Approach. *IEEE Internet Things J.* **2020**, *7*, 9517–9529. [CrossRef]

Disclaimer/Publisher's Note: The statements, opinions and data contained in all publications are solely those of the individual author(s) and contributor(s) and not of MDPI and/or the editor(s). MDPI and/or the editor(s) disclaim responsibility for any injury to people or property resulting from any ideas, methods, instructions or products referred to in the content.

Communication

Absorbing Material of Button Antenna with Directional Radiation of High Gain for P2V Communication

Yuanxu Fu *, Tao Shen, Jiangling Dou and Zhe Chen

Faculty of Information Engineering and Automation, Kunming University of Science and Technology, Kunming 650032, China; shentao@kust.edu.cn (T.S.); chenzhe@kust.edu.cn (Z.C.)
* Correspondence: 20171104001@stu.kust.edu.cn; Tel.: +86-150-871-56156

Abstract: Vehicular communication systems can be used to enhance the safety level of road users by exchanging safety/warning messages. In this paper, an absorbing material on a button antenna is proposed for pedestrian-to-vehicle (P2V) communication, which provides safety service to road workers on the highway or in a road environment. The button antenna is small in size and is easy to carry for carriers. This antenna is fabricated and tested in an anechoic chamber; it can achieve a maximum gain of 5.5 dBi and an absorption of 92% at 7.6 GHz. The maximum distance of measurement between the absorbing material of the button antenna and the test antenna is less than 150 m. The advantage of the button antenna is that the absorption surface is used in the radiation layer of the antenna so that the antenna can improve the radiation direction and gain. The absorption unit size is $15 \times 15 \times 5$ mm^3.

Keywords: NZIM; ME structure; flat gain; quality factor Q

1. Introduction

The cooperative intelligent transport systems (C-ITS) pilot deployment project, which has the objective of developing and experimenting with innovative road CITS solutions, offers pedestrian-to-vehicle (P2V) communication systems for vulnerable road user (VRU) protection [1–8]. To provide a good communication service to the P2V communication system, one should ensure that a minimum broadcast distance is achieved such that an adequate road safety level can be provided [9–15]. Therefore, directive antennas have been proposed to improve connectivity and security for vehicular communication and autonomous systems [16–18]. However, making the antenna smaller makes it easier to wear or carry. The button antenna is the most comfortable technology because it can be easily integrated into human clothes and provides a stable performance due to its rigidity [19,20]. Button antennas are used in many fields, such as in Internet of Things (IOT), medicine, Global Positioning System (GPS), and wireless body area network (WBAN) applications [21–23]. Researchers have focused on developing many types of button antennas. In [24], a wearable dual-port button antenna that excites pattern-diversity dual-polarized waves was proposed for ultra-wideband (UWB) applications. In [25], shorting vias are presented as standard components in planar technologies for textile button antennas. The high-profile button antenna was demonstrated for body-centric communication in [26], which resulted in a low gain value when tested on the body. A reconfigurable snap-on button antenna module with dedicated circuitry for wearable applications has been illustrated in [27]. This button antenna is an off-body antenna, starting with the radiating part, which is a monopole fed by a microstrip line for investigating the best structure based on a dielectric resonator [28]. In [29], a novel circularly polarized button antenna was proposed, covering the unlicensed national information infrastructure (U-NII) worldwide band. To date in the discussed button antennas, the antennas have been designed with large dimensions and less gain.

In this paper, an absorbing material for a button antenna is proposed. It mainly solves two problems of an antenna: antenna miniaturization and high gain directional radiation. The absorption rate of the button antenna can reach 92% at 7.6 GHz. It has an overall size of $15 \times 15 \times 5$ mm^3 and a measured gain of 5.5 dBi. This absorbing button antenna is suitable for short-distance communication. The structure of this paper is as follows. In the next section, the overall structure of the button antenna and its material composition are introduced. In Section 3, the absorption principle of near-zero-index metamaterials (NZIMs) is analyzed. To analyze the characteristics of the structure, Section 4 improves the radiation direction of the button antenna with metamaterials, and Section 5 analyzes the gain of the button antenna. Next, the SAR, S-parameters, efficiency, and gain of the measured and simulated results are assessed. Finally, the concluding remarks from this work are presented in Section 6.

2. Configuration

As shown in Figure 1, the overall size of the antenna is $15 \times 15 \times 5$ mm^3, where the grounding and feed isolation aperture is 2 mm and the coaxial feed diameter of the antenna is 1 mm. The antenna is composed of two substrates of glass fiber epoxy (FR-4) material. The FR-4 material has the advantages of a simple structure, easy fabrication, and good physical characteristics, meaning that the antenna radiation is not constrained by its own shape, allowing it to realize a stability of radiation. The permittivity of these substrates is 4.4, with a tangential loss of 0.027 and a thickness of 1.5 mm. A coaxial feed link is used to connect the parasitic patch and radiation patch to reduce coupling and impedance matching. The parasitic patch makes the antenna reach its maximum gain with a size of 13×13 mm^2.

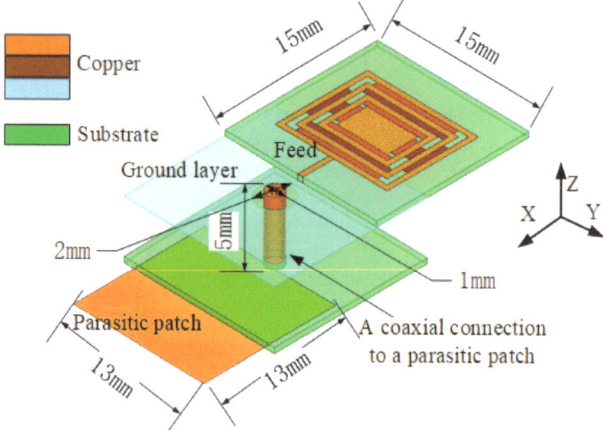

Figure 1. Button antenna using absorbing material configuration.

3. The Design and Analysis of Absorption Surface of the Button Antenna

This section mainly designs and analyzes the absorption material properties of the button antenna. Its main innovation comes from [30,31]. First, the near-zero-index metamaterial (NZIM) characteristics of a copper ring without embedded microstrip lines (CRO) and a copper ring with embedded microstrip lines (CRW) are analyzed. Second, the absorption rate and absorption angle of a copper ring with embedded microstrip lines (CRW) are analyzed.

Firstly, Figure 2a,c shows a copper ring without embedded microstrip lines (CRO) and a copper ring with embedded microstrip lines (CRW). The above two rectangular copper rings are symmetrical in structure, and the gap and microstrip line width are the same. Table 1 shows the parameter sizes of the two copper rings. Figure 2b,d shows the real and imaginary parts of the permittivity (ε), magnetic permeability (μ), and refractive index (*n*), which are approximately equal to zero by periodic electrical boundary (PEC) and periodic magnetic boundary (PMC) simulation results, which is obtained using Equations (1) and (2). The CRO and CRW belong to the NZIM. However, the coil without an embedded microstrip line (CRO) only belongs to the near-zero-index metamaterial at 7–7.6 GHz, while the inclusion of an embedded microstrip line (CRW) can improve the near-zero-index condition near 7.8 GHz.

$$z = \pm \sqrt{\frac{(1+S_{11})^2 - S_{21}^2}{(1-S_{11})^2 - S_{21}^2}} \quad (1)$$

$$n = \pm \frac{1}{kd}[\cos^{-1}(\frac{1 - S_{11}^2 + S_{21}^2}{2S_{21}})] \quad (2)$$

where S_{11} and S_{21} are reflection and transmission coefficients. *n* is the refractive index, *k* is the wave number, d is the length of the unit cell, and z is the sign of effective impedance. the permittivity is $\varepsilon = n/z$, and the magnetic permeability is $\mu = n \times z$ [32].

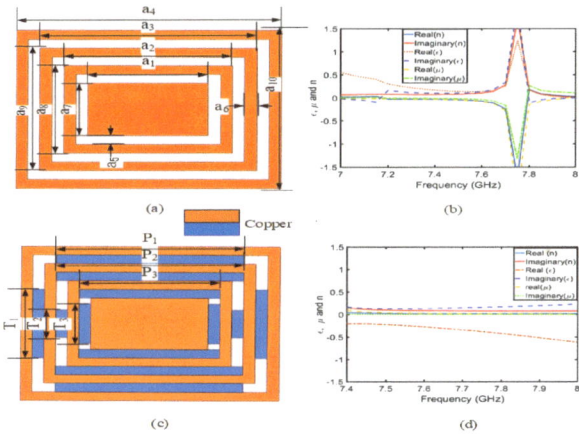

Figure 2. CRO and CRW models of NZIM: (**a**) CRO model, (**b**) CRO of ε, μ, and *n* parameters simulation, (**c**) CRW model, and (**d**) CRW of the ε, μ, and *n* parameters simulation.

Table 1. Dimensions of the CRO and CRW.

Parameter	Value (mm)	Parameter	Value (mm)
a_1	5	P_1	8
a_2	7	P_2	8
a_3	9	P_3	6
a_4	11	T_1	2.3
a_5	5	T_2	1.6
a_6	5	T_3	4
a_7	3	a_9	7
a_8	5	a_{10}	9

Secondly, Figure 3 shows the comparative absorption of CRO and CRW and the simulation model of CRW. The CRO and CRW is simulated by applying Floquet port and periodic boundary conditions in Ansys HFSS, and the absorptivity is obtained using Equation (3). In Figure 3b, the absorption of CRW is much better than that of CRO, and the absorption rate reaches 92% at 7.6 GHz. In the case of CRW, as the radiation surface of the button antenna has a certain slope, the slope absorption angle ((θ) and (ψ)) and absorption efficiency should be simulated in Figure 3a. Figure 4a,b shows that (θ) is the absorption rate after the change in the YZ slope angle, and (ψ) is the absorption rate after the change in the XY slope angle, which is the slope absorption rate of the CRW structure via PMC and PEC simulation. The slope angles vary from 0° to 60° with a step size angle of 15°. It can be observed that the proposed absorber material structure exhibits a different response at 7.6 GHz. The same absorptivity exists for different polarization angles (0°, 30°, 45°). However, when the incidence angle is 60°, the absorption rate begins to decrease, which is related to the resonant current at the resonant frequency. As shown in Figure 4c,d, the current density at an angle of 60° significantly decreases, where red indicates the strong concentration and blue indicates the weak one. Because of the change in the absorption current of the ψ = 60° angle, the absorption rate is in the complete absorption state.

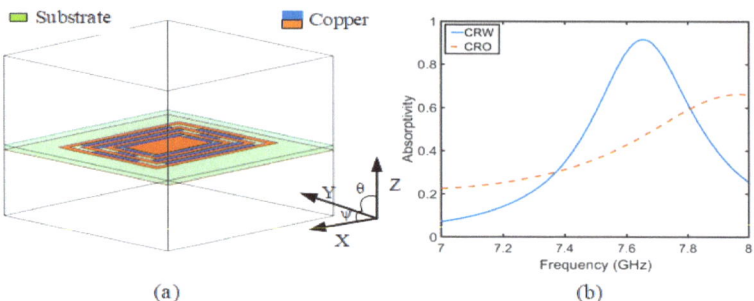

Figure 3. (a) Simulation model of CRW absorption and (b) the absorption rate of CRW and CRO.

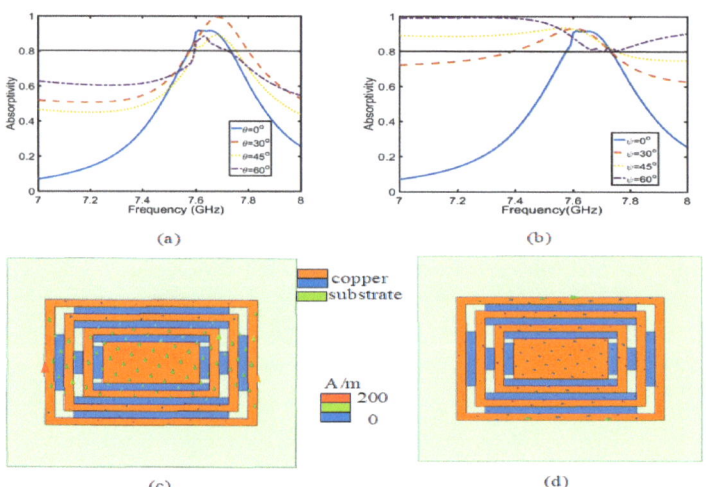

Figure 4. (a,b) are slope absorption rate of θ and ψ. (c,d) are the current distribution of θ = 60° and ψ = 60°, respectively.

$$A(\omega) = 1 - |S_{11}(\omega)|^2 - |S_{21}(\omega)|^2 \tag{3}$$

where $A(\omega)$ is the absorptivity and $|S_{11}(\omega)|^2$ and $|S_{21}(\omega)|^2$ represent the reflectivity and transitivity, respectively, with respect to frequency ω [33].

4. The Radiation Direction of Button Antenna Is Improved by the CRO Metamaterial

In this section, as shown in Figure 5, one design uses a CRO metamaterial as the radiation layer of the antenna while the other is a common patch antenna. By comparing the radiation, directivity factor (D), and quality factor (Q) of the two antennas, the CRO metamaterial can improve the radiation direction. First, the equivalent circuit method is used to analyze the relationship between the quality factor (Q) and directivity factor (D). Second, the quality factor Q is used to analyze the CRO structure, and the index parameter that has the greatest influence on Q is found by analyzing the CRO structure. In addition, it provides a theoretical basis for improving the gain in the next section.

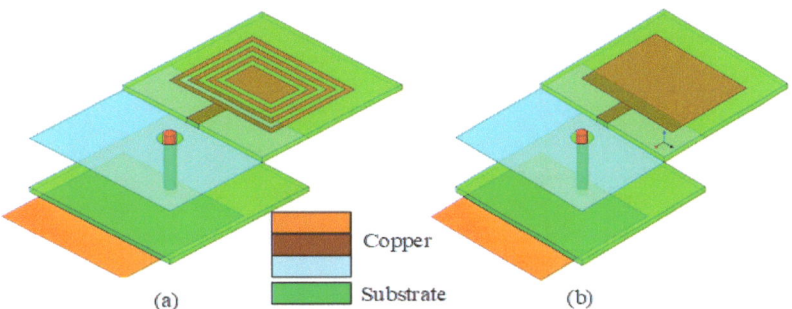

Figure 5. (a) CRO metamaterial antenna; (b) common patch antenna.

4.1. The Relationship between Q and D Is Analyzed According to the Equivalent Circuit Theory of Patch Antenna

In the equivalent circuit of the patch antenna, D is a directivity coefficient, which is defined as the ratio of the radiation intensity of the antenna in the maximum radiation direction to the average radiation intensity. The quality factor Q is a parameter that measures antenna bandwidth and gain. As shown in Equations (4) and (5), both are inversely proportional to the radiation conductance Gr; the other parameters are known numbers. Then, as shown in Figure 6, when the electromagnetic field model conforms to the TM_{01} mode and $2\lambda_0 \geq a \geq 0.35\lambda_0$ (a is the length of the antenna and λ_0 is the free space wavelength), Q and D become positive, as shown in Equation (6). The results are shown in Figure 7a,b. After 7.6 GHz, the values of D and Q of the CRO antenna increase relative to the patch antenna, wherein the radiation direction of the co-polarization changes throughout Figure 7c,d. The change in the Q value is greater than that observed in D, so it is very important to study Q.

$$D = \frac{2ab}{15\lambda_0 G_r} \quad (4)$$

$$Q = \frac{\varepsilon_r ab}{120\lambda_0 h G_r} \quad (5)$$

where a and b are the length and width of the antenna patch, respectively. G_r is the radiation conductance, ε_r is the relative dielectric constant, λ_0 is the free space wavelength, and h is the antenna height [34].

$$Q \propto D \quad (6)$$

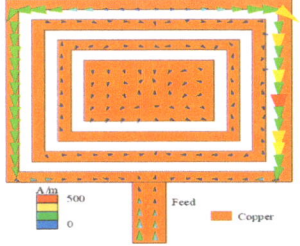

Figure 6. The surface current of the CRO metamaterial.

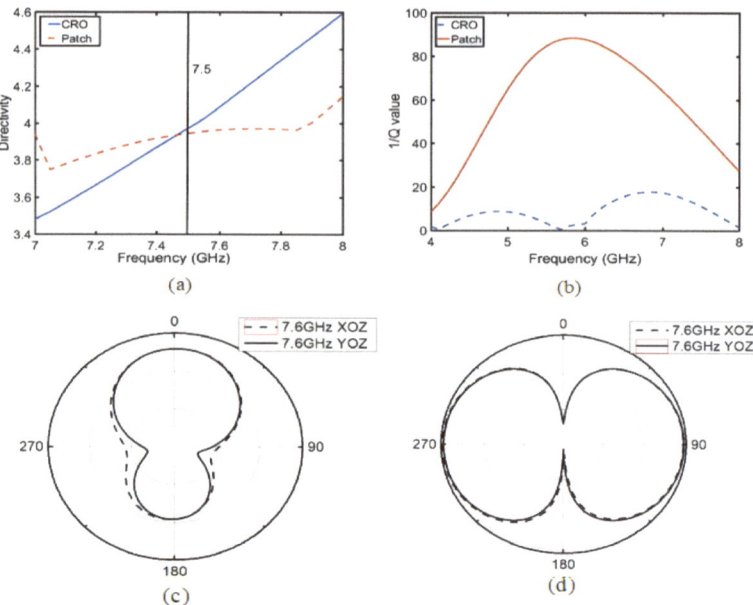

Figure 7. Comparison of antenna parameters of the CRO atennna and patch antenna: (**a**) is the directivity factor and (**b**) is the quality factor Q. At 7.6 GHz, (**c**) is the directional radiation pattern of the CRO metamaterial and (**d**) is the omnidirectional radiation pattern of the patch antenna.

4.2. The Quality Factor Q of CRO Antenna Is Analyzed by an Equivalent Circuit

The change in the Q value can affect the gain of the whole antenna. Therefore, to improve the gain of the CRO antenna, we should first analyze the factors in the CRO structure that can affect the Q value, determine which factors have the greatest influence on Q value, and finally adopt the methods that further improve the Q value to increase the antenna gain. As shown in Figure 8, it is assumed that the copper coil in the CRO structure is the real part of the impedance and that the coupling between the coils is the imaginary part. Equation (7) is used to calculate the relationship between the Q, copper coil, and coupling distance. Therefore, e_i is the coil width and d_i is the coil coupling distance. Equation (8) is established based on the principle of the microstrip line between the width of copper coil and the real part. Equation (9) is established according to the parallel relation of the real part in Figure 8, and Equation (10) is substituted into Equations (8) to obtain Equation (9); in this way, the relationship between the coil width e_i and the real part of impedance was obtained. When $e_i \geq h$ (h is the height of the coil), Z_{real} is proportional to the sum of e_i in the frequency range of 5–8 GHz, as shown in Figure 9a. Similarly, the relation between the coupling distance and imaginary part can be calculated using Equations (11) and (12). It can be concluded from Equation (12) that the imaginary part is proportional to the sum of the coupling distance d_i. The imaginary part and coupling distance shown in Figure 9b also conform to this relation. Substituting Equations (10) and (12) into Equation (7), we obtain Equation (13), which is the relationship between quality factor Q and the coupling distance d_i and coil width e_i. The value of Q is proportional to the sum of the coil widths and is inversely proportional to the coupling distance. Therefore, increasing the coil width e_i can further improve the Q value. As shown in Figure 10, the increase in the coil width e_i is much greater than the decrease in the coupling distance d_i.

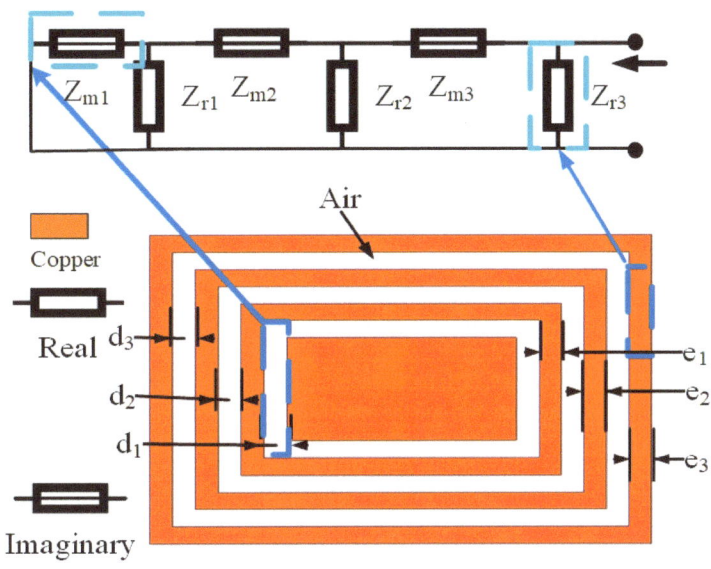

Figure 8. Equivalent circuit diagram of the CRO structure.

$$Q = \frac{Z_{real}}{Z_{im}} \qquad (7)$$

The Z_{im} is the imaginary part of the impedance and Z_{real} is the real part of the impedance [35].

$$Z_{ri}|_{i=(1,2,3)} = \frac{120\pi}{e_i/h + 2.42 - 0.44 \times h/e_i + (1 - h/e_i)^6} \qquad (8)$$

The h is the height of the coil; we are assuming that it is a constant. e_i (i = 1, 2, 3) is the width of coil [36].

$$Z_{real} = \frac{1}{Z_{r1}} + \frac{1}{Z_{r2}} + \frac{1}{Z_{r3}} \qquad (9)$$

$$Z_{real} = \frac{\sum_{i=1}^{3} e_i/h + 9.68 - 0.44 \times \sum_{i=1}^{3} h/e_i + \sum_{i=1}^{3} (1 - h/e_i)^6}{120\pi} \qquad (10)$$

$$Z_{mi}|_{i=(1,2,3)} = \frac{\varepsilon_{r1}\varepsilon_{r2}}{d_i} \qquad (11)$$

where ε_{r1} and ε_{r2} is the dielectric constant of the two materials. d_i (i = 1, 2, 3) is the coupling distance parameter of the coil, whose corresponding reactance is C_{mi} (i = 1, 2, 3). The effect of the coil area on the capacitance is ignored [37].

$$Z_{im} = \frac{1}{Z_{m1}} + \frac{1}{Z_{m2}} + \frac{1}{Z_{m3}} = \frac{\sum_{i=1}^{3} d_i}{\varepsilon_{r1}\varepsilon_{r2}} \qquad (12)$$

$$Q = \frac{\sum_{i=1}^{3} e_i}{\sum_{i=1}^{3} d_i} \qquad (13)$$

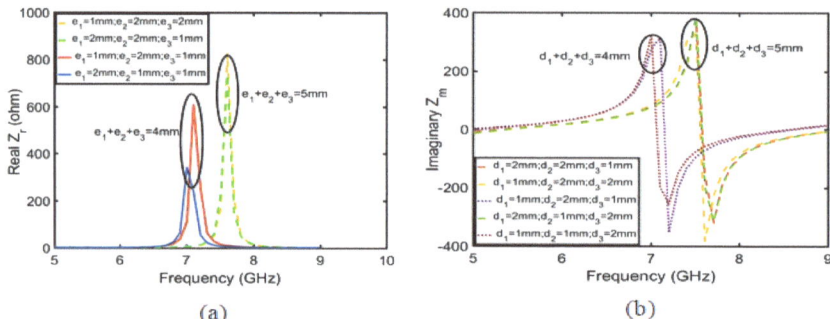

Figure 9. (**a**) is the relationship between the sum of e_i and the real part. (**b**) is the relationship between the sum of e_i and the imaginary part.

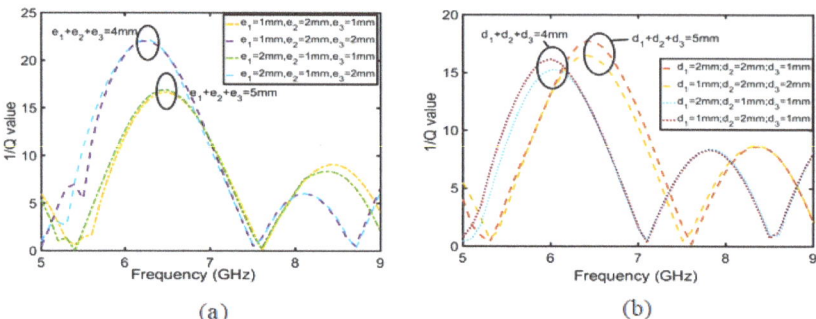

Figure 10. (**a**) is the relationship between the sum of e_i and $1/Q$. (**b**) is the relationship between the sum of d_i and $1/Q$.

5. Analysis of Gain of the CRW Structure of Button Antenna by Quality Factor Q

The main contents of this section are as follows: First, we analyze the improvement of the gain of the button antenna by quality factor Q. Second, the equivalent circuit method is used to analyze the embedded microstrip lines. Finally, the effect of its quality factor and CRW structure on the gain of the button antenna is analyzed. As seen in the previous section, the largest change in the CRO structure is the quality factor Q, as shown in Equation (14). In this way, the only parameter that affects antenna gain is the value Q. Therefore, it becomes an inevitable scheme to further improve the quality factor by embedding microstrip lines to improve the width of the copper coil of the CRO. The next step is to analyze the relationship between the T_1 length and quality factor Q and further analyze the change of the Q value after adding other embedded microstrip lines. According to Equation (7), Q is related to the real and imaginary parts of the impedance. Therefore, this chapter mainly analyzes the impedance after adding the T_1 symmetrical microstrip line. The first part is the real part analysis, and the second part is the imaginary part analysis; finally, all microstrip lines (T_2, T_3, P_1, P_2, P_3) are added to improve the antenna gain.

$$G = \frac{DQ}{Q_r} \tag{14}$$

where D is the antenna orientation coefficient; the D is constant is 6 when the antenna size is less than $0.35\lambda_0$ [38]. Q is the quality factor and Q_r is the radiation loss (Q_r is also a constant).

5.1. Analysis of Real and Imaginary Parts of Impedance with the Addition of T_1

As shown in Figure 11, the real part of the microstrip line is analyzed according to the principle of the equivalent circuit. According to the equivalent hypothesis in [39], the whole circuit is divided into the T-type circuit and π-type circuit so that the whole circuit is assumed to be a symmetrical homogeneity circuit; the change in the whole circuit after adding T_1 means adding impedance Z_{T2}. The process is as follows:

(1) The relationship between T_1 and the real part of the impedance of T-type and π-type is calculated using Equations (15)–(17);
(2) The impedance Equation (18) of the real part is listed by the circuit diagram in Figure 11;
(3) Taking the derivative of Equation (18), Equations (19) and (20) are obtained, from which it can be seen that the impedances Z_{T1} and Z_{T2} are directly proportional to the real part of the overall impedance, meaning that the length of T_1 is directly proportional to the overall impedance.

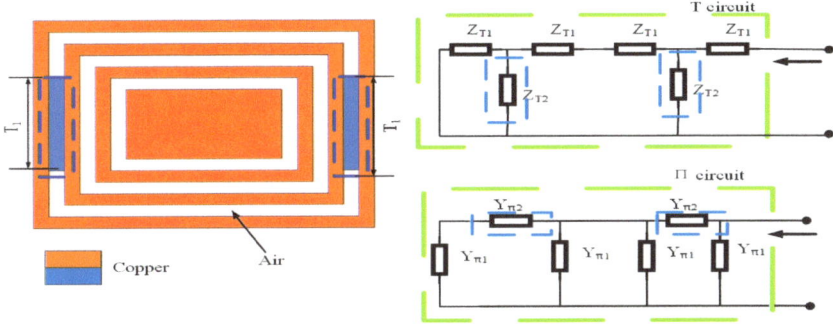

Figure 11. Real part analysis diagram of the rectangular metal ring structure after embedding T1 of the CRO.

After calculation, the following conclusion can be drawn: T_1 is inversely proportional to the real part of the impedance from 2 to 5.5 GHz in Figure 12a, which belongs to the π-type circuit; and T_1 is directly proportional to the real part of the impedance from 6.5 to 8.5 GHz in Figure 12b, which belongs to the T-type circuit. In the same way, the relation between the imaginary part of the impedance and the length of T_1 was calculated using Equations (21)–(23) and is shown in Figure 13. The conclusions are as follows: T_1 is proportional to the imaginary part impedance of the T-type and inversely proportional to the imaginary part impedance of the π-type. As shown in Figure 14, when T_1 continuously increases, the imaginary part of the impedance frequency from 2 to 5.5 GHz gradually decreases. Therefore, it is determined that this property belongs to a π-type circuit. In the same way, it belongs to a T-type circuit at 6–8.5 GHz. As shown in Figure 15, after adding T_1, Q is not improved at 2–5 GHz, whereas Q is greatly improved at 6–8 GHz. Therefore, the parameters of T_2, T_3, P_1, P_2, and P_3 are optimized based on a T-type circuit.

$$\alpha = \beta T_1 \tag{15}$$

where α is the phase, β is the attenuation coefficient of microstrip line, and T_1 is the load microstrip line length [40,41].

$$\begin{cases} Z_{T1} = Z_0 tan(\alpha/2) & Z_{T2} = Z_0[1/sin(\alpha)] \quad \text{T-type circuit} \\ Y_{\pi 1} = Y_0 tan(\alpha/2) & Y_{\pi 2} = Y_0[1/sin(\alpha)] \quad \pi\text{-type circuit} \end{cases} \tag{16}$$

The Z_0 is the microstrip line impedance constant, and Y_0 is the microstrip line admittance constant [42,43].

$$\begin{cases} Z_{T1} = Z_0 tan[(\beta T_1)/2] & Z_{T2} = Z_0[1/sin(\beta T_1)] \\ Y_{\pi 1} = Y_0 tan[(\beta T_1)/2] & Y_{\pi 2} = Y_0[1/sin(\beta T_1)] \end{cases} \tag{17}$$

$$\begin{cases} Z_{Treal} = (2Z_{T1}^2 Z_{T2} + 3Z_{T2}^2 Z_{T1})/(2Z_{T1}^2 + 4Z_{T1} + Z_{T2}^2) + Z_{T1} \\ Y_{\pi real} = (Y_{\pi 1}^3 + 4Y_{\pi 1}^2 Y_{\pi 2} + 2Y_{\pi 1} Y_{\pi 2}^2)/(4Y_{\pi}^4 + 6Y_{\pi 1} Y_{\pi 2} + 2Y_{\pi 2}^2) \end{cases} \quad (18)$$

$$\begin{cases} \partial Z_{Treal}/\partial Z_{T1} = 2Z_{T1}^2 Z_{T2}^2 + 3Z_{T2}^2 + 4Z_{T1} Z_{T2}^2 \\ \partial Z_{Treal}/\partial Z_{T2} = 4Z_{T1}^4 + 10Z_{T2}^2 Z_{T1}^2 + 12Z_{T1}^3 Z_{T2} \end{cases} \quad (19)$$

$$\begin{cases} \partial Y_{\pi real}/\partial Y_{\pi 1} = Y_{\pi 1}^4 + 12Y_{\pi 1}^3 Z_{\pi 2} + 36Y_{\pi 1}^2 Y_{\pi 2}^2 + 44Y_{\pi 1} Y_{\pi 2}^3 + 8Y_{\pi 2}^4 \\ \partial Y_{\pi real}/\partial Y_{\pi 2} = 10Y_{\pi 1}^4 + 14Y_{\pi 1}^2 Y_{\pi 2}^2 + 12Y_{\pi 1}^3 Y_{\pi 2} \end{cases} \quad (20)$$

$$\begin{cases} X_{L/2} = Z_0 tan(\alpha/2) & C_i = sin(\alpha)/Z_0 & \text{T-type circuit} \\ X_L = Z_0 sin(\alpha) & C_{i/2} = tan(\alpha/2)/Z_0 & \pi\text{-type circuit} \end{cases} \quad (21)$$

X_L is inductance of the imaginary part impedance, and C_i is the capacitance of the imaginary part impedance [44,45].

$$Z_{im} = \sqrt{\frac{X_L}{C_i}} \quad (22)$$

$$\begin{cases} Z_{imT} = Z_0/cos(\beta T_1/2) \\ Z_{im\pi} = Z_0 \cdot cos(\beta T_1/2) \end{cases} \quad (23)$$

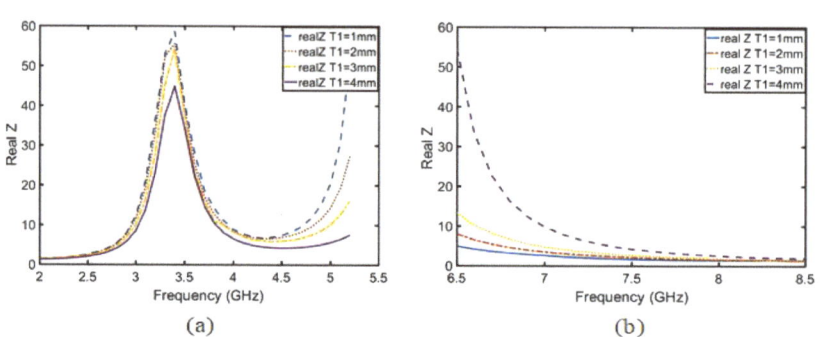

Figure 12. Real part analysis diagram of impedance after embedding T_1 at (**a**) 2–5.5 GHz and (**b**) 6.5–8.5 GHz.

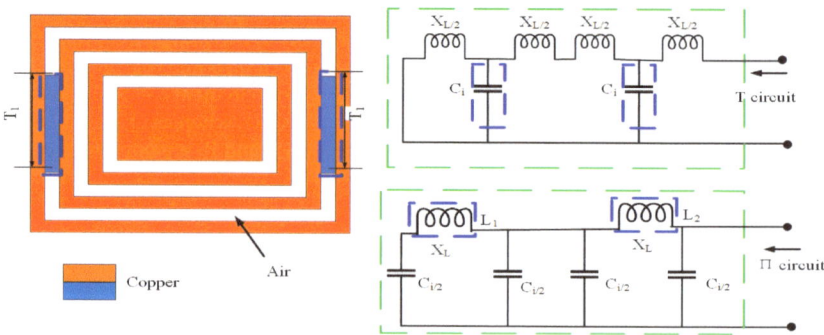

Figure 13. Imaginary part analysis diagram of the rectangular metal ring structure after embedding T_1 of the CRO.

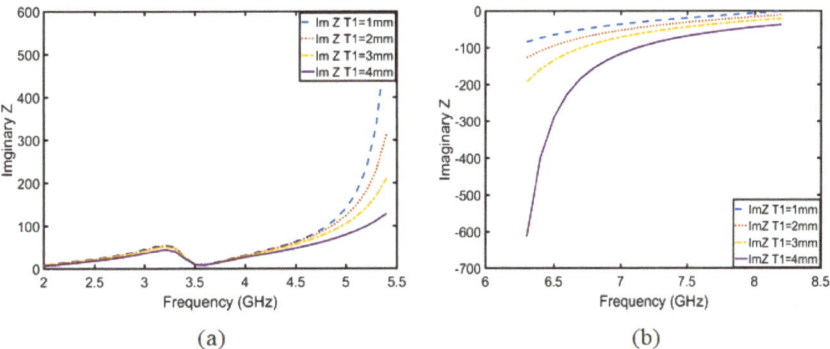

Figure 14. Imaginary part analysis diagram of impedance after embedding T_1 at (**a**) 2–5.5 GHz and (**b**) 6.5–8.5 GHz.

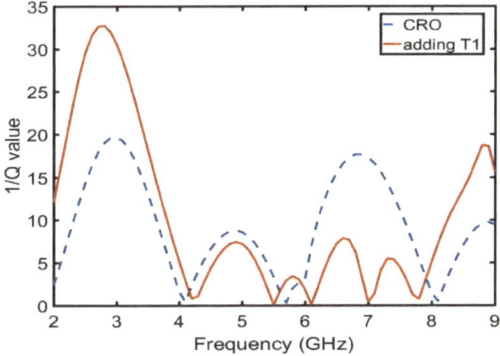

Figure 15. comparison of 1/Q values between embedding the T_1 of the CRO and CRO.

5.2. Final Result Analysis

To further improve the antenna gain, microstrip lines are continuously embedded. As shown in Figure 16, case 1, case 2, and case 3 show the current distribution. With the addition of embedded microstrip lines, the current density and intensity increase leads to an increase in antenna gain. As shown in Figure 16a, with the addition of microstrip lines T_1, T_2, T_3, P_1, P_2, and P_3, the value of 1/Q is progressively decreasing; therefore, we have to optimize for the smallest 1/Q, which is the maximum value of Q. HFSS software was adopted for the optimization, and the minimum value of 1/Q at 7.6 GHz is shown in Figure 16b. Figure 16c verifies that the maximum value of Q is the maximum value of the antenna gain by gain of the co-planar polarization.

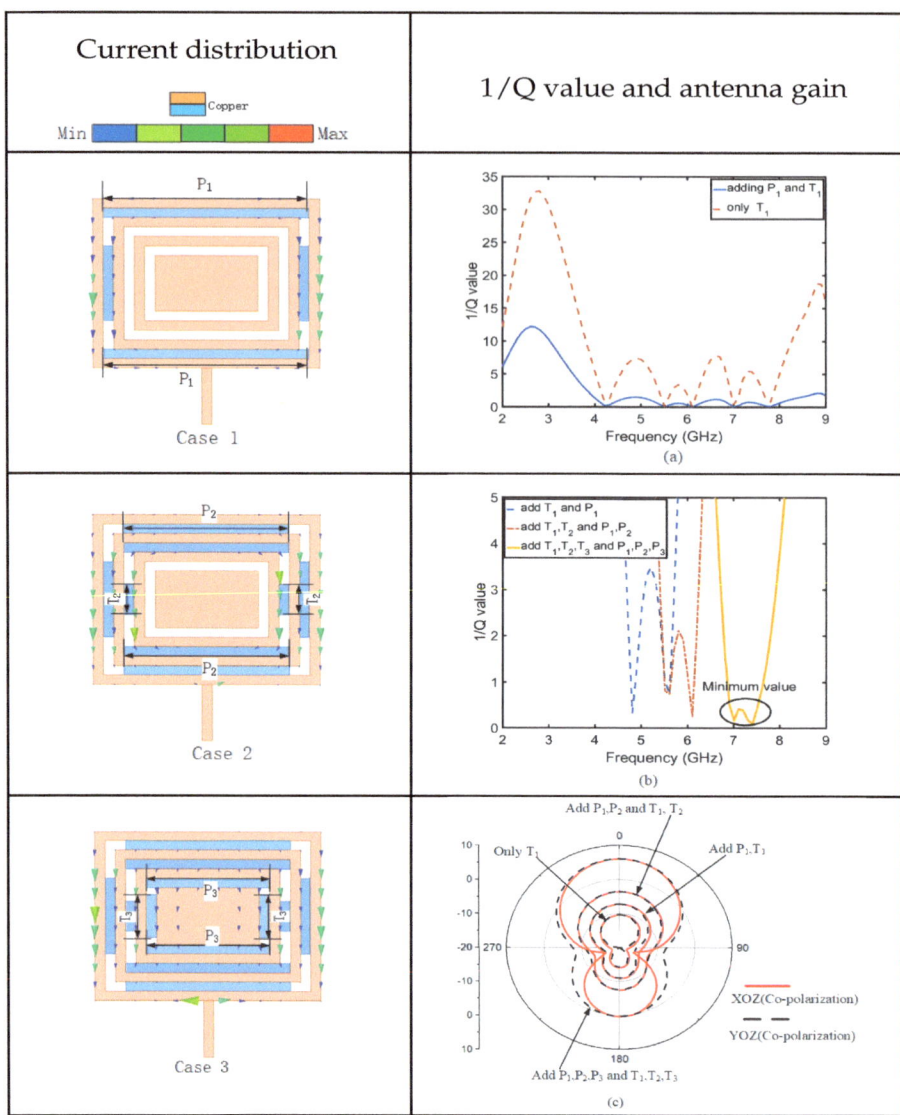

Figure 16. CRW analysis: adding the P_1 current distribution (case 1); adding the P_2 and T_2 current distribution (case 2); adding the P_3 and T_3 current distribution (case 3); comparison diagram of 1/Q after adding P_1 (**a**); the minimum value of 1/Q after CRW optimization (**b**); maximum gain after CRW optimization (**c**).

6. Measured and Simulated Results

In this section, the performance of the button antenna is evaluated on the human body. The absorptivity, S-parameters, efficiency, and radiation pattern are analyzed together with the SAR values. The scenario of the numerical model is shown in Figure 17. The cubic tissue model consists of a 1.7 mm thick skin layer, a 5 mm thick fat layer, a 20 mm thick muscle layer, and a 13 mm thick bone layer with a size of 150×150 mm^2. All of the material parameters of the tissue were obtained from the values available in the CST Studio human tissues library. A 5 mm air gap was set between the antenna and the surface of the skin to mimic any worn clothes. An R&S microwave signal generator was used to test the S_{11} parameters. The gain was measured using the antenna anechoic chamber with an signal generator R&S microwave signal generator, which has a maximum output power of 11 dBm. Figure 18 shows the physical antenna and measurement environment.

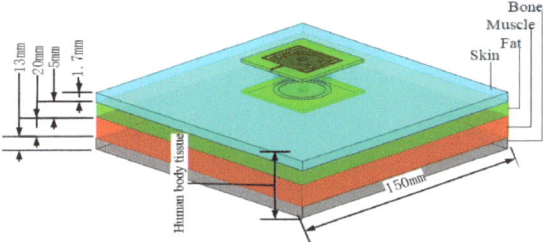

Figure 17. Four-layered human tissue.

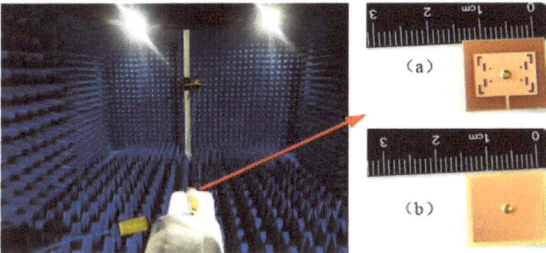

Figure 18. The (**a**) front and (**b**) back of the antenna object in test environment.

6.1. Specific Absorption Rate (SAR) and Absorptivity Property Analysis

For wearable applications, the SAR value is a critical parameter. The numerical SAR distributions of the proposed antennas are shown in Figure 19. The calculation is based on the IEEE C 95.3 standards, and the simulated power is 200 mW. According to the European standard, the SAR results should be under 2 W/kg averaged over 10 g of tissue. As shown in Figure 19, (a) the SAR peak value is 0.18 W/kg at 7.6 GHz and (b) the SAR peak value is 0.10 W/kg at 7.8 GHz, both of which meet the standard. The absorption efficiency is calculated by measuring the S-parameter of the antenna radiation surface. The absorption efficiency measurement is basically consistent with the simulation in Figure 20. In Table 2, compared with other button antennas, this button antenna has a smaller SAR and has less influence on body tissues.

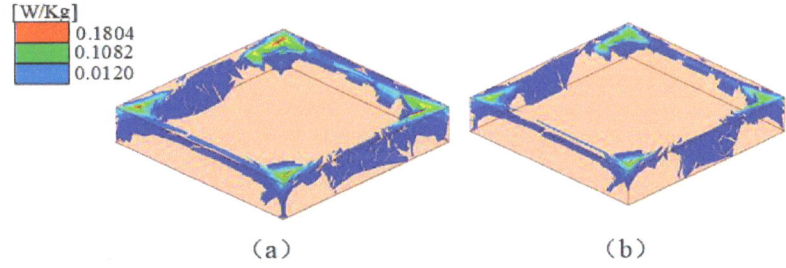

Figure 19. Simulated SAR of the button antenna on the human body at (**a**) 7.6 GHz and (**b**) 7.8 GHz.

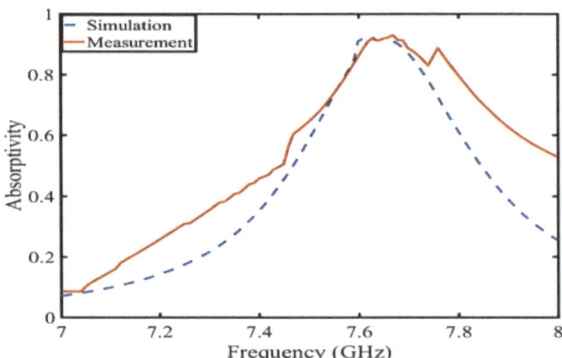

Figure 20. Simulation and measurement of absorptivity.

Table 2. Performance comparison with the literature.

Reference	$\lambda_0 \times \lambda_0 \times h$ mm³	BW (%)	Radiation Pattern	Gain (dBi)	SAR (W/kg)	Efficiency (%)
[20]	0.32×0.30×1	31	O	2.1	-	78
[21]	0.27×0.27	2.1	D	2.5	1.6	82
[22]	0.32×0.30×1.8	5	D	3.3	0.5	75
[23]	0.34×0.34×6.5	5	O	2	-	84
[24]	0.36×0.36×4.0	15	D	2	1.0	90
[25]	0.45×0.45×1.2	5.3	D	4.2	0.74	78
[26]	2.5×2.5×0.2	3	D	3.9	-	77
[27]	1.5×1.5×1.2	4	D	3.0	-	75
[28]	1.2×1.5×0.5	5	O	2.8	-	70
[29]	2.1×2.1×0.5	8	D	1.16	1.12	78
This work	0.24×0.24×5	4	D	5.5	0.18	92

O is the omnidirectional radiation and D is the directional radiation.

6.2. Analysis of S-Parameters, Efficiency, and Radiation Pattern on the Body, Free Space and Textile

As shown in Figure 21, (a) is the S-parameter value of the button antenna in different wearing substances, such as free space, arms of the human body, and cloth textiles. Judging from the S-parameter, this does not have much influence on the body tissue and clothing textile. The effect of the antenna radiation on the free space, human body, and cloth textiles is shown in Figure 21b–d. Although the human body and clothes improve antenna radiation and gain, the increase is 0.1 dBi. That is, it has no effect on the antenna radiation itself. As shown in Figure 22, other peak gains and the efficiency are unchanged, so wearing the button antenna with different substances has no influence on the performance and absorption. Compared with other button antennas, this button antenna has a small size, high gain, and good efficiency, as shown in Table 2. In other words, the antenna provides a good stable performance.

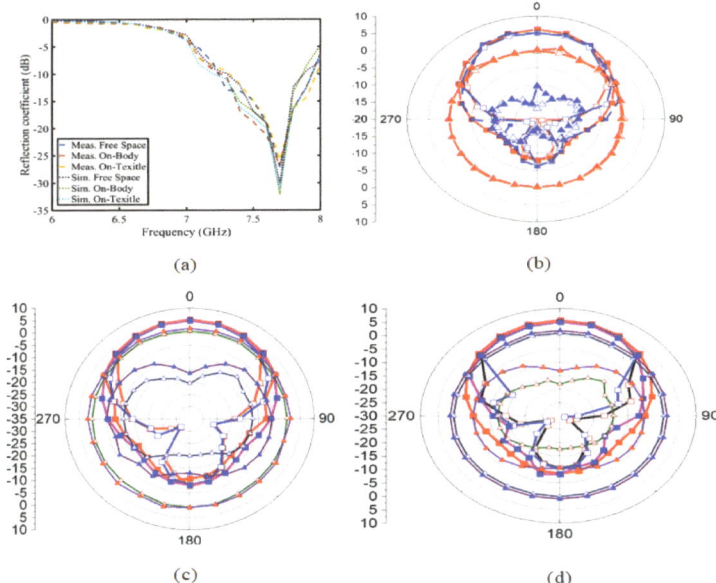

Figure 21. (**a**) is the S_{11} simulation and measurement diagram. (**b**–**d**) are the radiation simulation and measurement of the button antenna on the free space, on-body, and on-textile regions. (□ is the co-polarization simulation XOZ, ■ is the co-polarization measure XOZ, □ is the co-polarization simulation YOZ, ■ is the co-polarization measure YOZ, △ is the cross-polarization simulation XOZ, ▲ is the cross-polarization measure XOZ, △ is the cross-polarization simulation YOZ, and ▲ is the co-polarization measure YOZ).

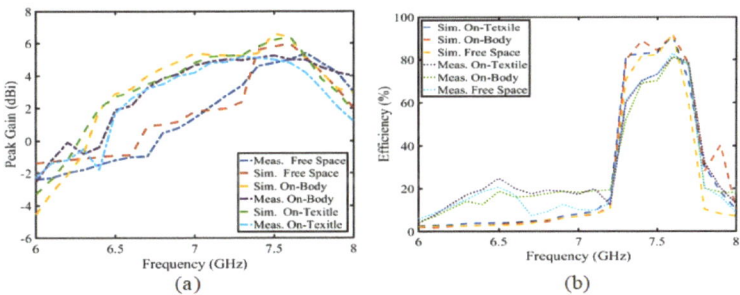

Figure 22. Simulation and measurement: (**a**) peak gain and (**b**) efficiency.

7. Conclusions

An absorbing material on a button antenna for P2V communication was proposed, fabricated, and measured, which provides safety service to road workers on highways or in a road environment. The maximum distance of measurement between the absorbing material of the button antenna and the test antenna is less than 150 m, which is a safe distance for traffic. This antenna can achieve a maximum gain of 5.5 dBi and absorption of 92% at 7.6 GHz. Compared with other similar antennas in Table 2, the proposed antenna is very compact in size and achieves an excellent gain and reasonable efficiency; the bandwidth is only 4%. The modified absorbing material structure and detail considerations for wearable scenarios make it a very suitable candidate for on-body applications.

Author Contributions: Conceptualization, Y.F.; methodology, Y.F.; investigation, T.S.; writing—original draft preparation, Y.F. and J.D.; writing—review and editing, Z.C. All authors have read and agreed to the published version of the manuscript.

Funding: This research was funded by Yunnan Fundamental Research project of fund grant number 202101AU070164, Yunnan Fundamental Research Projects, No. 202301AV07003; National Natural Science Foundation of China, No. 61971208; College personnel training project; Yunnan Young Top Talents of Ten Thousands Plan (Shen Tao, Yunren Social Development No. 2018 73); Yunnan Key Laboratory of Computer Technologies Application Open Fund.

Institutional Review Board Statement: Not applicable.

Informed Consent Statement: Not applicable.

Data Availability Statement: Not applicable.

Conflicts of Interest: The authors declare no conflict of interest.

References

1. Wu, Q.; Zhao, Y.; Fan, Q. Time-Dependent Performance Modeling for Platooning Communications at Intersection. *IEEE Internet Things J.* **2022**, *9*, 18500–18513. [CrossRef]
2. Wu, Q.; Zheng, J. Performance Modeling and Analysis of IEEE 802.11 DCF Based Fair Channel Access for Vehicle-to-Roadside Communication in a Non-Saturated State. *Wirel. Netw.* **2014**, *21*, 1–11. [CrossRef]
3. Wang, K.; Yu, F.R.; Wang, L.; Li, J.; Zhao, N.; Guan, Q.; Li, B.; Wu, Q. Interference Alignment With Adaptive Power Allocation in Full-Duplex-Enabled Small Cell Networks. *IEEE Trans. Veh. Technol.* **2019**, *68*, 3010–3015. [CrossRef]
4. Wu, Q.; Shi, S.; Wan, Z.; Fan, Q.; Fan, P.; Zhang, C. Towards V2I Age-aware Fairness Access: A DQN Based Intelligent Vehicular Node Training and Test Method. *Chin. J. Electron.* **2022**, *46*, 90–93.
5. Wu, Q.; Zhao, Y.; Fan, Q.; Fan, P.; Wang, J.; Zhang, C. Mobility-Aware Cooperative Caching in Vehicular Edge Computing Based on Asynchronous Federatedand Deep Reinforcement Learning. *IEEE J. Sel. Top. Signal Process.* **2022**, *17*, 66–81. [CrossRef]
6. Wu, Q.; Wang, X.; Fan, P.; Fan, Q.; Zhang, C.; Li, Z. High Stable and Accurate Vehicle Selection Scheme based on Federated Edge Learning in Vehicular Networks. *China Commun.* **2023**, *20*, 1–17. [CrossRef]
7. Wu, Q.; Ge, H.; Fan, P.; Wang, J.; Fan, Q.; Li, Z. Time-Dependent Performance Analysis of the 802.11 p-Based Platooning Communications Under Disturbance. *IEEE Trans. Veh. Technol.* **2020**, *69*, 15760–15773. [CrossRef]
8. Wu, Q.; Liu, H.; Zhang, C.; Fan, Q.; Li, Z.; Wang, K. Trajectory Protection Schemes Based on a Gravity Mobility Model in IoT. *Electronics* **2019**, *8*, 148. [CrossRef]
9. Wu, Q.; Xia, S.; Fan, P.; Fan, Q.; Li, Z. Velocity-Adaptive V2I Fair-Access Scheme Based on IEEE 802.11 DCF for Platooning Vehicles. *Sensors* **2018**, *18*, 4198. [CrossRef]
10. Wu, Q.; Zheng, J. Performance Modeling of IEEE 802.11 DCF Based Fair Channel Access for Vehicular-to-Roadside Communication in a Non-Saturated State. In Proceedings of the 2014 IEEE International Conference on Communications March, Sydney, NSW, Australia, 10–14 June 2014; pp. 10–14.
11. Wu, Q.; Zheng, J. Performance Modeling and Analysis of the ADHOC MAC Protocol for Vehicular Networks. *Wirel. Netw.* **2016**, *22*, 799–812. [CrossRef]
12. Wu, Q.; Zheng, J. Performance modeling and analysis of the ADHOC MAC protocol for VANETs. In Proceedings of the 2015 IEEE International Conference on Communications (ICC), London, UK, 8–12 June 2015; pp. 3646–3652.
13. Long, D.; Wu, Q.; Fan, Q.; Fan, P.; Li, Z.; Fan, J. A Power Allocation Scheme for MIMO-NOMA and D2D Vehicular Edge Computing Based on Decentralized DRL. *Sensors* **2023**, *23*, 3449. [CrossRef]
14. Jing, F.; Wu, Q.; Hao, J.F. Optimal deployment of wireless mesh sensor networks based on Delaunay triangulations. In Proceedings of the 2010 International Conference on Information, Networking and Automation (ICINA), Kunming, China, 18–19 October 2010; pp. 23–28.
15. Fan, J.; Yin, S.T.; Wu, Q.; Gao, F. Study on Refined Deployment of Wireless Mesh Sensor Network. In Proceedings of the 2010 6th International Conference on Wireless Communications Networking and Mobile Computing (WiCOM), Chengdu, China, 23–25 September 2010; pp. 1–5.
16. Duraj, D.; Rzymowski, M.; Nyka, K.; Kulas, L. Espar antenna for v2x applications in 802.11 p frequency band. In Proceedings of the 2019 13th European Conference on Antennas and Propagation (EuCAP), Krakow, Poland, 31 March–5 April 2019; pp. 1–4.
17. Maliatsos, K.; Marantis, L.; Bithas, P.S.; Kanatas, A.G. Hybrid multi-antenna techniques for v2x communications prototyping and experimentation. *Multidiscip. Digit. Publ. Inst.* **2020**, *1*, 80–95. [CrossRef]
18. Foysal, M.F.; Mahmud, S.; Baki, A. A novel high gain array antenna design for autonomous vehicles of 6g wireless systems. In Proceedings of the 2021 International Conference on Green Energy, Computing and Sustainable Technology (GECOST), Miri, Malaysia, 7–9 July 2021; pp. 1–5.

19. Xiaomu, H.; Yan, S.; Vandenbosch, G.A.E. Wearable Button Antenna for Dual-Band WLAN Applications with Combined on and off-Body Radiation Patterns. *IEEE Trans. Antennas Propag.* **2017**, *65*, 1384–1387. [CrossRef]
20. Yin, X.; Chen, S.J.; Fumeaux, C. Wearable Dual-Band Dual-Polarization Button Antenna for WBAN Applications. *IEEE Antennas Wirel. Propag. Lett.* **2020**, *19*, 2240–2244. [CrossRef]
21. Sambandam, P.; Kanagasabai, M.; Natarajan, R.; Alsath, M.G.N.; Palaniswamy, S. Miniaturized Button-Like WBAN Antenna for Off-Body Communication. *IEEE Trans. Antennas Propag.* **2020**, *68*, 5228–5235. [CrossRef]
22. Hu, X.; Yan, S.; Zhang, J.; Volski, V.; Vandenbosch, G.A.E. Omni-Directional Circularly Polarized Button Antenna for 5 GHz WBAN Applications. *IEEE Trans. Antennas Propag.* **2022**, *69*, 5054–5059. [CrossRef]
23. Saeidi, T.; Karamzadeh, S. A miniaturized multi-frequency wide-band leaky wave button antenna for ISM/5G communications and WBAN applications. *Radio Sci.* **2023**, *58*, 1–18. [CrossRef]
24. Le, T.T.; Kim, Y.-D.; Yun, T.-Y. Wearable Pattern-Diversity Dual-Polarized Button Antenna for Versatile On-/Off-Body Communications. *IEEE Access* **2022**, *10*, 98700–98711. [CrossRef]
25. Dang, Q.H.; Chen, S.J.; Zhu, B.; Fumeaux, C. Shorting Strategies for Wearable Textile Antennas: A review of four shorting methods. *IEEE Antennas Propag. Mag.* **2022**, *64*, 84–98. [CrossRef]
26. Ali, S.M.; Sovuthy, C.; Noghanian, S.; Abbasi, Q.H.; Asenova, T.; Derleth, P.; Casson, A.; Arslan, T.; Hussain, A. Low-profile Button Sensor Antenna Design for Wireless Medical Body Area Networks. In Proceedings of the Annual International Conference of the IEEE Engineering in Medicine & Biology Society (EMBC), Glasgow, UK, 11–15 July 2022; pp. 4618–4621.
27. Dang, Q.H.; Chen, S.J.; Fumeaux, C. Modular Wearable Textile Antenna with Pattern-Interchangeability using Snap-on Buttons. In Proceedings of the International Symposium on Antennas and Propagation (ISAP), Sydney, Australia, 31 October–3 November 2022; pp. 7–8.
28. Mersani, A.; Ribero, J.-M.; Osman, L. Small Button Antenna for Wearable applications. In Proceedings of the International Conference on Microelectronics (ICM), Casablanca, Morocco, 4–7 December 2022; pp. 78–81.
29. Hu, X.; Yan, S.; Vandenbosch, G.A.E. Compact Circularly Polarized Wearable Button Antenna with Broadside Pattern for U-NII Worldwide Band Applications. *IEEE Trans. Antennas Propag.* **2018**, *67*, 1341–1345. [CrossRef]
30. Coelho, C.A.T.; da Silva, M.W.B.; Magri, V.P.R.; de Matos, L.J. Application of thin dual-band electromagnetic absorbers to reduce the radar cross section of a microstrip patch antenna. In Proceedings of the 2020 IEEE MTT-S Latin America Microwave Conference (LAMC 2020), Cali, Colombia, 26–28 May 2021; pp. 1–4.
31. Chen, Z.; Shen, Z. Dual-Polarized Angle-Selective Surface Based on Three-Layer Frequency Selective Surfaces. In Proceedings of the 2022 IEEE International Symposium on Antennas and Propagation and USNC-URSI Radio Science Meeting (AP-S/URSI), Denver, CO, USA, 10–15 July 2022; pp. 910–911.
32. Deshmukh, R.; Marathe, D.; Kulat, K.D. Microstrip Patch Antenna Gain Enhancement using Near-zero Index Metamaterial Superstrate (NZIM Lens). In Proceedings of the 2019 10th International Conference on Computing, Communication and Networking Technologies, Kanpur, India, 6–8 July 2019; pp. 124–127.
33. Garg, P.; Jain, P. Isolation Improvement of MIMO Antenna Using a Novel Flower Shaped Metamaterial Absorber at 5.5 GHz WiMAX Band. *IEEE Trans. Circuits Syst. II Express Briefs* **2020**, *4*, 675–679. [CrossRef]
34. Xu, S.; Wang, Y.; Zhou, Q.; Cheng, L.; Cao, P.; Dou, J. A four-band wide-angle metamaterial absorber at microwave frequency. In Proceedings of the 2017 Sixth Asia-Pacific Conference on Antennas and Propagation (APCAP), Xi'an, China, 16–19 October 2017; pp. 1–3.
35. Luo, J.; Lai, Y. Photonic-doped Zero-index Media as Coherent Perfect Absorbers. In Proceedings of the 2018 Cross Strait Quad-Regional Radio Science and Wireless Technology Conference (CSQRWC), Xuzhou, China, 21–24 July 2018; pp. 1–2.
36. Luo, J.; Lu, W.; Hang, Z.; Chen, H.; Hou, B.; Lai, Y.; Chan, C.T. Arbitrary Control of Electromagnetic Flux in Inhomogeneous Anisotropic Media with Near-Zero Index. *Phys. Rev. Lett.* **2014**, *112*, 073903. [CrossRef] [PubMed]
37. Virushabadoss, N.; Henderson, R. Quality Factor of an Electrically Small Planar Slot Antenna with Different Matching Networks. In Proceedings of the 2019 IEEE Texas Symposium on Wireless and Microwave Circuits and Systems (WMCS), Waco, TX, USA, 28–29 March 2019; pp. 1–4.
38. Kaya, T.; Nesimoglu, T. Broadband Impedance Transformation by Defected Dielectric on Microstrip Lines. In Proceedings of the 2018 18th Mediterranean Microwave Symposium (MMS), Istanbul, Turkey, 31 October–2 November 2018; pp. 175–178.
39. Musa, S.M.; Sadiku, M.N.O.; Momoh, O.D. Finite element method for calculating capacitance and inductance of symmetrical coupled microstrip lines. In Proceedings of the 2012 Proceedings of IEEE Southeastcon, Orlando, FL, USA, 15–18 March 2012; pp. 1–4.
40. Vestenický, P. Analysis of Parasitic and Loss Elements Influence on RFID Loop Antenna Quality Factor. In Proceedings of the 2020 Cybernetics & Informatics (K&I), Velke Karlovice, Czech Republic, 29 January–1 February 2020; pp. 1–4.
41. Fu, S.; Zhao, X.; Li, C.; Wang, Z. Dual-band and omnidirectional miniaturized planar composite dipole antenna for WLAN applications. *Int. J. Microw. Comput.-Aided Eng.* **2021**, *31*, 22863. [CrossRef]
42. Fu, S.; Xiong, A.; Chen, W.; Fang, S. A compact planar Moxon-Yagi composite antenna with end-fire radiation for dual-band applications. *Microw. Opt. Technol. Lett.* **2022**, *62*, 2328–2334. [CrossRef]
43. Fu, S.; Kong, Q.; Fang, S.; Wang, Z. Broadband Circularly Polarized Microstrip Antenna with Coplanar Parasitic Ring Slot Patch for L-Band Satellite System Application. *IEEE Antennas Wirel. Propag. Lett.* **2014**, *13*, 943–946.

44. Fu, S.; Kong, Q.; Li, C. New design of helical antenna array for L-band land mobile satellite communications. In Proceedings of the 2014 3rd Asia-Pacific Conference on Antennas and Propagation, Harbin, China, 26–29 July 2014; pp. 37–39.
45. Fu, S.; Kong, Q.; Fang, S.J.; Wang, Z. Optimized design of helical antenna with parasitic patch for L-band satellite communications. *Prog. Electromagn. Res. Lett.* **2014**, *44*, 9–13. [CrossRef]

Disclaimer/Publisher's Note: The statements, opinions and data contained in all publications are solely those of the individual author(s) and contributor(s) and not of MDPI and/or the editor(s). MDPI and/or the editor(s) disclaim responsibility for any injury to people or property resulting from any ideas, methods, instructions or products referred to in the content.

Communication

Stereoscopic UWB Yagi–Uda Antenna with Stable Gain by Metamaterial for Vehicular 5G Communication

Yuanxu Fu *, Tao Shen, Jiangling Dou and Zhe Chen

Faculty of Information Engineering and Automation, Kunming University of Science and Technology, Kunming 650032, China; jianglingdou@kust.edu.cn (L.D.); chenzhe@kust.edu.cn (Z.C.)
* Correspondence: 20171104001@stu.kust.edu.cn; Tel.: +86-150-871-56156

Abstract: In this paper, a stereoscopic ultra-wideband (UWB) Yagi–Uda (SUY) antenna with stable gain by near-zero-index metamaterial (NZIM) has been proposed for vehicular 5G communication. The proposed antenna consists of magneto-electric (ME) dipole structure and coaxial feed patch antenna. The combination of patch antenna and ME structure allows the proposed antenna can work as a Yagi–Uda antenna, which enhances its gain and bandwidth. NZIM removes a pair of C-notches on the surface of the ME structure to make it absorb energy, which results in two radiation nulls on both sides of the gain passband. At the same time, the bandwidth can be enhanced effectively. In order to further improve the stable gain, impedance matching is achieved by removing the patch diagonally; thus, it is able to tune the antenna gain of the suppression boundary and open the possibility to reach the most important characteristic: a very stable gain in a wide frequency range. The SUY antenna is fabricated and measured, which has a measured −10 dBi impedance bandwidth of approximately 40% (3.5–5.5 GHz). Within it, the peak gain of the antenna reaches 8.5 dBi, and the flat in-band gain has a ripple lower than 0.5 dBi.

Keywords: NZIM; ME structure; flat gain; quality factor Q

Citation: Fu, Y.; Shen, T.; Dou, J.; Chen, Z. Stereoscopic UWB Yagi-Uda Antenna with Stable Gain by Metamaterial for Vehicular 5G Communication. *Sensors* **2023**, *23*, 4534. https://doi.org/10.3390/s23094534

Academic Editors: Peter Chong, Pingyi Fan and Qiong Wu

Received: 10 March 2023
Revised: 23 April 2023
Accepted: 30 April 2023
Published: 6 May 2023

Copyright: © 2023 by the authors. Licensee MDPI, Basel, Switzerland. This article is an open access article distributed under the terms and conditions of the Creative Commons Attribution (CC BY) license (https://creativecommons.org/licenses/by/4.0/).

1. Introduction

The 5G is the current generation of mobile communications expected to provide higher data rates and high-speed connectivity for multimedia applications [1–5]. In addition to mobile communication, there are more usages of the 5G communication systems such as the vehicle-to-everything (V2X) technique [6,7]. On one hand, the system integration of the terrestrial and satellite communication [8,9] can provide global seamless services. It is the development vision of the next generation wireless network [10,11]. On the other hand, the 5G communication system is convenient for communication between vehicles and base stations [12,13]. In order to reduce the interference between communication systems, an antenna with high gain and good direction is urgently needed [14,15]. Among them, the Yagi antenna is particularly prominent on account of its advantages [16], such as its light weight and simple structure [17–20]. However, it usually has a narrow operating band which cannot meet the multi-frequency operational requirements of wireless communication. Accordingly, many efforts have been made to enlarge the impedance bandwidth. In [21,22], impedance matching is adopted to improve the bandwidth of the antenna. In [23,24], this operational mode adopts a ground slot to solve the bandwidth problem of the bandwidth of antennas. In [25], bandwidth enhancement adopts a semielliptical monopole as a driven element, a half bow-tie-shaped truncated ground reflector, and elliptical director elements. The above structure can not enhance the gain while improving the bandwidth.

In this paper, the magneto-electric dipole (ME) structure is used to improve the bandwidth of the antenna by impedance matching and to enhance the antenna gain by electromagnetic superposition. In order to further improve the gain of the Yagi antenna, metamaterials are proposed below to improve the Yagi antenna gain. In [26], enhanced and stable gain are achieved by a wideband metematerial-loaded loop. In [20], to improve the in-band gain and enhance the stable gain capability, a new coplanar NZIM with wideband NZI characteristic and dual transmission notches in the lower and upper stopbands of the preliminary gain passband is introduced to the Yagi antenna. In [27],the four-step slotline along the center of the antenna is used to improve the impedance matching, and the metasurface is used to enhance the gain. The above method can improve the stable gain, but the bandwidth is narrow. The narrow bandwidth causes frequency shift, efficiency, and the voltage of the antenna receiving drop, which affects the communication quality.

In this paper, we propose the SUY antenna which has ultra-wideband and stable gain for the RF energy harvesting to obtain a stable output voltage and conversion efficiency. The following three steps are used to achieve the characteristics of the ultra-wideband and stable gain: First, the ME structure is used to achieve the reinforcement of broadband and gain. Second, a pair of C-notches is cut out of the surface of the ME structure to make a near-zero-index metamaterial (NZIM). This metamaterial structure not only produces out-of-band radiation suppression by energy absorption but also increases the bandwidth of the antenna. Finally, in order to further improve the stable gain, the director and driven element achieve impedance matching for enhancement of the gain of the suppression boundary. To show the advantages of the SUY antenna stable gain, as shown in Table 1, the advantages of the SUY antenna are stable and high gain, while the disadvantage is that the three-dimensional dimension is larger. Compared with wideband Yagi antenna, it has high gain characteristics, and compared with stable gain antenna, it has wide bandwidth and high gain. We use $\triangle G$ to signify the difference value of gain and BW to signify bandwidth. $\triangle G/BW$ is defined as the standard of the stable gain of the antenna [27].

Table 1. Performance Comparison with the Literature.

Reference	$\lambda_0 \times \lambda_0 \times h$ (mm^2)	BW(GHz)& $\triangle$BW	G& $\triangle$G (dBi)	$\triangle$G/$\triangle$BW (dBi/GHz)
[24]	0.5×0.63 $^{(p)}$	5.0–7.5 & 3.5	3.5–5.5 & 2	-
[25]	0.64×0.64 $^{(p)}$	0.8–3.0 & 2.2	4–5.2 & 1.2	-
[26]	$0.48 \times 0.69 \times 1$ $^{(s)}$	1.8–2.8 & 1	6.5–9 & 2.5	2.5
[20]	1.2×1.1 $^{(p)}$	8.5–9.5 & 1	5.0–8.3 & 3.3	2
[27]	0.9×1.1 $^{(p)}$	8.5–10 & 1.5	5.5–7 & 1.5	1
This work	$0.7 \times 0.7 \times 0.05$ $^{(s)}$	3.5–5.5 & 2.0	7.7–8.7 & 1.0	0.5

(p) is plane structure and (s) is the stereoscopic structure.

2. Configuration

Figure 1 shows the design configuration of the stereoscopic UWB Yagi–Uda antenna (length: 60 mm; width: 60 mm; height: 8.5 mm). The antenna consists of four parts: the driven element, director layer, ME structure, and coaxial feed. The ME structure is the reflection layer. Patch 1 and patch 2 are the driven element and director of the SUY antenna, respectively. They are the same size by impedance optimization. The two substrates are made of glass-fiber epoxy resin copper-clad laminate (FR4-epoxy). The permittivity of these substrates is 4.4, with a tangential loss of 0.0027 and a thickness of 1.5 mm. As shown in Figure 2, the ME structure is made of copper material. An F4B Rogers PCB was used for the ground substrate, which has a dielectric permittivity of 2.65, tangential loss of 0.001, and thickness of 2 mm. Table 2 shows the specific component dimension of the SUY antenna. As can be seen from Figure 2, its structure is relatively complex, but the raw materials are economical and popular. So this antenna is used in car wireless networks.

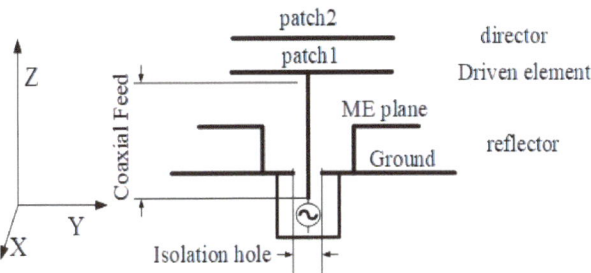

Figure 1. Design configuration of stereoscopic UWB Yagi–Uda.

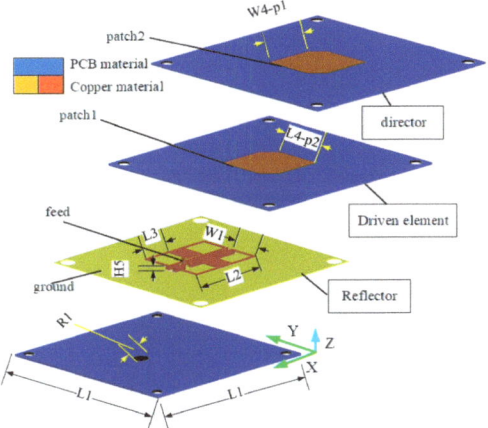

Figure 2. Processing drawing of the vertical plane of the stereoscopic UWB Yagi–Uda antenna.

Table 2. Dimensions of the proposed SUY antenna.

Parameter	Value (mm)	Parameter	Value (mm)
L1	60	L2	32
L3	11	L4	23
W4	23	H5	3
W1	13	p1	5
p2	5	R1	2

3. Antenna Design and Analysis

To analyze the SUY antenna more clearly, the SUY antenna is divided into the A-type antenna, B-type antenna, C-type antenna, and D-type antenna in Figure 3. In order to further understand the four antennas, as shown in Figure 4, the gain and reflection coefficient reflect the characteristics of each antenna. The specific characteristics are as follows: The A-type antenna is an original patch antenna with coaxial feed. The B-type antenna adopts ME structure to realize the double increase of bandwidth and gain, which appears as a zero radiation point due to current convection.

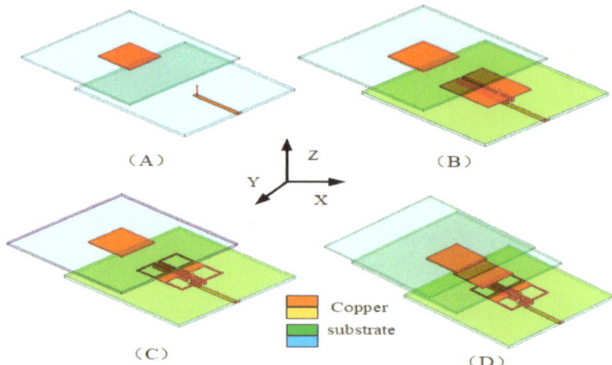

Figure 3. A-type (**A**), B-type (**B**), C-type (**C**), and D-type (**D**) antenna structures.

The C-type antenna realizes the filtering of out of band suppression by NZIM, which also increases the antenna bandwidth. The D-type antenna increases and flattens the gain of the upper suppression boundary at 4.9–5.5 GHz by adding director layer. The SUY antenna achieves stable the gain of lower suppression boundary from 3.5 to 4 GHz by cutting out the diagonal of patch 1 and patch 2. The four kinds of antennas are analyzed in the following sections.

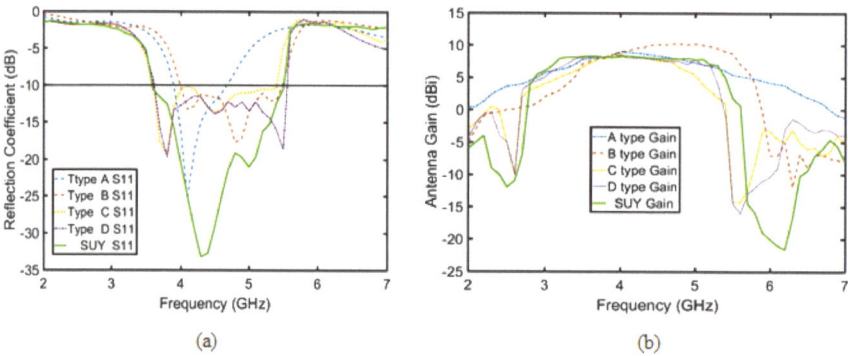

Figure 4. Four-antenna types and the SUY antenna: (**a**) reflection coefficient and (**b**) gain.

3.1. ME Characteristic Analysis

The ME structure has two characteristics: one is to improve the antenna gain by the electromagnetic superposition principle and the other is its impedance matching to improve the bandwidth as shown in Equations (1) and (2). As shown in Figure 5, the B-type antenna is made by adding the ME structure to the A-type antenna, which has increased bandwidth and gain simultaneously. The gain of the B-type antenna exhibits upper stopband suppression.

First, the reason for the electromagnetic superposition at 4.7 GHz is illustrated to explain that the gain of the B-type antenna can reach more than 9 dBi. As shown in Figure 6, according to the Simth circle diagram, impedance matching is best when the real part of impedance is 50 ohms and the imaginary part is 0. When the imaginary part is capacitive and the real part is greater than 50 ohms, the reflected induced current is in the same direction as the induced current of the driven element. Their phase difference is 0 degrees, which leads to high gain by electromagnetic superposition. In contrast, when the imaginary part is inductive and real part is 50 ohms smaller, the induced current in the reflection layer is opposite to that in the driven element layer with a phase difference of 180 degrees. Only the driven element current exists, so the gain at 3–4 GHz is not high gain. To demonstrate this phenomenon, the current distributions of 3.4 GHz and 4.7 GHz are shown in Figure 7. Secondly, the bandwidth of the antenna in Equation (1) is related to VSWR and 1/Q. Figure 8 shows the calculation result of 1/Q and VSWR. From the above results, it is concluded that the bandwidth of the B-type antenna is determined by Q. The change of Q value mainly depends on the impedance optimization in ME structures in Equation (2). Finally, as shown in Figure 7b, because the opposite current direction and equal current density produce zero radiation (Null 1) at 6 GHz. However, its filtering selectivity is not adequate, especially in the lower stopband suppression. We apply dual-C-notch NZIM to improve lower suppression selectivity.

$$BW = \frac{s-1}{\sqrt{sQ}} \quad (1)$$

BW is the bandwidth, s is the voltage standing wave ratio (VSWR), and Q is the quality factors [28].

$$\frac{1}{Q} = \frac{Z_{im}}{Z_{real}} \quad (2)$$

Z_{im} is the imaginary part of the impedance, and Z_{real} is the real part of the impedance [29].

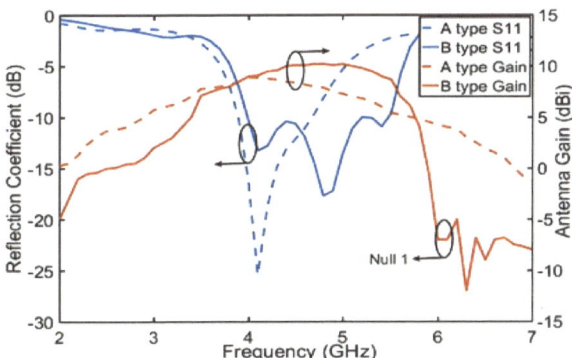

Figure 5. Gain and reflection coefficient simulation of A-type and B-type antennas.

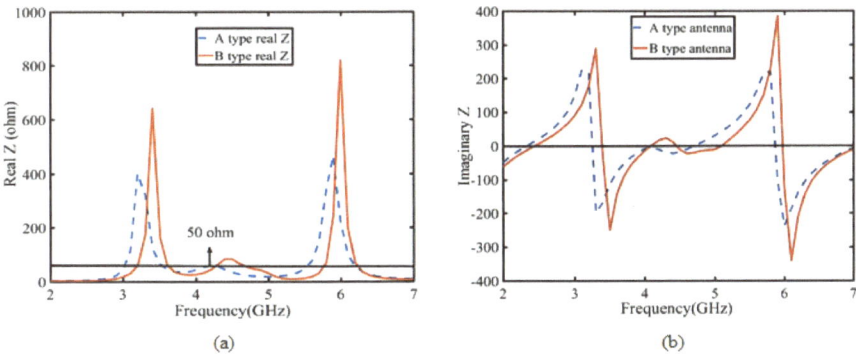

Figure 6. Impedance simulation of A-type and B-type antennas: (**a**) real Z and (**b**) imaginary Z.

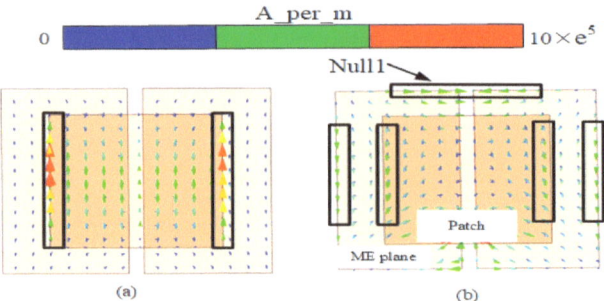

Figure 7. Current distribution at (**a**) 3.4 GHz and (**b**) 4.7 GHz.

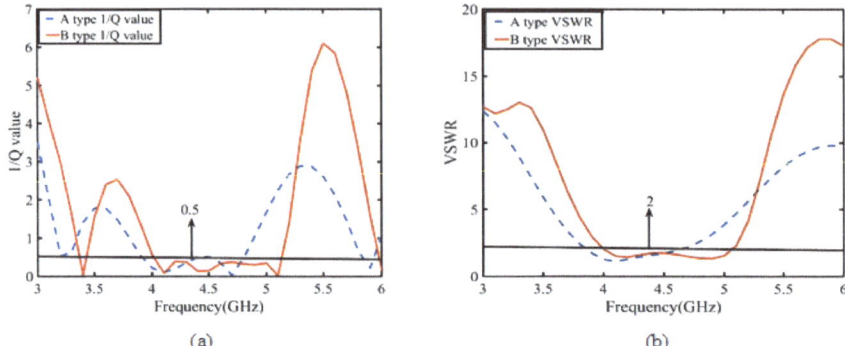

Figure 8. Simulation of the 1/Q value and VSWR parameter between A-type and B-type antennas: (**a**) 1/Q value and (**b**) VSWR parameter.

3.2. Design of the Dual-C-Notch NZIM

In this section, the dual-C-notch NZIM structure is designed to achieve two targets: one is to improve the out-of-band suppression by introducing the radiation nulls of the lower and upper gain stopbands by energy absorption and other is to enhance the bandwidth by impedance matching. This design idea of dual-C-notch NZIM is derived from absorptive-branch-loaded structure and dual-notch NZIM structure [30,31]. For convenience of expression, the dual-C-notch NZIM structure is defined as a C-type antenna (this C-type antenna is cut into dual-C-notch on the surface of the ME structure of the B-type antenna.). The dual-C-notch NZIM structure enhances the bandwidth and improves the out-of-band suppression by absorption.

The absorption function of NZIM is verified from two aspects: first, it is proved that the dual-C-notch belongs to NZIM at 2.6 GHz and 5.5 GHz; second, it has absorption function at 2.6 GHz and 5.5 GHz by electric field simulation. First, as shown in Figure 9, the dual-C-notch structure with periodic boundary condition is modeled to evaluate the scattering performance. Periodic electrical boundaries (PEC) are assigned along the Y-axis direction and periodic magnetic boundaries (PMC) are excited perpendicularly to the dielectric substrate. The input/output port of waveports is set along the Z-axis direction. The real and imaginary parts of the permittivity (ϵ), the magnetic permeability (μ), and the refractive index (n) by Formula (1) and the calculation method of the equivalent medium theoretical model are shown in Figure 10. It can be seen that the dual-C-notch structure conforms to the NZIM theorem because almost all parameters are near zero in the figure at 2.6 GHz and 5.5 GHz. Second, As shown in Figure 11, the absorptivity values of 2.6 GHz and 5.5 GHz frequencies are 40% and 60% respectively, and such energy absorption results in radiation attenuation. Figure 12 shows that the 2.6 GHz energy absorption position is an intermediate coupled with the dual-C-notch, while the 5.5 GHz energy absorption position is at one end of the direction loop of the dual-C-notch. The above proves the existence of zero radiation (null2 and null3), as shown in Figure 13. At the same time, the bandwidth of the C-type antenna can be shown to be increased.

$$n = \pm \frac{1}{kd}[\cos^{-1}(\frac{1 - S_{11}^2 + S_{21}^2}{2S_{21}})] \quad (3)$$

where S_{11} and S_{21} are reflection and transmission coefficients. The n is the refractive index, k is wave number, d is the length of the unit cell [32].

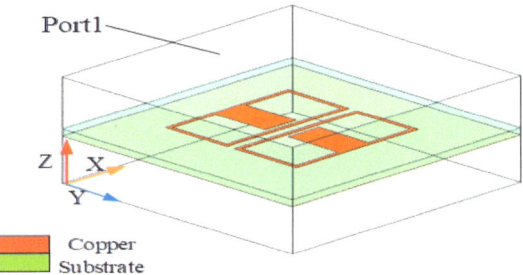

Figure 9. NZIM simulation: electromagnetic field boundaries.

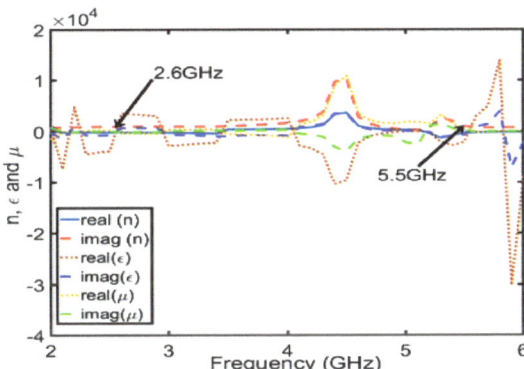

Figure 10. Real part and imaginary part of the permittivity (ϵ), the magnetic permeability (μ), and the refractive index (n).

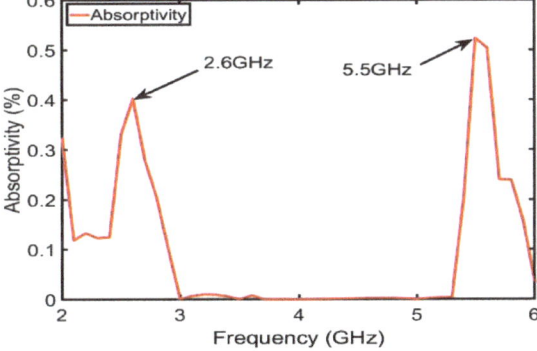

Figure 11. Absorption simulation.

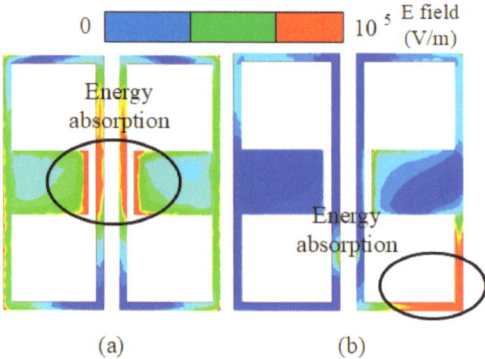

Figure 12. E field simulation at (**a**) 2.6 GHz and (**b**) 5.5GHz.

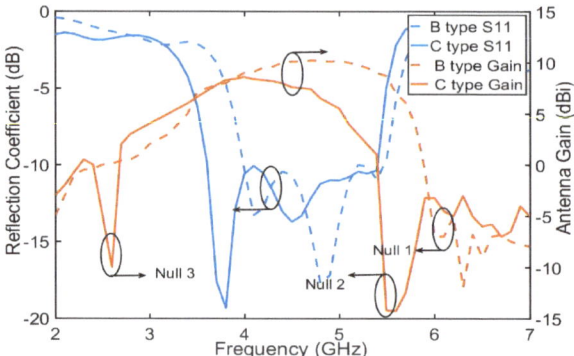

Figure 13. Simulation of gain and reflection coefficient of the B-type and C-type antennas.

3.3. Enhancement Suppression Boundary of Gain

In order to further improve the stable gain, the gain of the suppression boundary is enhanced to achieve two targets: one is to enhance the gain of upper suppression boundary by addition of the director layer from 4.9 to 5.5 GHz and the other is to improve the gain of lower suppression boundary by the diagonal cut of the patch from 3.5 to 4 GHz.

Firstly, according to the Yagi antenna principle, the director layer is added to the C-type antenna to improve antenna gain. The antenna is defined as a D-type antenna. As shown in Figure 14, after adding the director layer, the real part of the D-type antenna is close to 50 ohms, and the imaginary part is close to 0 in the range of 4.9–5.5 GHz. According to the Smith circle diagram, the gain is improved by the optimal matching. According to Equations (2) and (4), the gain increases with the increase of the quality factor (Q) after impedance matching in Figure 15a. So, the D-type antenna achieves enhancement gain of the upper suppression boundary by adding director to realize impedance matching.

Secondly, there are three ways to improve the gain of upper suppression boundary: one is to add another director layer, another is to change the length of patch 1 and 2, and the last is to change the shape of patch 1 and 2. The first method reduces the overall gain of the antenna, so it is not used. The second and third methods are verified in terms of which method is suitable to improve the gain of upper suppression boundary by the quality factor (Q). According to Equations (4)–(7), where radiation loss (Q_r) is constant, only direction coefficient (D) is related to length of patch 1 and 2, but patch 1 and 2 must be square. As shown in Figure 16b, the direction coefficient (D) is less than or equal to 0.03. This means that increasing the patch length will not change the antenna gain, and only changing its shape can improve the antenna gain. The method of diagonal cut of the patch 1 and 2 increase the gain of lower suppression boundary. As shown in Figure 17, when p1 = p2 = 5 mm, impedance matching reaches the best point. Finally, the Q reaches its maximum value in Figure 15b. The enhanced suppression boundary gain is to achieve the peak gain reaching 8.5 dBi, and the flat in-band gain has a ripple lower than 0.5 dBi, as shown in Figure 16a.

$$G = \frac{DQ}{Q_r} \quad (4)$$

G is the antenna gain, D is the antenna radiation direction coefficient, and Q_r is radiation loss [33].

$$Q_r = \frac{3\lambda_0 \varepsilon_r}{8h} \quad (5)$$

where λ_0 is resonant frequency wavelength, a is the length of the patch, ε_r is dielectric constant of the dielectric plate and h is dielectric plate thickness [34].

$$D = 2x^2 \left[xsi(x) + cosx - 2 + \frac{sinx}{x} \right]^{-1} \quad (6)$$

$$x = 2\pi a \lambda_0 \quad (7)$$

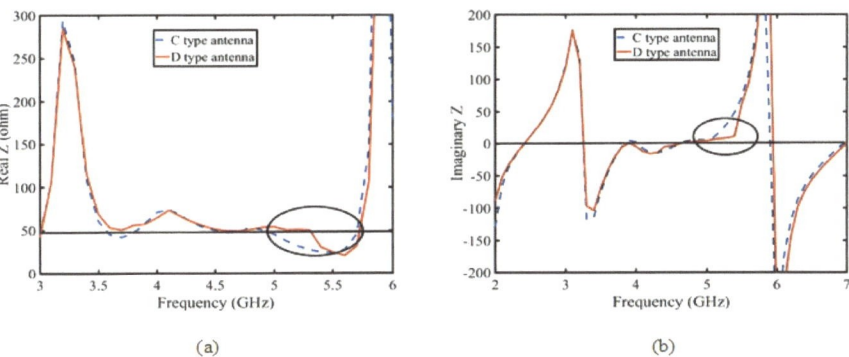

Figure 14. Impedance of the C-type and D-type antennas: (a) real Z; (b) imaginary Z.

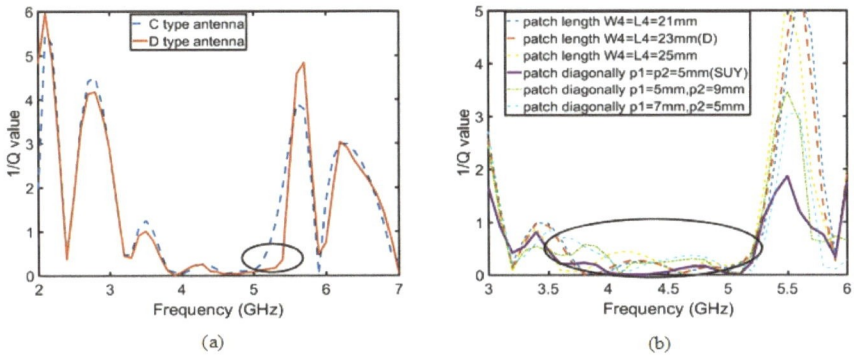

Figure 15. Analysis of the 1/Q value of flat gain. (a) The Q value of the D-type antenna is more stable than that of the C-type antenna. (b) The Q value of patch 1 and patch 2 diagonal cuts is more stable than that their variation in length.

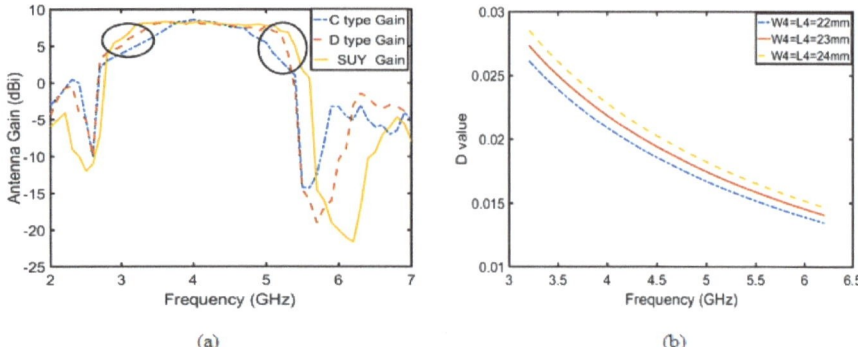

Figure 16. Flat gain of the SUY antenna is analyzed according to Equation (6). (**a**) Gain of the C-type, D-type, and SUY antenna and (**b**) relation between the radiation direction (D) and frequency.

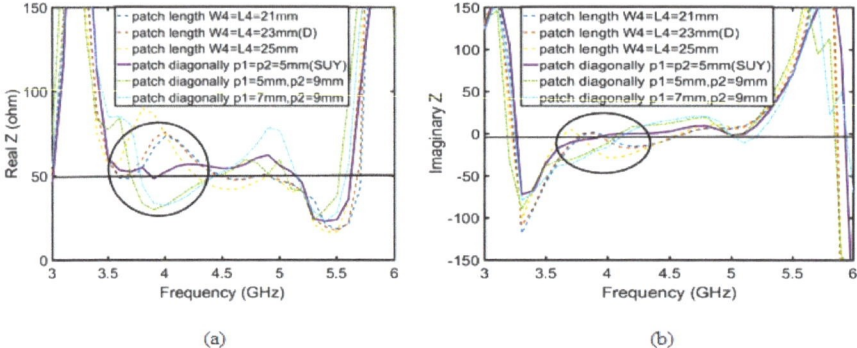

Figure 17. Contrast between the length variation impedance of patch 1 and patch 2 and their diagonal excision impedance: (**a**) real Z; (**b**) imaginary Z.

4. Measured and Simulated Results

To verify the validity of our design concept, a prototype of the proposed SUY antenna is fabricated and measured. The experimental setup in the anechoic chamber and the SUY antenna fabricated prototype are shown in Figure 18. The port of the antenna is welded to a 50 ohm SMA. HFSS software is used to simulate the SUY antenna. An R&S ZNB20 vector network analyzer is used to measure the reflection coefficient of the antenna, for which the insertion loss is less than 0.05 dB. Figure 19a shows the comparison between the measured results of the gain and reflection coefficients and the simulation results. The maximum measured filtering suppression reaches 14.6 dBi. The flat gain is measured at 0.6 dBi. Figures 19b and 20 show the simulated and measured values of efficiency and radiation patterns of the SUY antenna. The data were measured in an antenna anechoic chamber, where the signal generator is an R&S microwave signal generator. The output frequency range of this signal generator is from 100 kHz to 20 GHz, and the maximum output power is +11 dBm from 50 MHz to 20 GHz. A step of 1 cm is used to test the radiation and efficiency of the SUY antenna, for which a test distance of approximately 3–5 wavelengths is best. With the adjustment of the step size, the measurement efficiency of the antenna at 3.5–5.5 GHz is more than 90%, as shown in Figure 19b. The radiation angle of the SUY antenna is measured by $\pi/24$ step rotations. Figure 20 shows the co-polarization and cross-polarization of 4 GHz and 5 GHz. In manufacturing process of the NZIM structure, welding increases the impedance value, so the copper tape with a thickness of 0.001 mm is used for winding binding. In the test is a certain tilt angle caused by a certain cross-planning error. In the measurement, the small discrepancy may be a result of the manual installation and the error of the measurement system itself.

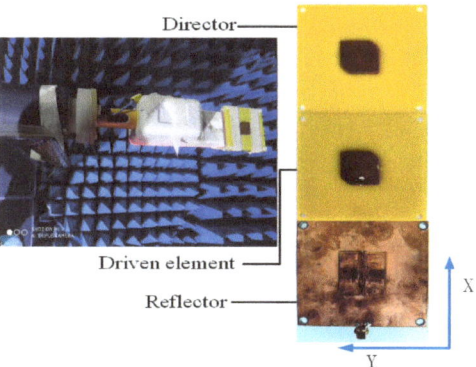

Figure 18. Experimental setup in an anechoic chamber and SUY antenna fabricated prototype.

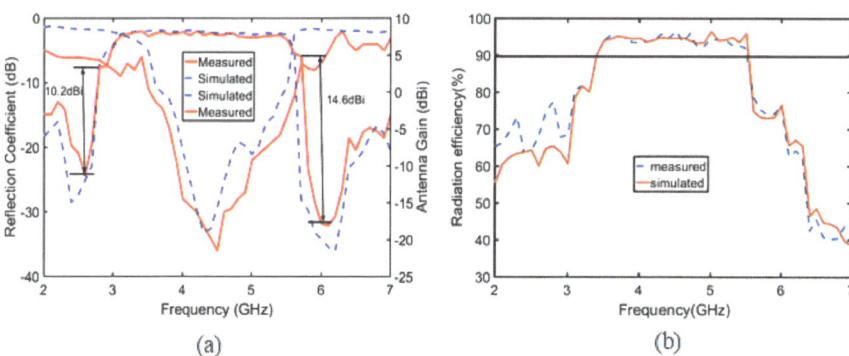

Figure 19. Simulated and measured (**a**) reflection coefficient, gain, and (**b**) efficiency of the SUY antenna.

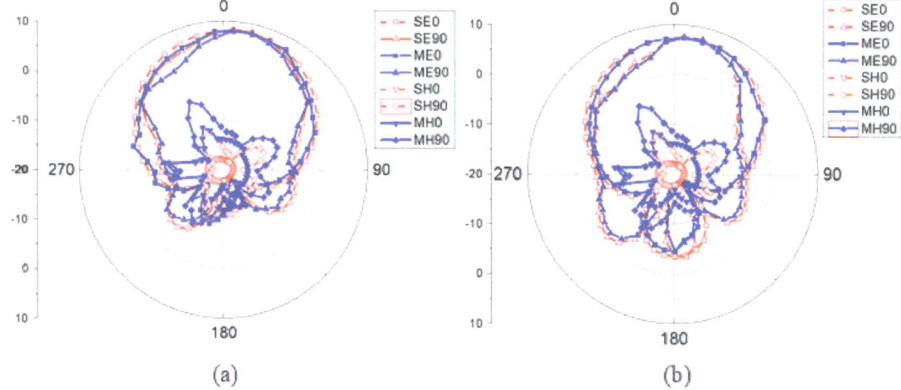

Figure 20. Co-polarization and cross-polarization radiation of (**a**) 4 GHz and (**b**) 5 GHz. SE0 and SE90 are polarized simulations; SH0 and SH90 are cross-polarized simulations; ME0 and ME90 are co-polarized measurements; MH0 and MH90 are cross-polarized measurements.

5. Conclusions

In this paper, stereoscopic ultra-wideband (UWB) Yagi–Uda (SUY) antenna with stable gain by NZIM has been discussed and analyzed in detail. Its advantages are high gain, wide bandwidth, and stable gain. The disadvantage is that the three-dimensional size is larger. The combination of coaxial feed and ME structure using electromagnetic superposition and the impedance matching principle achieve high gain and excellent broadband performance. The NZIM with two radiation nulls as well as flat in-band radiation gain is realized in a poof-of-concept prototype of SUY antenna. In order to further improve the stable gain, adjusting the director and driven element impedance enhance the gain of the suppression boundary. The demonstration antenna has been designed and implemented, and the simulated and measured results with good accordance have been presented.

Author Contributions: Conceptualization, Y.F.; Methodology, Y.F.; Investigation, T.S.; Writing—Original Draft Preparation, Y.F. and J.D.; Writing—Review & Editing, Z.C. All authors have read and agreed to the published version of the manuscript.

Funding: This research was funded by Yunnan Fundamental Research project of fund grant number 202101AU070164, Yunnan Fundamental Research Projects number: 202301AV07003 and National Natural Science Foundation of China number: 61971208.

Institutional Review Board Statement: Not applicable.

Informed Consent Statement: Not applicable.

Data Availability Statement: Not applicable.

Conflicts of Interest: The authors declare no conflict of interest.

References

1. Wu, Q.; Zhao, Y.; Fan, Q. Time-Dependent Performance Modeling for Platooning Communications at In-tersection. *IEEE Internet Things J.* **2022**, *9*, 18500–18513. [CrossRef]
2. Wu, Q.; Zheng, J. Performance modeling and analysis of IEEE 802.11 DCF based fair channel access for vehicle-to-roadside communication in a non-saturated state. *Wirel. Netw.* **2014**, *21*, 1–11. [CrossRef]
3. Wang, K.; Yu, F.R.; Wang, L.; Li, J.; Zhao, N.; Guan, Q.; Li, B.; Wu, Q. Interference Alignment with Adaptive Power Allocation in Full-Duplex-Enabled Small Cell Networks. *IEEE Trans. Veh. Technol.* **2019**, *3*, 3010–3015. [CrossRef]
4. Wu, Q.; Shi, S.; Wan, Z.; Fan, Q.; Fan, P.; Zhang, C. Towards V2I Age-aware Fairness Access: A DQN Based Intelligent Vehicular Node Training and Test Method. *Chin. J. Electron.* **2022**, *2*, 90–93.
5. Wu, Q.; Zhao, Y.; Fan, Q.; Fan, P.; Wang, J.; Zhang, C. Mobility-Aware Cooperative Caching in Vehicular Edge Computing Based on Asynchronous Federatedand Deep Reinforcement Learning. *IEEE J. Sel. Top. Signal Process.* **2022**, *1*, 66–81.
6. Wu, Q.; Wang, X.; Fan, Q.; Fan, P.; Zhang, C.; Li, Z. High stable and accurate vehicle selection scheme based on federated edge learning in vehicular networks. *China Commun.* **2023**, *3*, 1–17. [CrossRef]
7. Wu, Q.; Ge, H.; Fan, P.; Wang, J.; Fan, Q.; Li, Z. Time-Dependent Performance Analysis of the 802.11p-Based Platooning Communications Under Disturbance. *IEEE Trans. Veh. Technol.* **2020**, *69*, 15760–15773. [CrossRef]
8. Wu, Q.; Liu, H.; Zhang, C.; Fan, Q.; Li, Z.; Wang, K. Trajectory Protection Schemes Based on a Gravity Mobility Model in I. *Electronics* **2019**, *8*, 148. [CrossRef]
9. Wu, Q.; Xia, S.; Fan, P.; Fan, Q.; Li, Z. Velocity-Adaptive V2I Fair-Access Scheme Based on IEEE 802.11 DCF for Platooning Vehicles. *Sensors* **2018**, *18*, 4198. [CrossRef]
10. Wu, Q.; Zheng, J. Performance modeling and analysis of the ADHOC MAC protocol for VANETs. *IEEE Int. Conf. Commun.* **2015**, *2015*, 3646–3652.
11. Long, D.; Wu, Q.; Fan, Q.; Fan, P.; Li, Z.; Fan, J. A Power Allocation Scheme for MIMO-NOMA and D2D Vehicular Edge Computing Based on Decentralized DRL. *Sensors* **2023**, *23*, 3449. [CrossRef]
12. Jing, F.; Wu, Q.; Hao, J.F. Optimal deployment of wireless mesh sensor networks based on Delaunay tri-angulations. *Int. Conf. Inf. Netw. Autom.* **2010**, *1*, 23–28.
13. Fan, J.; Yin, S.T.; Wu, Q.; Gao, F. Study on Refined Deployment of Wireless Mesh Sensor Network. In Proceedings of the International Conference on Wireless Communications Networking & Mobile Computing, Chengdu, China, 23–25 September 2010; pp. 1–5.
14. Asci, Y. Wideband and Stable Gain Cavity Backed Slot Antenna with Inner Cavity Walls and Baffle for X and Ku-Band Applications. *IEEE Trans. Antennas Propag.* **2023**, *2*, 391–394. [CrossRef]

15. Nouri, M.; Behroozi, H.; Jafarieh, A.; Aghdam, S.A.; Piran, M.J.; Mallat, N.K. A Learning-based Dipole Yagi-Uda Antenna and Phased Array Antenna for mmWave Precoding and V2V Communication in 5G Systems. *IEEE Trans. Veh. Technol.* **2022**, *72*, 2789–2803. [CrossRef]
16. Chen, Z.; Hu, Z.; Zhang, J.; Zhang, G.; Guo, C. Compact Normal-Mode Hybrid-Helix Antenna and Its Ap-plication to Circularly Polarized Yagi Array. *IEEE Trans. Antennas Propag.* **2021**, *69*, 95–99. [CrossRef]
17. Wang, Z.; Ning, Y.; Dong, Y. Compact Shared Aperture Quasi-Yagi Antenna with Pattern Diversity for 5G-NR Applications. *IEEE Trans. Antennas Propag.* **2021**, *7*, 4178–4183. [CrossRef]
18. Ahmed, S.; Le, D.; Sydänheimo, L.; Ukkonen, L.; Björninen, T. Wearable Metasurface-Enabled Qua-si-Yagi Antenna for UHF RFID Reader with End-Fire Radiation Along the Forearm. *IEEE Access* **2021**, *9*, 77229–77238. [CrossRef]
19. Jia, W.-Q.; Lu, W.-J. Dual-Resonant Wideband Yagi-Uda Antennas Using Full-Wavelength Sectorial Di-pole. In Proceedings of the 2021 International Conference on Microwave and Millimeter Wave Technology (ICMMT), Nanjing, China, 23–26 May 2021; Volume 1, pp. 1–3.
20. Zhang, Y.; Wu, S. A Wideband Quasi-Yagi Antenna Using Four-Step Slotline with Metasurface for Sub-6G Applications. In Proceedings of the 2021 IEEE 4th International Conference on Electronic Infor-mation and Communication Technology (ICEICT), Xi'an, China, 18–20 August 2021; Volume 1, pp. 156–158.
21. Jia, W.-Q.; Ji, F.-Y.; Lu, W.-J.; Pan, C.-X.; Zhu, L. Dual-Resonant High-Gain Wideband Yagi-Uda Antenna Using Full-Wavelength Sectorial Dipoles. *IEEE Open J. Antennas Propag.* **2021**, *2*, 872–881. [CrossRef]
22. Chaudhari, A.D.; Ray, K.P. Design of Ultra-Wide Bandwidth Printed Quasi-Yagi Antenna with Semi-Elliptical Monopole Driver. In Proceedings of the 2020 IEEE International Symposium on Antennas and Propagation and North American Radio Science Meeting, Montreal, QC, Canada, 5–10 July 2020; Volume 2, pp. 219–220.
23. Xu, C.; Wang, Z.; Wang, Y.; Wang, P.; Gao, S. A Polarization-Reconfigurable Wideband High-Gain An-tenna Using Liquid Metal Tuning. *IEEE Trans. Antennas Propag.* **2021**, *2*, 5835–5841.
24. Yang, K.; Ye, L.; Zhu, W.; Li, J.; Wu, D. A Novel Wideband Quasi-Yagi Antenna for Base-station Appli-cations. In Proceedings of the Photonics & Electromagnetics Research Symposium (PIERS), Hangzhou, China, 21–25 November 2021; Volume 2, pp. 1830–1834.
25. Rezaeieh, S.A.; Antoniades, M.A.; Abbosh, A.M. Gain Enhancement of Wideband Metamaterial-Loaded Loop Antenna with Tightly Coupled Arc-Shaped Directors. *IEEE Trans. Antennas Propag.* **2018**, *4*, 2090–2095.
26. Ke, Y.-H.; Yang, L.-L.; Zhu, Y.-Y.; Wang, J.; Chen, J.-X. Filtering Quasi-Yagi Strip-Loaded DRR Antenna with Enhanced Gain and Selectivity by Metamaterial. *IEEE Access* **2021**, *9*, 31755–31761. [CrossRef]
27. Tao, J.; Feng, Q.; Vandenbosch, G.A.E.; Volskiy, V. Director-Loaded Magneto-Electric Dipole Antenna with Wideband Flat Gain. *IEEE Trans. Antennas Propag.* **2019**, *11*, 6761–6769. [CrossRef]
28. Virushabadoss, N.; Henderson, R. Quality Factor of an Electrically Small Planar Slot Antenna with Dif-ferent Matching Networks. In Proceedings of the 2019 IEEE Texas Symposium on Wireless and Micro-wave Circuits and Systems (WMCS), Waco, TX, USA, 28–29 March 2019; Volume 11, pp. 1–4.
29. Best, S.R. A discussion on the quality factor of impedance matched electrically small wire antennas. *IEEE Trans. Antennas Propag.* **2015**, *9*, 502–508.
30. EEsmail, B.A.F.; Koziel, S.; Golunski, L.; Majid, H.B.A.; Barik, R.K. Overview of Metamaterials-Integrated Antennas for Beam Manipulation Applications: The Two Decades of Progress. *IEEE Access* **2022**, *9*, 67096–67116. [CrossRef]
31. Zhang, S.; Zeng, Q.; Feng, N.; Hu, N.; Xie, W.; Denidni, T. A High Gain Filtering Dielectric Resonator Antenna Based on Metamaterial. In Proceedings of the 2020 International Conference on Microwave and Millimeter Wave Technology (ICMMT), Shanghai, China, 20–23 September 2020; Volume 1, pp. 1–3.
32. Küçükvural, K.; Uçar, M.H.B.; Çakir, G. CPW-Fed Microstrip Monopole Antenna Design with 5.5 GHz Notch-band Filtering Characteristic for Ultra-Wideband Communications. In Proceedings of the 2021 29th Signal Processing and Communications Applications Conference (SIU), Istanbul, Turkey, 9–11 June 2021; Volume 1, pp. 1–4.
33. Zhu, H.; Zhang, Y.; Ye, L.; Li, Y.; Dang, Z.; Xu, R.; Yan, B. A High Q-factor Metamaterial Absorber and Its Refractive Index Sensing Characteristics. *IEEE Trans. Microw. Theory Tech.* **2021**, *12*, 5383–5391. [CrossRef]
34. Hansen, T.V.; Kim, O.S.; Breinbjerg, O. Quality Factor and Radiation Efficiency of Dual-Mode Self-Resonant Spherical Antennas with Lossy Magnetodielectric Cores. *IEEE Trans. Antennas Propag.* **2014**, *62*, 467–470. [CrossRef]

Disclaimer/Publisher's Note: The statements, opinions and data contained in all publications are solely those of the individual author(s) and contributor(s) and not of MDPI and/or the editor(s). MDPI and/or the editor(s) disclaim responsibility for any injury to people or property resulting from any ideas, methods, instructions or products referred to in the content.

MDPI AG
Grosspeteranlage 5
4052 Basel
Switzerland
Tel.: +41 61 683 77 34

Sensors Editorial Office
E-mail: sensors@mdpi.com
www.mdpi.com/journal/sensors

Disclaimer/Publisher's Note: The title and front matter of this reprint are at the discretion of the Guest Editors. The publisher is not responsible for their content or any associated concerns. The statements, opinions and data contained in all individual articles are solely those of the individual Editors and contributors and not of MDPI. MDPI disclaims responsibility for any injury to people or property resulting from any ideas, methods, instructions or products referred to in the content.

www.ingramcontent.com/pod-product-compliance
Lightning Source LLC
LaVergne TN
LVHW072337090526
838202LV00019B/2435